T0135680

Asymptotic Change-Point Analysis of the Dependencies in Time Series

Inaugural-Dissertation

zur
Erlangung des Doktorgrades
der Mathematisch-Naturwissenschaftlichen Fakultät
der Universität zu Köln

vorgelegt von
Christoph Heuser
aus Düren

Köln 2016

Berichterstatter: Prof. Dr. Josef G. Steinebach, Universität zu Köln
Prof. Dr. Wolfgang Wefelmeyer, Universität zu Köln

Tag der letzten mündlichen Prüfung: 23. Mai 2016

Bibliographic information published by the Deutsche Nationalbibliothek

The Deutsche Nationalbibliothek lists this publication in the Deutsche Nationalbibliografie; detailed bibliographic data are available on the Internet at http://dnb.d-nb.de .

©Copyright Logos Verlag Berlin GmbH 2016
All rights reserved.

ISBN 978-3-8325-4314-3

Logos Verlag Berlin GmbH
Comeniushof, Gubener Str. 47,
10243 Berlin
Tel.: +49 (0)30 42 85 10 90
Fax: +49 (0)30 42 85 10 92
INTERNET: http://www.logos-verlag.de

Abstract

In recent papers, Wied and his coauthors have introduced change-point procedures to detect and estimate structural breaks in the correlation between time series. To prove the asymptotic distribution of the test statistic and stopping time as well as the change-point estimation rate, they use an extended functional Delta method and assume nearly constant expectations and variances of the time series.
In this thesis, we allow asymptotically infinitely many structural breaks in the means and variances of the time series. For this setting, we present test statistics and stopping times which are used to determine whether or not the correlation between two time series is and stays constant, respectively. Additionally, we consider estimates for change-points in the correlations. The employed nonparametric statistics depend on the means and variances. These (nuisance) parameters are replaced by estimates in the course of this thesis. We avoid assuming a fixed form of these estimates but rather we use 'blackbox' estimates, i.e. we derive results under assumptions that these estimates fulfill. These results are supplement with examples.
This thesis is organized in seven sections. In Section 1, we motivate the issue and present the mathematical model. In Section 2, we consider a posteriori and sequential testing procedures, and investigate convergence rates for change-point estimation, always assuming that the means and the variances of the time series are known. In the following sections, the assumptions of known means and variances are relaxed.
In Section 3, we present the assumptions for the mean and variance estimates that we will use for the mean in Section 4, for the variance in Section 5, and for both parameters in Section 6. Finally, in Section 7, a simulation study illustrates the finite sample behaviors of some testing procedures and estimates.

Zusammenfassung

Die vorliegende Arbeit befasst sich mit der Strukturbruchanalyse in den Korrelationen zwischen zwei Zeitreihen. In kürzlich erschienenen wissenschaftlichen Arbeiten haben sich Wied und seine Koautoren sowohl mit a-posteriori und sequenziellen Testverfahren als auch mit Schätzmethoden für Strukturbrüche in den Korrelationen befasst. Hierbei haben die Autoren nahezu konstante Erwartungswerte und Varianzen für die beiden Zeitreihen angenommen.
In der vorliegenden Arbeit präsentieren wir das asymptotische Verhalten von Testverfahren (sowohl für a-posteriori als auch für sequenzielle Testprobleme) und Schätzern für die Strukturbrüche in den Korrelationen zwischen zwei Zeitreihen, bei denen jeweils Strukturbrüche in den Erwartungswerten und Varianzen erlaubt sind. Dabei ist die Arbeit wie folgt aufgebaut: In Kapitel 1 motivieren wir die Problematik und erläutern das mathematische Modell. In Kapitel 2 stellen wir die oben genannten Verfahren und Schätzmethoden vor, die die exakten Parameter, d.h. die Erwartungswerte und die Varianzen, verwenden. Diese Verfahren werden in den folgenden Kapiteln erweitert, um mit unbekannten Erwartungswerten und Varianzen umgehen zu können. In Kapitel 3 präsentieren wir die Annahmen der Parameterschätzung, die in den nachfolgenden Kapiteln zugrunde gelegt werden.
In Kapitel 4, 5 und 6 präsentieren wir Verfahren, in denen ein Austauch der exakten Parameter durch ihre Schätzer stattfindet. Dabei werden in Kapitel 4 die (unbekannten) exakten Erwartungswerte durch ihre Schätzer ersetzt; in Kapitel 5 ist dies der Austausch der Varianzen und in Kapitel 6 ersetzen wir beide der vorgenannten Parameter durch ihre Schätzer. In Kapitel 7 beleuchten wir schließlich das Verhalten der verschiedenen Methoden bei endlicher Beobachtungszahl mit Hilfe von Simulationen und wenden die neuen Verfahren auf Finanzdaten an.

Acknowledgments

This thesis has been developed during my employment as a scientific assistant at the Mathematical Institute at the University of Cologne. First of all, I want to thank my supervisor, Prof. Dr. Josef G. Steinebach. I started as his last PhD student before his retirement. His support, experience and knowledge have been key factors to finish this thesis. I have always been able to count on his advice even after he retired last year. He has always given me calm and confidence.
I also want to thank Prof. Dr. Wolfgang Wefelmeyer for being my second examiner and particularly for all the valuable and lively discussions. In addition, I want to thank the whole working group for a unique and wonderful time. In particular, I want to thank Béatrice Bucchia for the productive joint research work, Dr. Maren Schmeck and Leonid Torgovitski for the marvelous time we had together. I thank Prof. Dr. Dr. Hanspeter Schmidli, Prof. Dr. Alexander Drewitz, and Dr. Markus Schulz for the many interesting, vivid, and helpful discussions during lunchtime, and Heidi Anderka and Vera Sausen for their help with all the paper work involved. I also want to express my sincere thanks to Prof. Dr. Martin Wendler for his suggestions, remarks, and the many discussions about my thesis during as well as after our time together in Cologne, and to Dr. Sandra Kliem for her help to reduce my mistakes concerning English grammar and spelling.
Finally, I want to thank my girlfriend Julia and both our families for their support and encouragement. I also want to thank many of my friends, particularly Alexander, Christoph, and Martin for their endless willingness to read the countless pages of formulas, tables, and figures.

Contents

0 Notation and Convention

$Z_{l,n}$	$Z_{1,n} = X_n$ And $= Z_{2,n} = Y_n$	p. 4
$\mu_{l,i}$, $\sigma^2_{l,i}$ and ρ_i	$\mathbb{E}[Z_{l,i}] = \mu_{l,i}$, $\mathrm{Var}[Z_{l,i}] = \sigma^2_{l,i}$, $\mathrm{Corr}(X_i,Y_i) = \rho_i$ for $i = 1,2,\dots$	p. 4
k^*, k^*_i and Δ_ρ, $\Delta_{\rho,i}$	change-point(s) and change size(s) in the correlations	p. 4ff.
$\underline{\Delta}_{k^*}$ and $\underline{\Delta}_\rho$	$\min_r \lvert k^*_r - k^*_{r-1}\rvert$ and $\min_r \lvert\Delta_{\rho,r}\rvert$	p. 5
$k^*_{\xi,i}$ and $\Delta_{\xi,i}$	change-point(s) and change size(s) in the parameter $\xi \in \{\mu_1,\mu_2,\sigma^2_1,\sigma^2_2\}$	p. 86
$T_n^{\iota,\psi,l,\gamma}$	general detector which is equal to $f_\iota(B_n^{\psi,l,\gamma}(\cdot))$	p. 8
$\tau^{(c)}_{n,\iota,\psi,l,\gamma}$ and $\tau^{(o)}_{n,\iota,\psi,l,\gamma}$	general closed-end and open-end stopping times	p. 8f.
D_l and $\hat{D}_l$	long-run variance (LRV) and its estimate	p. 8
$\tilde{Q}_n^{(\psi)}(s,t)$	detector for the change-points in an epidemic change-point setting	p. 19
$Q_n^{(\psi)}(\cdot)$	detector for the change-points in a general (multiple) change-point setting	p. 19
$g_\rho(\cdot)$, $g_\mu(\cdot)$, and $g_{\sigma^2}(\cdot)$	bounded and integrable functions represent the changes in the correlations, means, and variances	p. 4, 90, and 94
$m_{\mu,l,n}$, $m_{\sigma,l,n}$, and $m_{l,n}$	number of change-points in the means, variances	p. 48
$c_{\alpha,\cdot}$	critical value for the significance level α	p. 7

n,i,k,l	frequently used natural numbers
x,z,s	frequently used real numbers
$W(\cdot)$ and $B(\cdot)$	standard Brownian motion and bridge
A^T	transpose of a matrix A
$[x]$	integer part of $1+x$ for $x \in \mathbb{R}$
$\lambda(\cdot)$, $\#(\cdot)$	Lebesgue measure and cardinality
$\max A$ and $\min A$	maximum and minimum of a set $A \subset \mathbb{R}$ with $\max\emptyset = -\infty$ and $\min\emptyset = \infty$
$\lVert\cdot\rVert_p$, $\lVert\cdot\rVert_A$, and $\lVert\cdot\rVert$	L_p-norm, uniform norm on a set $A \subset \mathbb{R}$, and uniform norm on a corresponding set
$\xrightarrow{D}$, $\xrightarrow{\mathcal{D}[0,1]}$, $\xrightarrow{P}$	convergence in distribution of a random variable, of a random càdlàg function, convergence in probability
$a_n \sim b_n$	the real valued sequences a_n and b_n fulfill $\lim_{n\to\infty} \frac{a_n}{b_n} = c \in \mathbb{R}_{\neq 0}$
$a_n \ll b_n$	$a_n/b_n = o(1)$ as $n \to \infty$ for real valued sequences
$Y_n = X_n + o_P(1)$	the S-valued random sequences $\{X_n\}$ and $\{Y_n\}$ fulfill $\lVert X_n - Y_n\rVert_S \xrightarrow{P} 0$
$X_n \leq Y_n + o_P(1)$	for random sequences $\{X_n\}$ and $\{Y_n\}$ a random sequence $r_n = o_P(1)$ exists such that $X_n \leq Y_n + r_n$ almost surely (a.s.)
$a \wedge b$ and $a \vee b$	the minimum and maximum of $a,b \in \mathbb{R}$
$a_n \equiv a$	$a_n = a$ for all $n \in \mathbb{N}$
$\overline{x}$ and $\overline{x}_k$,	sample means of a set $\{x_1,\dots,x_n\}$, i.e., $\overline{x} = n^{-1}\sum_{i=1}^n x_i$ and $\overline{x} = k^{-1}\sum_{i=1}^k x_i$
$\overline{x}_j^k$	sample means of a set $\{x_{l+1},\dots,x_k\}$ for $j,k \in \mathbb{N}$ and $j < k$, i.e., $\overline{x}_j^k = (k-j)^{-1}(k\overline{x}_k - j\overline{x}_j)$

1 Introduction

1.1 Introduction

In our modern society the amount of collected data per year is higher than ever before. For this reason it is necessary to provide suitable big data storage and highly efficient search algorithms. Examples of where such procedures are required to draw conclusions from enormous quantities are predictions in financial markets, early warning systems for natural disasters, and buying patterns of internet users. In each of these fields certain parameters structure the data set and reveal the required information concerning future development. Sometimes it is important to know whether these parameters have undergone gradual changes, have suddenly changed significantly, or have remained more or less constant, e.g. when observing the changes of the global mean surface temperature, of the risk of some stock price, of the linear correlation between two groups etc. One possibility to investigate whether there is a change of a parameter or not is the change-point analysis. For a survey on the change-point analyses we refer to Müller and Siegmund (1994), Aue and Horváth (2013), and Brodsky and Darkhovsky (2013).

This thesis investigates the issue whether there may be a change in the correlation between two time series. Among many other things, this is motivated by questions such as: Does the risk of a portfolio of many stock prices remain constant or not? Is the correlation between sunshine duration and the electricity rates traded at the market stable?

Testing whether there is a change in a correlation or not is not new and has been treated by some research studies. Firstly, Aue et al. (2009) investigated the related issue of break detection in the covariance structure of multivariate time series. Then, Wied (2010) treated structural breaks in the correlation and published the main results in Wied et al. (2012). They presented a cumulative sum (CUSUM) test where the asymptotic property of test statistic, the convergence in distribution, was proved by a functional Delta method. Under certain assumptions these results allow to test a posteriori whether there has been a change in the correlation between two given time series. Testing whether there is a sudden change in the correlations while the observation of the data set is incomplete was considered by Wied and Galeano (2013). Again the testing procedure is based on CUSUM detectors and the convergence in distribution was proved by the same functional Delta method as before. Recently, Wied and Galeano (2014) introduced a multiple break detection method for the correlations.

Each of the main results in these papers possesses some special kind of assumptions. Technically, the authors' proofs are based on the functional Delta method for which a little bit more than the fourth moments of the two considered time series have to be bounded. Furthermore, they assume that the mean and the variance of both time series do not change significantly. In our approach to analyze a change of the correlations we present different CUSUM test statistics, which allow changes in the means and the variances in the two time series.

This thesis is organized in seven sections. Section 1 illustrates the mathematical models and the motivation for the approach. In Section 2, we consider the case where it is assumed that the means and variances of the time series are known. Under this assumption we demonstrate an asymptotic a posteriori change-point procedure under a functional central limit theorem (FCLT). For normalization of the limit process we present long-run variance (LRV) estimates. Additionally, we investigate a change-point estimate and consider open-end and closed-end sequential change-point procedures. In Section 3, we introduce the mean and variance estimate assumption. In contrast to a special type of estimate we present some sufficient conditions for the estimation errors so that the main results of Section 2 are still satisfactory. Section 4 contains the same procedures and change-point estimates as Section 2, but we replace the expectations by some unspecified mean estimates which fulfill the assumptions of the third section. In Section 5, we concentrate on the same procedures as in Section 4, now assuming that the expectations are once again known and the variances will be replaced by some estimates. In Section 6, we consider the already known procedures with both parameters replaced by their estimates. Finally, in Section 7, we illustrate the asymptotic behavior of the testing procedures and change-point estimates by some simulation studies.

1.2 Main Model and its Assumptions

This subsection presents the main models and their assumptions. The following box contains the basic assumptions of the mathematical model:

Let $X_1,\ldots,X_n$ and $Y_1,\ldots,Y_n$ be a part of two real-valued time series $\{X_n\}_{n\in\mathbb{N}}$ and $\{Y_n\}_{n\in\mathbb{N}}$ on a probability space $(\Omega,\mathcal{F},P)$, which will have the following general design throughout the whole thesis:

$$\begin{pmatrix} X_i \\ Y_i \end{pmatrix} = \begin{pmatrix} Z_{1,i} \\ Z_{2,i} \end{pmatrix} = \begin{pmatrix} \mu_{1i} \\ \mu_{2i} \end{pmatrix} + B_i \begin{pmatrix} \tilde{\epsilon}_{1,i} \\ \tilde{\epsilon}_{2,i} \end{pmatrix} \quad \text{for } i \in \mathbb{N}, \tag{1.2.1}$$

where $\{\tilde{\epsilon}_{1,n}\}_{n\in\mathbb{N}}$ and $\{\tilde{\epsilon}_{2,n}\}_{n\in\mathbb{N}}$ are two centered and normalized random sequences with

$$\operatorname{Corr}(\tilde{\epsilon}_{1,i},\tilde{\epsilon}_{2,i}) = 0 \qquad \text{and} \qquad \Sigma_{XY,i} = B_i B_i^T = \begin{pmatrix} \sigma_{1,i}^2 & \sigma_{1,i}\sigma_{2,i}\rho_i \\ \sigma_{1,i}\sigma_{2,i}\rho_i & \sigma_{2,i}^2 \end{pmatrix} \tag{1.2.2}$$

for all $i \in \mathbb{N}$, where $\{\mu_{1,n}\}_{n\in\mathbb{N}}$, $\{\sigma_{1,n}^2\}_{n\in\mathbb{N}}$, $\{\mu_{2,n}\}_{n\in\mathbb{N}}$, $\{\sigma_{2,n}^2\}_{n\in\mathbb{N}}$, and $\{\rho_n\}_{n\in\mathbb{N}}$ are deterministic, uniformly bounded sequences, which represent the mean, variance and correlation of $\{X_n\}_{n\in\mathbb{N}}$ and $\{Y_n\}_{n\in\mathbb{N}}$, respectively. Furthermore, we assume that $\inf_n \sigma_{l,n} > 0$ for $l = 1,2$ and for all $n \in \mathbb{N}$.

Remark 1.2.1. *1. We use $Z_{1,n}$ and $Z_{2,n}$ instead of X_n and Y_n, respectively, to avoid repetitions if something holds for both.*

2. It holds that $\mu_{l,i} = \mathbb{E}[Z_{l,i}]$, $\sigma_{l,i}^2 = \operatorname{Var}[Z_{l,i}]$, and $\rho_i = \operatorname{Corr}(X_i,Y_i)$ for all $i \in \mathbb{N}$.

3. With $Z_i^{(0)} = (X_i - \mu_{1,i})(Y_i - \mu_{2,i})/(\sigma_{1,i}\sigma_{2,i})$ we obtain that $\rho_i = \mathbb{E}\left[Z_i^{(0)}\right]$. The variance of $Z_i^{(0)}$ depends on the matrix B_i, which is not unique. Notably for each $i \in \mathbb{N}$ there are infinitely many B_i fulfilling the conditions of the same model. This could provide structural breaks in the second moments of $Z_i^{(0)}$ and could influence our methods such that the fluctuation of B_i will be implicitly restricted by further assumptions.

A Posteriori Analysis One aim of this thesis is to present asymptotic tests and to prove their asymptotic distributions, where the tests decide for given samples whether the null hypothesis

$$H_0: \quad \rho_1 = \ldots = \rho_n \tag{H_0}$$

cannot be rejected or whether it is rejected in favor of an alternative. Firstly, we investigate a local alternative of the form:

Assumption $\mathbf{H_{LA}}$. *Suppose $\{\rho_n\}$ fulfills*

$$\rho_i = \rho_{i,n} = \rho_0 + \frac{1}{\sqrt{n}} g_\rho\left(\frac{i}{n}\right), \tag{1.2.3}$$

where $g_\rho : [0,1] \to \mathbb{R}$ is a bounded integrable function with

$$\sup_{z\in[0,1]} \left| n^{-1} \sum_{i=1}^{[nz]} g_\rho\left(\frac{i}{n}\right) - \int_0^z g_\rho(x)dx \right| = o(1), \quad \text{as } n \to \infty. \tag{1.2.4}$$

Secondly, we analyze the following (local) epidemic change-point setting:

Assumption $\mathbf{H_A}$. *Set $R_{k^*} = (k_1^*,k_2^*] \subset [1,n)$ and suppose $\{\rho_n\}$ fulfills for $i = 1,\ldots,n$*

$$\rho_i = \rho_{i,n} = \begin{cases} \rho, & i \notin R_{k^*}, \\ \rho + \Delta_\rho, & i \in R_{k^*}, \end{cases} \tag{1.2.5}$$

where $|\Delta_\rho| \lambda(R_{k^*}) \lambda(R_{k^*}^c) n^{-3/2} \to \infty$ as $n \to \infty$.

Remark 1.2.2. *1. The so-called change-points k_1^* and k_2^* depend on the sample size n, whereas the change size $\Delta_\rho \neq 0$ could depend on n. Furthermore, we denote $(k_1^*,k_2^*]$ to be the change-set. We will see that the assumed divergence of the combination between the change size and change location is necessary to obtain an asymptotic power one under $\mathbf{H_A}$. Heuristically, the change-sets must not be too small or too large compared with $[1,n]$.*

*2. The epidemic change-point setting goes back to **Levin and Kline (1985)** and is a generalization of the at most one change (AMOC) alternative, $\rho_1 = \ldots = \rho_{k^*} \neq \rho_{k^*+1} = \ldots = \rho_n$. If $k_1^* = 1$, we would get an AMOC alternative.*

After we have decided that the null hypothesis can be rejected, we want to know where the change-points are. Hence, the estimation of the change-points is another aim of this work. Moreover, we want to estimate the numbers and the locations of the unknown change-points in a multiple change-point setting.

Assumption $\mathbf{H_A^{(M)}}$. *Suppose $\{\rho_n\}$ fulfills for $i = 1,\ldots,n$*

$$\rho_i = \rho_0 + \sum_{j=1}^{R^*} \mathbb{1}_{\{k_j^*<i\}} \Delta_{\rho,j,n}, \tag{1.2.6}$$

where $R^* \in \mathbb{N}$, $1 = k_0^* < k_1^* < \ldots < k_{R^*}^* < k_{R^*+1}^* = n$, $\underline{\Delta}_\rho = \inf_{1\leq j} |\Delta_{\rho,j,n}| > 0$ for all $n \in \mathbb{N}$, and $k_j^*/n \to \theta_{\rho,j} \in [0,1]$ as $n \to \infty$ for all $j = 1,\ldots,R$. Furthermore, we set $\Delta_{k^*,r,n} = k_r^* - k_{r-1}^*$ and $\underline{\Delta}_{k^*,n} = \min_{1\leq r\leq R+1} \Delta_{k^*,r,n}$.

Remark 1.2.3. *We refer to k_j^* and $\theta_{\rho,j}$, $j = 1,\ldots,R$, as **change-points**.*

Summing up, we will deal with asymptotic tests which decide whether there is or is not a change in the correlation for two given time series. Additionally, we will present some change-point estimates. The following Figure 1 contains one example under the null hypothesis, H_0, and one for each presented alternatives above, $\mathbf{H_{LA}}$, $\mathbf{H_A}$, and $\mathbf{H_A^{(M)}}$. The figure shows cases where a change in the correlation is not really visible on the basis of the plotted processes X and Y. To illustrate the correlation between the two time series we added the exact correlations (in green) and the sliding window sample correlations (in black).

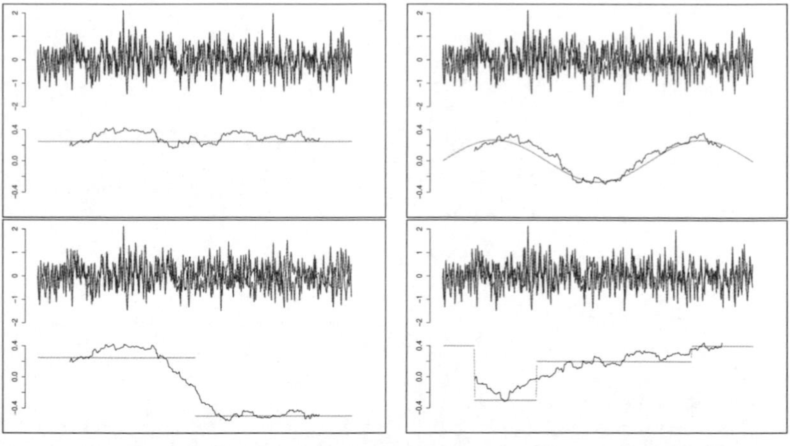

Figure 1: Each box contains a figure showing two graphs of a Moving-average (MA) process path, one red- and one blue-colored, and a second figure showing a green graph of the correlations between the two MA processes. The black graph shows the path of the sample correlation coefficient based on a sliding window method with a forward–backward window size of 50–50.

Sequential Analysis The previous analyses and estimations assume a given sample of the two time series $\{X_n\}$ and $\{Y_n\}$. But if they are not completely given and we receive the samples one by one, we want to test sequentially whether the correlation remains constant or not, i.e., we want to decide whether

$$H_0^{(2)}: \quad \rho_1 = \ldots = \rho_n = \rho_{n+1} = \ldots \tag{1.2.7}$$

cannot be rejected or whether the alternative

$$H_1^{(2)}: \quad \rho_1 = \ldots = \rho_n = \ldots = \rho_{k^*} \neq \rho_{k^*+1} = \ldots \tag{1.2.8}$$

is accepted. This procedure is called sequential change-point analysis. Here, we distinguish between **closed-end**, i.e., the observation will be finished after $n + [mn]$ observations with $m > 0$, and **open-end**, i.e., the observation will only be stopped if a change-point comes up.

Again, we are interested in a local setting:

Assumption $\mathbf{H_{LA}^{(c)}}$. *Suppose* $\{\rho_n\}$ *fulfills*

$$\rho_0 = \rho_1 = \ldots = \rho_n \quad \text{and} \quad \rho_i = \rho_{i,n} = \rho_0 + \frac{1}{\sqrt{n}} g_\rho\left(\frac{i}{n}\right) \mathbb{1}_{\{i>n\}}, \tag{1.2.9}$$

where g_ρ *is a bounded, integrable function on* $[1,1+m]$ *with*

$$\sup_{z\in[0,m]} \left| n^{-1} \sum_{i=n+1}^{n+[nz]} g_\rho\left(\frac{i}{n}\right) - \int_1^{1+z} g_\rho(x)dx \right| = o(1), \quad \text{as } n \to \infty. \tag{1.2.10}$$

In an AMOC model we assume:

Assumption $\mathbf{H_{A}^{(c)}}$. *Suppose* $\{\rho_n\}$ *fulfills*

$$\rho_i = \rho_{i,n} = \begin{cases} \rho, & i \in \{1, \ldots, n + [\theta_\rho^* n]\}, \\ \rho + \Delta_\rho, & i \in \{n + [\theta_\rho^* n], \ldots, n + [nm]\}, \end{cases} \tag{1.2.11}$$

where $0 < \theta_\rho^* < m$ *and* $\Delta_\rho \neq 0$.

Additionally, we consider the open-end local setting and an open-end AMOC model:

Assumption $\mathbf{H_{LA}^{(o)}}$. *Suppose* $\{\rho_n\}$ *fulfills*

$$\rho_i = \rho_{i,n} = \rho_0 + \frac{1}{\sqrt{n}} g_\rho\left(\frac{i}{n}\right) \mathbb{1}_{\{i>n\}}, \tag{1.2.12}$$

where g_ρ *is a bounded, integrable function on* $[1,\infty)$ *with*

$$\sup_{z\in[0,\infty)} \left| n^{-1} \sum_{i=n+1}^{n+[nz]} g_\rho\left(\frac{i}{n}\right) - \int_1^{1+z} g_\rho(x)dx \right| = o(1), \quad \text{as } n \to \infty. \tag{1.2.13}$$

Assumption $\mathbf{H_{A}^{(o)}}$. *Suppose* $\{\rho_n\}$ *fulfills*

$$\rho_i = \rho_{i,n} = \begin{cases} \rho, & i \in \{1, \ldots, n + [\theta^* n]\}, \\ \rho + \Delta_\rho, & i \in \{n + [\theta^* n], \ldots\} \end{cases} \tag{1.2.14}$$

with $0 < \theta^* < \infty$ *and* $\Delta_\rho \neq 0$.

In this setting we continue the observation until we detect a possible change-point. Mathematically, this procedure can be described by a stopping time. Hence, we are interested in the asymptotic behavior of stopping times.

Figure 2 illustrates the path of two MA processes where the observations until $n (= 200)$ are already observed and do not reject $H_0^{(2)}$. The data are analyzed to determine whether the correlation is changing or not. On the one hand, in a closed-end procedure we stop the analysis at $n + mn (= 500)$ if no change is detected in the correlation. On the other hand, an open-end procedure we only stop the analysis until we detect a change.

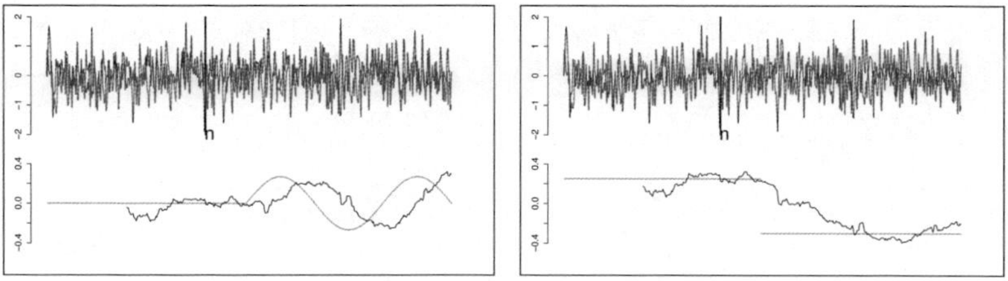

Figure 2: Each box contains a figure showing two graphs of a path of a Moving-average (MA) process, one red- and one blue-colored, and a second figure showing a green graph of the correlation coefficient between the two MA processes. The black graph shows the path of the sample correlation coefficient based on a sliding window method with a backward window of size 100.

1.3 Test Statistics and Stopping Times

For the **a posteriori** setting we construct asymptotic tests to decide whether there is a change in the correlation between two given time series $X_1,\dots,X_n$ and $Y_1,\dots,Y_n$ or not. This means that if some suitable detectors are larger than a critical value, we will reject the null hypothesis, otherwise we will not. Hence, we consider the following test

$$\phi^{\gamma}_{\iota,\psi,l}(X,Y) = \begin{cases} 1, & T_n^{\iota,\psi,l,\gamma} > c_{\alpha,\iota,\psi,l}, \\ 0, & T_n^{\iota,\psi,l,\gamma} \le c_{\alpha,\iota,\psi,l}, \end{cases} \tag{1.3.1}$$

where we call $T_n^{\iota,\psi,l,\gamma}$ a detector and $c_{\alpha,\iota,\psi,l}$ the critical value. The other indices will be specified later on. The test $\phi^{\gamma}_{\iota,\psi,l}(X,Y) = \phi^{\gamma}_{\iota,\psi,l}(X_1,\dots,X_n;Y_1,\dots,Y_n)$ is called consistent if it has asymptotic power one under an alternative, i.e., $\phi^{\gamma}_{\iota,\psi,l}(X_1,\dots,X_n;Y_1,\dots,Y_n) \xrightarrow{P} 1$, as $n \to \infty$.

Firstly, we consider the mean change model for the random variables $Z_1,\dots,Z_n$. For this testing problem the classical approach is based on the maximum-likelihood statistic, which depends on the distribution of the random variables. If we assume that the innovations are independent, identically N(0,1)–distributed, and that the change-point is known to be located at k^*, then the log-likelihood ratio will yield the detector

$$w(k^*,n)|\overline{Z}_{k^*} - \overline{Z}_n|,$$

where $\overline{Z}_k$ is the sample mean of the observations $Z_1,\dots,Z_k$ and $w(k^*,n)$ is a suitable weighting function. Since k^* is not known in general, we take the maximum over all possible k^*'s. Alternatively, we can transfer the previous form into a functional one, i.e., consider $w([n\cdot],n)|\overline{Z}_{[n\cdot]} - \overline{Z}_n|$ and apply some other continuous functional mapping from $D[0,1]$ on $\mathbb{R}$, where $D[0,1]$ is the Skorokhod space, i.e., the set of all càdlàg functions from $[0,1]$ to $\mathbb{R}$. More generally, we may consider a weighting map of the distance between two estimates, calculated by the observations $Z_1,\dots,Z_{[\cdot n]}$ and $Z_1,\dots,Z_n$. This concept of splitting the samples into the groups with indices from 1 to k, or $[n\cdot]$, and from $k+1$, or $[n\cdot]+1$, to n to compare the estimates will be used throughout the thesis.

If we adapt this detector to our change-point problem in the correlation, we get a similar detector of the form:

$$T_n = f\left(w([[n\cdot],n)|\hat{\rho}_{[n\cdot]} - \hat{\rho}_n|\right),$$

where $\hat{\rho}_k$ is some estimate for the correlation and $f : D[0,1] \to \mathbb{R}$ any suitable, continuous function. Wied et al. (2012) considered

$$f(g(\cdot)) = \sup_{z\in[0,1]} |g(z)|, \quad w(k,n) = \frac{k}{\sqrt{n}} D^{-\frac{1}{2}}, \quad \hat{\rho}_k = \begin{cases} 0, & \text{if } k < 2, \\ \frac{\sum_{i=1}^k (X_i - \overline{X}_k)(Y_i - \overline{Y}_k)}{\sqrt{\sum_{i=1}^k (X_i - \overline{X}_k)^2 \sum_{i=1}^k (Y_i - \overline{Y}_k)^2}}, & \text{if } k \ge 2, \end{cases}$$

where $D^{-1/2}$ is a certain unknown weighting factor which can be estimated and $\hat{\rho}_n$ is the well-known sample correlation coefficient. They proved that the test statistic converges under certain assumptions in distribution towards the maximum of a Brownian bridge: In particular, they assumed that the five dimensional vector $(X_i^2,Y_i^2,X_i,Y_i,X_iY_i)$ is L_2 near epoch dependent (see Def. on p. 41) with uniformly bounded rth moments ($r > 2$) and nearly constant first and second moments. They relied on a new functional Delta method and came up with the estimate $\hat{D}^{-\frac{1}{2}}$.
In Heuser (2013), we considered another weighting function, namely $w(k,n) = w_\gamma(k,n) = \frac{k}{\sqrt{n}}D^{-1/2}(k(n-k)/n^2)^{-\gamma}$ with some $\gamma \in [0,1/2)$.

In this thesis we consider the general statistics in an a posteriori model of the following form

$$T_n^{\iota,\psi,l,\gamma} = f_\iota(B_n^{\psi,l,\gamma}(\cdot)) \tag{1.3.2}$$

under more general assumptions where the basic assumptions are the following:

- ι is an index to distinguish functions $f_\iota : D[0,1] \to \mathbb{R}$ which are continuous with respect to $\|\cdot\|_{[0,1]}$ and fulfill the property that $f_\iota(g_1) \geq f_\iota(g_2)$ for all $g_1,g_2 \in D[0,1]$ with $|g_1(z)| \geq |g_2(z)|, \forall z \in [0,1]$. Here, we have the functions $f_1(g) = \sup_{z\in[0,1]}|g(z)|$ and $f_2(g) = \int_0^1 |g(z)|dz$ in mind;
- $B_n^{\psi,l,\gamma}(\cdot)$ is a process defined by

$$B_n^{\psi,l,\gamma}(z) = \hat{D}_l^{-1/2} w_\gamma\left(\frac{[zn]}{n}\right)\frac{[zn]}{\sqrt{n}}\left(\hat{\rho}_{\psi,[zn]} - \hat{\rho}_{\psi,n}\right); \tag{1.3.3}$$

- n is the sample size, which tends towards infinity;
- l is the index to distinguish the different real valued long–run variance estimates $\hat{D}$, where $l = 0$ denotes some suitable consistent LRV estimate;
- γ is a constant to specify a weighting function w_γ which allows us to highlight some special areas where we suppose a change-point;
- ψ is the index to distinguish the different correlation estimates $\hat{\rho}$.

In the **sequential** model we construct a stopping time to decide sequentially whether there is or is not a sudden change in the correlation between observations $X_1,\ldots,X_n$, $X_{n+1},\ldots$ and $Y_1,\ldots,Y_n,Y_{n+1},\ldots$ that come in bit by bit. Wied and Galeano (2013) have already considered the closed-end stopping time

$$\tau_n = \inf\left\{1 \leq k \leq nm \;:\; \hat{D}^{-1/2}w(k,n)\frac{k}{\sqrt{n}}|\hat{\rho}_{n+1}^{n+k} - \hat{\rho}_1^n| > c_{\alpha,\iota,\psi,l}\right\}$$

with

$$w(k,n) = \left(\max\left\{\epsilon,(1+k/n)\left(\frac{k}{n+k}\right)^\gamma\right\}\right)^{-1}, \quad \gamma \in [0,1/2),\ \epsilon > 0,$$

where $\hat{\rho}_i^k$, $k \geq 1+i$, is Pearson's correlation coefficient calculated by the observations $(X_i,Y_i),\ldots,(X_k,Y_k)$.

Relying on the same idea we get the following closed-end and open-end stopping times

$$\tau_{n,\iota,\psi,l,\gamma}^{(c)} = \inf\left\{0 \leq x \leq nm \;:\; u_\iota(\mathbb{1}_{\{\cdot\leq m\}}\tilde{B}_n^{\psi,l,\gamma}(\cdot))(x/n) > c_{\alpha,\iota,\psi,l}\right\} \tag{1.3.4}$$

and

$$\tau_{n,\iota,\psi,l,\gamma}^{(o)} = \inf\left\{0 \leq x \;:\; u_\iota(\tilde{B}_n^{\psi,l,\gamma}(\cdot))(x/n) > c_{\alpha,\iota,\psi,l}\right\}, \tag{1.3.5}$$

where $\inf\emptyset = \infty$ and

- ι is an index to distinguish functions $u_\iota : D[0,\infty) \to D[0,\infty)$ which are continuous with respect to $\|\cdot\|_{(0,\infty)}$ and fulfill
  1. $u_\iota(g_1)(z) \geq u_\iota(g_2)(z)$ for all $z \in [0,\infty)$ and all $g_1,g_2 \in D[0,\infty)$ with $|g_1(x)| \geq |g_2(x)|, \forall x \in [0,\infty)$;
  2. $u_\iota(\mathbb{1}_{\{\cdot \leq z\}} g(\cdot))(x) = \mathbb{1}_{\{x \leq z\}} u_\iota(g(\cdot))(x)$ for all $x \in [0,\infty)$, $g \in D[0,\infty)$ and $z \in \mathbb{R}_+$.

  Here, we have the functions $u_1(g)(x) = |g(x)|$ and $u_2(g)(x) = x^{-1}\int_1^{1+x} |g(z)|dz$ in mind;
- $\tilde{B}_n^{\psi,l,\gamma}(\cdot)$ is a process defined by

$$\tilde{B}_n^{\psi,l,\gamma}(z) = \begin{cases} \hat{D}_l^{-1/2} w_\gamma\left(\frac{[nz]}{n}\right) \frac{n}{n+[x]} \frac{[nz]}{\sqrt{n}} \left(\hat{\rho}_{\psi,[nz],n+1}^{n+[nz]} - \hat{\rho}_{\psi,[nz],1}^{n}\right), & \text{if } [nz] \geq 1, \\ 0, & \text{else;} \end{cases} \tag{1.3.6}$$

- w_γ, $\hat{\rho}$ and the indices ψ, l, and γ have the same meaning as in the a posteriori setting.

Remark 1.3.1. *With the different functions f_ι and u_ι we can regulate the sensitivity of the test, as well as the robustness. The above f_1 and u_1 produce a quite sensitive test, whereas the above f_2 and u_2 produce a test statistic which is rather robust against single outliers of $B_n^{\psi,l,\gamma}(\cdot)$ and $\tilde{B}_n^{\psi,l,\gamma}(\cdot)$, respectively. (Some other robust test are discussed on p. 8 of Dehling et al. (2015))*
Essentially, we will prove under the null hypotheses H_0 and $H_0^{(2)}$, as well as under the alternatives $\mathbf{H_{LA}}$, $\mathbf{H_{LA}^{(c)}}$, and $\mathbf{H_{LA}^{(o)}}$ that $B_n^{\psi,l,\gamma}(\cdot)$ and $\tilde{B}_n^{\psi,l,\gamma}([\cdot n])$ converge in distribution towards Gaussian processes on some functional space and apply the CMT to obtain the asymptotic distribution of $f_\iota(B_n^{\psi,l,\gamma})$ and $u_\iota(\tilde{B}_n^{\psi,l,\gamma})([\cdot n])$. However, it is sometimes possible to relax the assumptions of our model if we use functions f_ι and u_ι which results in a more robust method. For example, if we use $f_2(g) = \int_0^1 |g(x)|dx$, the convergence of $f_2(B_n^{\psi,l,\gamma} + \hat{R})$ will be independent of the error function $\hat{R}([n\cdot]/n)$ in case of $\sum_{i=1}^n |R(i/n)| = o_P(n)$, whereas for the case $f_1(g) = \sup_{x\in[0,1]} |g(x)|$, the error function must vanish uniformly in probability so that the error has no influence on the limit. However, in the case of f_2 we get that $B_n^{\psi,l,\gamma}(\cdot) + \hat{R}([n\cdot]/n)$ does not have to converge towards the limit of $B_n^{\psi,l,\gamma}(\cdot)$ in order to guarantee that the limits of the transformed terms are equal.

As a result, we consider many different test statistics, different stopping times and investigate their asymptotic behaviors. We focus our attention on the effect of the location parameters $\{\mu_{1,n}\}$ and $\{\mu_{2,n}\}$, as well as the effect of the variation parameters $\{\sigma_{1,n}^2\}$ and $\{\sigma_{2,n}^2\}$ of the observed time series $\{X_n\}$ and $\{Y_n\}$.

Remark 1.3.2. *In this thesis we focus on the one–dimensional change-point setting, i.e., we consider a change in the one–dimensional correlation. Naturally, it is possible to consider a d-dimensional random sequence $\{X_n\}$ with $d > 2$ and investigate whether there is a change in the $d\times d$-dimensional sequence of correlation matrices*

$$\Sigma_{X,\rho,n} = \begin{pmatrix} 1 & \rho_{12,n} & \cdots & \rho_{1d,n} \\ \rho_{21,n} & 1 & \cdots & \rho_{2d,n} \\ \vdots & & \ddots & \\ \rho_{d1,n} & \cdots & \rho_{dd-1,n} & 1 \end{pmatrix}.$$

Many of the following testing procedures can be extended without much effort by replacing the functional central limit theorem by a multidimensional one or the absolute value by supremum norm. Naturally, the assumptions of the alternatives $\mathbf{H_{LA}}$ and $\mathbf{H_A}$ have to be adapted, e.g. under $\mathbf{H_{LA}}$ the change function g_ρ maps from $[0,1]$ to $\mathbb{R}^{d(d-1)/2}$ and in equation (1.2.3) the absolute value is replaced by the Euclidean or the supremum norm. Additionally, the test statistics are based on functions f_ι mapping from $\mathcal{D}[0,1]^{d(d-1)/2}$ to $\mathbb{R}$.

2 Change-Point Analysis of the Correlations under Known Means and Variances

In addition to the main model, we assume in this section that the parameters μ_{1i}, μ_{2i}, σ_{1i}, and σ_{2i} are known. Then our estimate for the correlation in (1.3.3) is defined for $n \in \mathbb{N}$ as

$$\hat{\rho}_{0,n} = \frac{1}{n}\sum_{i=1}^{n} Z_i^{(0)} = \frac{1}{n}\sum_{i=1}^{n} \frac{(X_i - \mu_{i1})(Y_i - \mu_{i2})}{\sigma_{1i}\sigma_{2i}}, \tag{2.0.7}$$

and the sequential correlation estimates in (1.3.6) are defined as

$$\hat{\rho}_{0,k,l+1}^{l+k} = \frac{l+k}{k}\hat{\rho}_{0,l+k} - \frac{l}{k}\hat{\rho}_{0,l} \quad \text{for } l = 1,\ldots,n-1, \ k = 1,\ldots,n-l.$$

This section is divided into three subsections, where we consider well-known general results of the change-point analysis. Firstly, we focus on the asymptotic behavior of the test statistics with estimates for the change-points and for the long-run variance. Secondly, we consider the asymptotic behavior of the stopping times. Afterwards, we present some examples.

2.1 A Posteriori Analysis under General Dependency Framework

2.1.1 Testing under a Functional Central Limit Theorem

It is well-known that the asymptotic behavior of CUSUM test statistics is based on FCLTs, cf. Aue and Horváth (2013). In the following theorem we summarize the asymptotic behavior which we are interested in.

Theorem 2.1.1. *Let a Brownian motion $W(\cdot)$ and a standard Brownian bridge $B(\cdot)$ on $[0,1]$, $D > 0$, and*

$$\frac{1}{\sqrt{n}}\sum_{i=1}^{[n\cdot]}(Z_i^{(0)} - \rho_i) \overset{\mathcal{D}[0,1]}{\longrightarrow} D^{1/2}W(\cdot) \tag{2.1.1}$$

be given. Then, it holds
(i) under H_0 and $|\hat{D}_0 - D| = o_P(1)$ that

$$T_n^{0,0,0,0} \overset{\mathcal{D}}{\longrightarrow} f_0(B(\cdot)); \tag{2.1.2}$$

(ii) under Assumption $\mathbf{H_{LA}}$ and $|\hat{D}_0 - D| = o_P(1)$ that

$$T_n^{0,0,0,0} \overset{\mathcal{D}}{\longrightarrow} f_0(B(\cdot) + h(\cdot)), \tag{2.1.3}$$

where $h(z) = D^{-1/2}\left(\int_0^z g_\rho(x)dx - z\int_0^1 g_\rho(x)dx\right)$;
(iii) under Assumption $\mathbf{H_A}$ and for each finite $c \in \mathbb{R}^+$, each continuous $f_0 : \mathcal{D}[0,1] \to \mathbb{R}$ with $\lim_{\|x\|\to\infty} f_0(x) = \infty$, $|\hat{D}_0^{1/2}| = o_P(a_n|\Delta_\rho|\lambda(R_{k^*})\lambda(R_{k^*}^c)n^{-2})$, and $a_n \to \infty$ that

$$P\left(f_0\left(\frac{a_n}{\sqrt{n}}B_n^{0,0,0}\right) \geq c\right) \to 1. \tag{2.1.4}$$

Proof. See appendix (p. 177). □

Remark 2.1.2. *1. We obtain the convergence in (i) even if $\sum_{i=1}^n |\rho_i| = o(\sqrt{n})$.*

2. Choosing $f_0(g) = \sup_{h<z<1-h}|\frac{g(z+h)+g(z-h)-2g(z)}{h}|$, $h \in (0,\frac{1}{2})$ provides the well-known procedure of sliding windows with bandwidth $[nh]$. To avoid the search of an optimal bandwidth choice we could use $f_0(g) = \sup_z \int_\epsilon^{z\wedge(1-z)} |\frac{g(z+h)+g(z-h)-2g(z)}{h}|dh$ for an arbitrarily small $\epsilon > 0$. Furthermore, it is possible to choose $\epsilon = 0$ without any additional assumptions, for which the law of the iterated logarithm comes into account.

3. *Under Assumption* $\mathbf{H_A}$ *the conditions* $\lim_{\|x\|\to\infty} f_0(x) = \infty$ *and*

$$|\hat{D}^{1/2}| = o_P(a_n|\Delta_\rho|\lambda(R_{k^*})\lambda(R_{k^*}^c)n^{-2})$$

can be replaced by the conditions that f_0 *satisfies the triangle inequality and that*

$$f_0\left(\hat{D}^{-1/2}a_n\Delta_\rho\sqrt{n}\frac{n\lambda(R_{k^*}\cap(0,[n\cdot])) - [n\cdot]\lambda(R_{k^*})}{n^2}\right) \xrightarrow{P} \infty.$$

4. *More generally, assuming that* f_0 *fulfills the property of the triangle inequality,* H_0 *will be rejected with power one if*

$$\left\| f_0\left(\hat{D}_n^{-1/2}\frac{[n\cdot]}{\sqrt{n}}(\overline{\rho}_{[n\cdot]} - \overline{\rho}_n)\right)\right\| \xrightarrow{P} \infty, \qquad as\ n\to\infty.$$

5. *The proof makes it clear that each of the three claims holds true if we replace* $B_n^{0,0,0}$ *by* $B_n^{0,0,0} + o_P(\hat{D}_0)$. *In the preceding theorem, we indirectly assume that the variances of* $\{Z_n^{(0)}\}$ *are nearly constant such that the LRV* D *exists. This assumption can be weakened if we replace* $Z_i^{(0)}$ *by* $Z_i^{(0)}/\sigma_{Z^{(0)},i}$, *where* $\sigma^2_{Z^{(0)},i} = \operatorname{Var}\left[Z_i^{(0)}\right]$ *is not necessarily constant. Since this parameter is usually unknown, we have to estimate it. Then, the claimed convergences of Theorem 2.1.1 hold true under the replacement of* $\sigma_{Z^{(0)},i}$ *by* $\hat{\sigma}_{Z^{(0)},i}$ *if*

$$\left\|\sum_{i=1}^{[n\cdot]}\left(\frac{\sigma_{Z^{(0)},i}}{\hat{\sigma}_{Z^{(0)},i}} - 1\right)(Z_i^{(0)} - \rho_i)\right\| = o_P(n^{1/2}).$$

In the proof of Theorem 4.1.10 below, we will consider similar equations. There, we will see sufficient conditions for this equation to hold.

6. *The convergence of the test statistic in a multidimensional change-point setting can similarly be proven if we replace the one-dimensional FCLT by a multidimensional one.*

We are also interested in weighted test statistics, which allow to be more sensitive in areas where we expect a change. Assuming the weighting function $w(\cdot)$ to be continuous on $[0,1]$, we can just apply the CMT to obtain the weighted convergence. Thus, the limit would be $f_0(w_\gamma(\cdot)B(\cdot))$ or $f_0(w_\gamma(\cdot)(B(\cdot) + h(\cdot)))$. But to highlight the observation's start or end more prominently we may drop the continuity of w_γ in 0 or 1. To get weighted asymptotic results in this case, we pursue various approaches. One is to assume a FCLT rate, as Csörgö and Horváth (1997) do in their proof of Theorem 4.1.2. (ii). Another approach is to impose some additional assumptions on the random sequence, over which the partial sum is taken, e.g. the fulfillment of Kolmogorov-type inequalities. We consider three settings, where $\{S_n\}$ is the sequence of the partial sums of $\{Z_n\}$ with $S_0 = 0$ (a.s.):

Definition. *We say that a random sequence* $\{Z_n\}_{n\in\mathbb{N}}$ *fulfills the first Kolmogorov-type inequality for* $r > 0$ *if a constant* $C \in \mathbb{R}$ *exists such that for each* $n \in \mathbb{N}$, *and* $\eta > 0$

$$P\left(\max_{1\le k\le n}|S_k| \ge \eta\right) \le \frac{C}{\eta^r}\sum_{i=1}^{n}\alpha_i \qquad (\mathcal{K}_{\mathbf{r}}^{(\mathbf{1})})$$

holds true. Here $\{\alpha_n\}_{n\in\mathbb{N}}$ *is a non-negative, uniformly bounded sequence.*

Definition. *We say that a random sequence* $\{Z_n\}_{n\in\mathbb{N}}$ *fulfills the second Kolmogorov-type inequality for* $r > 0$ *if a constant* $C \in \mathbb{R}$ *exists such that for each* $n \in \mathbb{N}$, $m \in \{1,\ldots,n\}$, *and* $\eta > 0$

$$P\left(\max_{1\le k\le m}|S_n - S_{n-k}| \ge \eta\right) \le \frac{C}{\eta^r}\sum_{i=1}^{m}\alpha_{n,i} \qquad (\mathcal{K}_{\mathbf{r}}^{(\mathbf{2})})$$

holds true. Here $\{\alpha_{n,i}\}_{n\in\mathbb{N},i\in\{1,\ldots,n\}}$ *is a non-negative, uniformly bounded array.*

Definition. *We say that a random sequence* $\{Z_n\}_{n\in\mathbb{N}}$ *fulfills the third (shifted) Kolmogorov-type inequality for* $r>0$ *if a constant* $C\in\mathbb{R}$ *exists such that for each* $n,v\in\mathbb{N}$ *and* $\eta>0$

$$P\left(\max_{1\leq k\leq n}|S_{v+k}-S_v|\geq\eta\right)\leq\frac{C}{\eta^r}\sum_{i=1}^{n}\alpha_{i,v} \tag{$\mathcal{K}_\mathbf{r}^{(\mathbf{3})}$}$$

holds true. Here $\{\alpha_{n,v}\}_{n,v\in\mathbb{N}}$ *is a non-negative, uniformly bounded array.*

Remark 2.1.3. *1. Usually, the first Kolmogorov-type inequality is defined without the assumption of* α_n *being uniformly bounded. We add this assumption since we will only use the Kolmogorov-type inequality in this context. In the following, we will denote them* $(\mathcal{K}_\mathbf{r}^{(\mathbf{1})})$, $(\mathcal{K}_\mathbf{r}^{(\mathbf{2})})$, *and* $(\mathcal{K}_\mathbf{r}^{(\mathbf{3})})$.

2. *Tomacs and Líbor (2006) showed that* $(\mathcal{K}_\mathbf{r}^{(\mathbf{1})})$ *is equivalent to the Hájek–Rényi inequality*

$$P\left(\max_{1\leq k\leq m}\left|\frac{S_k}{\beta_k}\right|\geq\eta\right)\leq\frac{\tilde{c}}{\eta^r}\sum_{i=1}^{m}\frac{\alpha_i}{\beta_i^r} \tag{2.1.5}$$

for each $m\in\mathbb{N}$, *where* $\{\beta_n\}$ *is a non-decreasing sequence of positive constants,* $\eta,r,c,\tilde{c}>0$. *This statement is traceable to Fazekas and Klesov (2001), Theorem 1.1.*

3. *Similarly, it is possible to show that* $(\mathcal{K}_\mathbf{r}^{(\mathbf{2})})$ *is equivalent to*

$$P\left(\max_{1\leq k\leq m}\left|\frac{S_n-S_{n-k}}{\beta_k}\right|\geq\eta\right)\leq\frac{\tilde{c}}{\eta^r}\sum_{i=1}^{m}\frac{\alpha_{n,i}}{\beta_i^r} \tag{2.1.6}$$

where β_n *is a non-decreasing sequence of positive constants,* $\eta,r,c,\tilde{c}>0$ *and* $n-1\geq m$. *Moreover,* $(\mathcal{K}_\mathbf{r}^{(\mathbf{3})})$ *can also be extended to a Hájek–Rényi-type inequality.*

4. *When stating that "a* d*-dimensional random vector* $\{Z_n\}$ *fulfills a Kolmogorov-type inequality" we mean that each of the* d *components fulfills the Kolmogorov-type inequality.*

As described above, we are interested in weighted convergence with weighting functions which are noncontinuous in 0 and 1. The following sets of weighting functions are useful to point out which additional condition is sufficient in the various steps of the proof. Later, we will only use the last set, i.e., (2.1.9), of the following four sets of weighting functions for the a posteriori analysis. Basically, we assume that each weighting function is an element of

$$\mathcal{WF}=\Big\{w:(0,1)\to\mathbb{R}_{>0}\ :\ w\text{ is continuous, non-increasing in a neighborhood of zero}$$
$$\text{and increasing in a neighborhood of one}\Big\}.$$

The next set contains continuous weighting functions so that the maximum over $(0,1)$ of a weighted Brownian bridge, $w(\cdot)B(\cdot)$, exists a.s.

$$\widetilde{\mathcal{WF}}^{(A)}=\Big\{w\in\mathcal{WF}\ :\ w(t)=o(1/\sqrt{t\log\log(1/t)})=w(1-t),\text{as }t\to0\Big\}. \tag{2.1.7}$$

The next set of weighting functions contains those which enable us to work with the Hájek–Rényi-type inequality

$$\widetilde{\mathcal{WF}}^{(\gamma,r)}=\Big\{w:(0,1)\to\mathbb{R}_{>0}\ :\ w\text{ continuous, non-increasing in a neighborhood}$$
$$\text{of zero, increasing in a neighborhood of one, and it holds that}$$
$$n^{-r/2}\sum_{i=1}^{[nt]}w(i/n)^r\to0,\ n^{-r/2}\sum_{i=1}^{[nt]}w(1-i/n)^r\to0,\text{as }n\to\infty,t\to0\Big\}. \tag{2.1.8}$$

The last set contains those weighting functions which we will focus on, that is

$$\mathcal{WF}^{(\gamma)} = \Big\{ w_\gamma : (0,1) \to \mathbb{R}_{>0} \ : \ w_\gamma \text{ is continuous so that} \\ w_\gamma(t) = O(t^{-\gamma}) = w_\gamma(1-t), \text{as } t \to 0, \text{for some } \gamma \in \left[0,\frac{1}{2}\right) \Big\}. \tag{2.1.9}$$

It is obvious that $\mathcal{WF}^{(\gamma)} \subset \mathcal{WF}^{(\gamma,2)} \subset \widetilde{\mathcal{WF}}^{(A)}$ holds true. We will see that it is sufficient that, additionally, the Kolmogorov-type inequalities are fulfilled for $r \geq 2$ and $w_\gamma \in \mathcal{WF}^{(\gamma,2)}$ to ensure that the weighted test statistics converge towards a non-degenerate distribution.

Theorem 2.1.4. *Let* $w_\gamma \in \mathcal{WF}^{(\gamma)}$ *for some fixed* $\gamma \in [0,\frac{1}{2})$, *and let* $\{Z_n^{(0)} - \rho_n\}$ *fulfill* $(\mathcal{K}_{\mathbf{r}}^{(\mathbf{1})})$ *and* $(\mathcal{K}_{\mathbf{r}}^{(\mathbf{2})})$ *for* $r_z = 2$. *Then, under the assumption of Theorem 2.1.1 it holds*

(i) under H_0 *and* $|\hat{D}_0 - D| = o_P(1)$ *that*

$$T_n^{0,0,0,\gamma} \xrightarrow{\mathcal{D}} f_0(w_\gamma(\cdot)B(\cdot)); \tag{2.1.10}$$

(ii) under Assumption $\mathbf{H_{LA}}$ *and* $|\hat{D}_0 - D| = o_P(1)$ *that*

$$T_n^{0,0,0,\gamma}(\cdot) \xrightarrow{\mathcal{D}} f_0(w_\gamma(\cdot)(B(\cdot) + h(\cdot))) \tag{2.1.11}$$

with $h(z) = D^{-1/2}\left(\int_0^z g_\rho(x)dx - z\int_0^1 g_\rho(x)dx\right)$;

(iii) under Assumption $\mathbf{H_A}$ *and with*

$$|\hat{D}_0^{1/2}| = o_P\left(a_n|\Delta_\rho|\frac{\lambda(R_{k^*})}{n}\left(\left(\frac{k_1^*}{n}\right)^{1-\gamma}\left(\frac{n}{n-k_1^*}\right)^{\gamma} \vee \left(\frac{n-k_2^*}{n}\right)^{1-\gamma}\left(\frac{n}{k_2^*}\right)^{\gamma}\right)\right),$$

where $a_n \to \infty$, *that for each finite* $c \in \mathbb{R}^+$ *and each continuous* $f_0 : \mathcal{D}[0,1] \to \mathbb{R}$ *with* $\lim_{\|x\| \to \infty} f_0(x) = \infty$

$$P\left(f_0\left(\frac{a_n}{\sqrt{n}}B_n^{0,0,\gamma}\right) \geq c\right) \to 1.$$

Remark 2.1.5. *1. The remarks on Theorem 2.1.1 can be adapted to Theorem 2.1.4.*

2. The assumption that $\{Z_n^{(0)} - \rho_n\}$ *fulfills* $(\mathcal{K}_{\mathbf{r}}^{(\mathbf{1})})$ *and* $(\mathcal{K}_{\mathbf{r}}^{(\mathbf{2})})$ *can be reduced to:* $\{Z_1^{(0)}, \ldots, Z_{[n\epsilon]}^{(0)}\}$ *and* $\{Z_{n-[n\epsilon]}^{(0)}, \ldots, Z_n^{(0)}\}$ *fulfill* $(\mathcal{K}_{\mathbf{r}}^{(\mathbf{1})})$ *and* $(\mathcal{K}_{\mathbf{r}}^{(\mathbf{2})})$ *for an arbitrarily small* $\epsilon > 0$, *respectively. This allows for a stronger dependency in* $\{Z_{[n\epsilon]+1}^{(0)}, \ldots, Z_{n-[n\epsilon]}^{(0)}\}$.

We split the proof into several lemmas. In the following, we will call (2.1.14) the weighted functional convergence (WFC).

Lemma 2.1.6. *Suppose* $g \in \widetilde{\mathcal{WF}}^{(A)}$ *and let* $\{Z_n\}$ *fulfill the FCLT*

$$\frac{1}{\sqrt{n}}\sum_{i=1}^{[n\cdot]} Z_i \xrightarrow{D[0,1]} D^{1/2}W(\cdot). \tag{2.1.12}$$

If and only if

$$\lim_{\epsilon \to 0}\lim_{n \to \infty} P\left(\max_{\{k \leq n\epsilon\} \cup \{k \geq n - n\epsilon\}} \left|g(k/n)\frac{k}{\sqrt{n}}(\overline{Z}_k - \overline{Z}_n)\right| \geq \eta\right) = 0 \tag{2.1.13}$$

holds true for every $\eta > 0$, *we obtain*

$$g([n\cdot]/n)\frac{[n\cdot]}{\sqrt{n}}(\overline{Z}_{[n\cdot]} - \overline{Z}_n) \xrightarrow{D[0,1]} g(\cdot)\, D^{1/2}B(\cdot). \tag{2.1.14}$$

Proof. "$\Rightarrow$": The CMT yields that the process $B_n(\cdot) = \frac{[n\cdot]}{\sqrt{n}}(\overline{Z}_{[n\cdot]} - \overline{Z}_n)$ converges in distribution towards a Brownian bridge as $n \to \infty$. Since g is continuous on $[\epsilon,1-\epsilon]$ for every $\epsilon > 0$, the weighted process $g_n B_n = g([n\cdot]/n)B_n$ converges in distribution towards a weighted Brownian bridge $g(\cdot)B(\cdot)$ on $[\epsilon,1-\epsilon]$ by applying the CMT again. Define $B_n^\epsilon = g_n B_n \mathbb{1}_{[\epsilon,1-\epsilon]}$ and $B^\epsilon = gB\mathbb{1}_{[\epsilon,1-\epsilon]}$. Then, Billingsley (1968, Th. 4.2) completes this part of the proof since (2.1.13) holds and B^ϵ converges in probability towards the weighted Brownian bridge gB on $[0,1]$ as $\epsilon \to 0$. Here, we also use the law of iterated logarithm and $w(t) = o(1/\sqrt{t \log\log(1/t)}) = w(1-t)$ as $t \to 0$.
"$\Leftarrow$": Suppose (2.1.14) holds true, the CMT implies that for every $\eta > 0$

$$\lim_{\epsilon\to 0}\lim_{n\to\infty} P\left(\sup_{z\in[0,\epsilon]}\left|g([nz]/n)\frac{[nz]}{\sqrt{n}}(\overline{Z}_{[nz]} - \overline{Z}_n)\right| \geq \eta\right) \tag{2.1.15}$$

$$= \lim_{\epsilon\to 0} P\left(\sup_{z\in[0,\epsilon]}|g(z)B(z)| \geq \eta\right) = 0, \tag{2.1.16}$$

where the last equality follows from the law of the iterated logarithm. The convergence on $[1-\epsilon,1]$ can be shown in the same way as on $[0,\epsilon]$. □

Lemma 2.1.7. *Let the assumptions of Lemma 2.1.6 hold true and $g \in \mathcal{WF}^{(\gamma)}$ for a $\gamma \in [0,\frac{1}{2})$. Then the equation (2.1.13) can be reduced to*

$$\left\|\left(\frac{n}{[n\cdot]}\right)^\gamma \frac{1}{\sqrt{n}}\sum_{i=1}^{[n\cdot]} Z_i\right\|_{[0,\epsilon]} = o_p(1) \quad \text{and} \quad \left\|\left(\frac{n}{n-[n\cdot]}\right)^\gamma \frac{1}{\sqrt{n}}\sum_{i=1+[n\cdot]}^{n} Z_i\right\|_{[1-\epsilon,1]} = o_p(1) \tag{2.1.17}$$

as $n \to \infty$, followed by $\epsilon \to 0$.

Proof. Firstly, we obtain that

$$\begin{aligned}
&\max_{k\leq n\epsilon \cup k\geq n-n\epsilon}\left|g(k/n)\frac{k}{\sqrt{n}}(\overline{Z}_k - \overline{Z}_n)\right| \\
&\leq \max_{k\leq n\epsilon}\left|g(k/n)\frac{k}{\sqrt{n}}(\overline{Z}_k - \overline{Z}_n)\right| + \max_{k\geq n-n\epsilon}\left|g(k/n)\frac{k}{\sqrt{n}}(\overline{Z}_k - \overline{Z}_n)\right| \\
&\leq \max_{k\leq n\epsilon}\left|g(k/n)\frac{k}{n}\frac{1}{\sqrt{n}}\sum_{i=1+k}^{n} Z_i\right| + \max_{k\leq n\epsilon}\left|g(k/n)\frac{n-k}{n}\frac{1}{\sqrt{n}}\sum_{i=1}^{k} Z_i\right| \\
&\quad + \max_{k\geq n-n\epsilon}\left|g(k/n)\frac{k}{n}\frac{1}{\sqrt{n}}\sum_{i=1+k}^{n} Z_i\right| + \max_{k\geq n-n\epsilon}\left|g(k/n)\frac{n-k}{n}\frac{1}{\sqrt{n}}\sum_{i=1}^{k} Z_i\right|.
\end{aligned}$$

Since

$$\max_{k\leq n\epsilon}\left|g(k/n)\frac{k}{n}\right| = o(1), \qquad \max_{k\leq n\epsilon}\left|\frac{1}{\sqrt{n}}\sum_{i=1+k}^{n} Z_i\right| = O_P(1), \qquad \max_{n\geq k\geq n-n\epsilon}\left|g(k/n)\frac{n-k}{n}\right| = o(1),$$

and $\max_{n\geq k\geq n-n\epsilon}\left|\frac{1}{\sqrt{n}}\sum_{i=1}^{k} Z_i\right| = O_P(1)$ the first and the fourth summand is equal to $o_P(1)$ as $n \to \infty$, followed by $\epsilon \to 0$. The uniform boundedness of $g(k/n)(n-k)/n(k/n)^\gamma$ and $g(k/n)(k/n)(n/(n-k))^\gamma$ holds for each $k \leq n\epsilon$ and for $k \geq n - n\epsilon$, such that (2.1.13) holds if and only if (2.1.17) is fulfilled. □

Remark 2.1.8. *In the same way it can be proven that under the assumptions of Lemma 2.1.6 the equation (2.1.13) can be reduced to*

$$\left\|\frac{1}{\sqrt{\log(\log(n/[n\cdot]))}}\frac{1}{\sqrt{[n\cdot]}}\sum_{i=1}^{[n\cdot]} Z_i\right\|_{[0,\epsilon]} = O_p(1) \tag{2.1.18}$$

and

$$\left\| \frac{1}{\sqrt{\log(\log(1-n/[n\cdot]))}} \frac{1}{\sqrt{n-[n\cdot]}} \sum_{i=1+[n\cdot]}^{n} Z_i \right\|_{[1-\epsilon,1]} = O_p(1) \tag{2.1.19}$$

as first $n \to \infty$, *followed by* $\epsilon \to 0$.

Lemma 2.1.9. *Under the assumptions of Lemma 2.1.7, the equations in* (2.1.17) *are satisfied if* $(\mathcal{K}_\mathbf{r}^{(1)})$ *and* $(\mathcal{K}_\mathbf{r}^{(2)})$ *hold for some* $r \geq 2$.

Proof. Since the first Kolmogorov-type inequality $(\mathcal{K}_\mathbf{r}^{(1)})$ is satisfied, it follows from the Hájek–Rényi-type inequality (2.1.5), which equivalent to $(\mathcal{K}_\mathbf{r}^{(1)})$, that

$$P\left(\left\| \left(\frac{n}{[n\cdot]}\right)^{\gamma} \frac{1}{\sqrt{n}} \sum_{i=1}^{[n\cdot]} Z_i \right\|_{[0,\epsilon]} \geq \eta \right) = P\left(\max_{k\in[1,[n\epsilon]]} \frac{1}{k^\gamma} \left| \sum_{i=1}^{k} Z_i \right| \geq \eta n^{\frac{1}{2}-\gamma} \right)$$

$$\leq \frac{C}{\eta^r n^{r(\frac{1}{2}-\gamma)}} \sum_{i=1}^{[n\epsilon]} \frac{\alpha_i}{i^{\gamma r}} \leq \frac{C_1}{\eta^r n^{r(\frac{1}{2}-\gamma)}} \sum_{i=1}^{[n\epsilon]} \frac{1}{i^{\gamma r}} = \begin{cases} O(n^{-r(\frac{1}{2}-\gamma)}), & \text{for } \gamma r > 1, \\ O(n^{-r(\frac{1}{2}-\gamma)} \log(n\epsilon)), & \text{for } \gamma r = 1, \\ O(\epsilon^{1-\gamma r} n^{1-r\frac{1}{2}}), & \text{for } \gamma r < 1. \end{cases}$$

This implies that $\lim_{\epsilon\to 0} \lim_{n\to\infty} P\left(\left\| \left(\frac{n}{[n\cdot]}\right)^{\gamma} \frac{1}{\sqrt{n}} \sum_{i=1}^{[n\cdot]} Z_i \right\|_{[0,\epsilon]} \geq \eta \right) = 0$.
Since

$$P\left(\left\| \left(\frac{n}{n-[n\cdot]}\right)^{\gamma} \frac{1}{\sqrt{n}} \sum_{i=1+[n\cdot]}^{n} Z_i \right\|_{[1-\epsilon,1]} \geq \eta \right) = P\left(\max_{1\leq k\leq n-[n\epsilon]} \left| \frac{S_n - S_{n-k}}{k^\gamma} \right| \geq \eta n^{\frac{1}{2}-\gamma} \right),$$

it similarly follows that $\lim_{\epsilon\to 0} \lim_{n\to\infty} P\left(\left\| \left(\frac{n}{n-[n\cdot]}\right)^{\gamma} \frac{1}{\sqrt{n}} \sum_{i=1+[n\cdot]}^{n} Z_i \right\|_{[1-\epsilon,1]} \geq \eta \right) = 0$. □

Remark 2.1.10. *1. Under the assumptions of Lemma 2.1.6 the equations (2.1.18) and (2.1.19) are satisfied if* $(\mathcal{K}_\mathbf{r}^{(1)})$ *and* $(\mathcal{K}_\mathbf{r}^{(2)})$ *hold for* $r > 2$. *Note that this assumption contradicts the one of the FCLT. Hence, we will not consider* $\widetilde{\mathcal{WF}}^{(A)}$ *anymore.*

2. Analogously, we can prove

$$\max_{1\leq k\leq[n\epsilon]} n^{-1/2} w(k/n) |\sum_{i=1}^{k} Z_i| = O_P\left(n^{-1/2} \left(\sum_{i=1}^{[n\epsilon]} w(i/n)^r \right)^{1/r} \right) = o_P(1),$$

$$\max_{1\leq k\leq[n\epsilon]} \frac{w(1-k/n)}{\sqrt{n}} \left| \sum_{i=n-k}^{n} Z_i \right| = O_P\left(n^{-1/2} \left(\sum_{i=1}^{[n\epsilon]} w(1-i/n)^r \right)^{1/r} \right) = o_P(1)$$

as $n \to \infty$, *followed by* $\epsilon \to 0$ *for all* $w \in \widetilde{\mathcal{WF}}^{(\gamma,r)}$.

Proof of Theorem 2.1.4. The convergence under H_0 directly follows from the combination of Lemma 2.1.6, Lemma 2.1.7, and Lemma 2.1.9. Furthermore, it holds that $w_\gamma([n\cdot]/n)$ and $\frac{1}{n}\sum_{i=1}^{[n\cdot]} g_\rho([n\cdot]/n)$ converge towards $w_\gamma(\cdot)$ and $\int_0^{\cdot} g_\rho(x)dx$ uniformly on $[\epsilon,1-\epsilon]$, $\epsilon > 0$, respectively. In addition, $w_\gamma([n\cdot]/n)R_{n,\rho}(\cdot)$ converges uniformly towards zero on $[0,\epsilon]$ and $[1-\epsilon,1]$, as $n \to \infty$ and $\epsilon \to 0$. Hence, $w_\gamma([nz]/n)R_{n,\rho}(z)$ converges uniformly towards $w_\gamma(z)\left(\int_0^z g_\rho(x)dx - z\int_0^1 g_\rho(x)dx\right)$ on $[0,1]$ and we get the second result. In the third claim, we obtain that

$$\|w_\gamma([n\cdot]/n)R_{n,\rho}(\cdot)\| \geq |\Delta_\rho| \frac{\lambda(R_{k^*})}{\sqrt{n}} \left(\left(\frac{k_1^*}{n}\right)^{1-\gamma} \left(\frac{n}{n-k_1^*}\right)^{\gamma} \vee \left(\frac{n-k_2^*}{n}\right)^{1-\gamma} \left(\frac{n}{k_2^*}\right)^{\gamma} \right).$$

Thus, the assertion follows from using the same arguments. □

Remark 2.1.11. *Obviously, the asymptotic distribution of a test statistic in a multidimensional setting can similarly be proven.*

Misspecification of the Parameters So far, we have assumed that the exact values of all parameters $\mu_{1,n}$, $\mu_{2,n}$, $\sigma^2_{1,n}$, and $\sigma^2_{1,n}$ are known. But the question is, what happens if information about the true parameters is incorrect? The next lemma shows that there is some tolerance for misinformation in these parameters.

Lemma 2.1.12. *Let the assumptions of Theorem 2.1.1 hold true and set*

$$\mu_{l,i} = m_{l,i} + d_{\mu,l,i} \qquad \text{and} \qquad \sigma_{l,i} = s_{l,i} + d_{l,\sigma,i},$$

where $\inf_{i\in\mathbb{N}} s_{l,i} > \epsilon > 0$ *and* $\sum_{i=1}^n |d_i| = o(\sqrt{n})$ *for each* $\{d_i\} \in \{\{d_{1,\mu,i}\}, \{d_{2,\mu,i}\}, \{d_{1,\sigma,i}\}, \{d_{2,\sigma,i}\}\}$. *Then,* $Z_i = (X_i - m_{1,i})(Y_i - m_{2,i})(s_{1,i}s_{2,i})^{-1} - \rho_i$ *fulfills a FCLT. Thus, if we replace* $Z_i^{(0)}$ *by* Z_i, *the convergences of Theorem 2.1.1 hold true.*

Proof. Firstly, we note that

$$\begin{aligned}\frac{1}{\sqrt{n}}\sum_{i=1}^{[\cdot n]}\left(\frac{(X_i - m_{1,i})(Y_i - m_{2,i})}{s_{1,i}s_{2,i}} - \rho_i\right) &= \frac{1}{\sqrt{n}}\sum_{i=1}^{[\cdot n]}\left(Z_i^{(0)} - \rho_i\right)\\ &+ \frac{1}{\sqrt{n}}\sum_{i=1}^{[\cdot n]}\frac{d_{1,\mu,i}(Y_i - m_{2,i})}{s_{1,i}s_{2,i}} + \frac{1}{\sqrt{n}}\sum_{i=1}^{[\cdot n]}\frac{d_{2,\mu,i}(X_i - \mu_{1,i})}{s_{1,i}s_{2,i}}\\ &+ \frac{1}{\sqrt{n}}\sum_{i=1}^{[\cdot n]}\left(Z_i^{(0)} - \rho_i\right)\left[\left(\frac{\sigma_{1,i}}{s_{1,i}} - 1\right)\left(\frac{\sigma_{2,i}}{s_{2,i}} - 1\right) + \left(\frac{\sigma_{1,i}}{s_{1,i}} - 1\right) + \left(\frac{\sigma_{2,i}}{s_{2,i}} - 1\right)\right].\end{aligned}$$

Now, we show that each of the last three summands is of order $o_P(1)$. We use Markov's inequality and the uniform boundedness of the first moments:

$$\begin{aligned}P\left(\left\|\frac{1}{\sqrt{n}}\sum_{i=1}^{[\cdot n]}\frac{d_{1,\mu,i}(Y_i - m_{2,i})}{s_{1,i}s_{2,i}}\right\| \geq \eta\right) &\leq \frac{1}{\eta}\mathbb{E}\left[\left\|\frac{1}{\sqrt{n}}\sum_{i=1}^{[\cdot n]}\frac{d_{1,\mu,i}(Y_i - m_{2,i})}{s_{1,i}s_{2,i}}\right\|\right]\\ &\leq \frac{1}{\eta\sqrt{n}}\sum_{i=1}^{n}\frac{|d_{1,\mu,i}|\mathbb{E}\left[|Y_i - m_{2,i}|\right]}{s_{1,i}s_{2,i}} = o(1),\end{aligned}$$

as $n \to \infty$. Analogously, both other summands are of order $o_P(1)$. So the claim follows. □

Remark 2.1.13. *1. The errors* $\{d_i\}$ *have no effect on the asymptotic of the unweighted test statistics.*

2. *To get a WFC* ($\gamma > 0$), *we need to impose further dependence and moment assumptions on* $\{(X_n,Y_n)\}$ *or assumptions on the* $\{d_{l,i}\}$. *One option is to assume that* $\sum_{i=1}^n i^{-\gamma}|d_i| = o(n^{1/2-\gamma})$ *so that we can estimate the weighted error terms in the following way:*

$$\left\|\left(\frac{n}{[\cdot n]}\right)^{\gamma}\frac{1}{\sqrt{n}}\sum_{i=1}^{[\cdot n]}\frac{d_{1,\mu,i}(Y_i - m_{2,i})}{s_{1,i}s_{2,i}}\right\| \leq n^{\gamma-\frac{1}{2}}\sum_{i=1}^{n}\left|\frac{d_{1,\mu,i}(Y_i - m_{2,i})}{i^{\gamma}s_{1,i}s_{2,i}}\right| = o_P(1).$$

The first step follows from the triangle inequality. In the second step we use Markov's inequality and the uniform boundedness of the moments and of the parameter sequences.

Another option is to assume that the sequences $d_{l,\cdot,i}$ have no influence on the fulfillment of the Kolmogorov-type inequalities:

Lemma 2.1.14. *Let the assumptions of Theorem 2.1.4 and Lemma 2.1.12 hold true. Moreover, for each* $\{d_iZ_i\}$ *with* $\{Z_i\} \in \{\{\epsilon_{1,i}\},\{\epsilon_{2,i}\},\{Z_i^{(0)} - \rho_i\}\}$ *and some sequence* $\{d_i\}$ *with* $\sum_{i=1}^n |d_i| = o(\sqrt{n})$ *let* $(\mathcal{K}_{\mathbf{r}}^{(\mathbf{1})})$ *and* $(\mathcal{K}_{\mathbf{r}}^{(\mathbf{2})})$ *be satisfied for* $r = 2$. *Then, the convergences of Theorem 2.1.4 hold true.*

Proof. It is clear that the three additive error summands of the proof of Lemma 2.1.12 weighted by $(n/[\cdot n])^\gamma$ are of the order $o_P(1)$. Each of these can be divided into sums of type $\frac{1}{C}\left(\frac{n}{[n\cdot]}\right)^\gamma \frac{1}{\sqrt{n}}\sum_{i=1}^{[\cdot n]} d_i Z_i$, where $\{Z_i\} \in \{\{\epsilon_{1,i}\},\{\epsilon_{2,i}\},\{Z_i^{(0)}-\rho_i\}\}$, $C \in \{\sigma_1,\sigma_2,\sigma_1\sigma_2\}$, and

$$\{d_i\} \in \{\{d_{1,\mu,i}\},\{d_{2,\mu,i}\},\{d_{1,\sigma,i}\},\{d_{2,\sigma,i}\},\{d_{1,\mu,i}d_{2,\mu,i}\},\{d_{1,\sigma,i}d_{2,\sigma,i}\},\{d_{1,\sigma,i}d_{2,\mu,i}\},\{d_{1,\mu,i}d_{2,\sigma,i}\}\}.$$

Hence, it is sufficient to verify that for each combination it holds, on the one hand, that

$$\left\|\frac{1}{C}\left(\frac{n}{[n\cdot]}\right)^\gamma \frac{1}{\sqrt{n}}\sum_{i=1}^{[\cdot n]} d_i Z_i\right\|_{[0,\epsilon]} = o_P(1)$$

and, on the other hand, that

$$\left\|\frac{1}{C}\left(\frac{n}{n-[n\cdot]}\right)^\gamma \frac{1}{\sqrt{n}}\sum_{i=[\cdot n]+1}^{n} d_i Z_i\right\|_{[1-\epsilon,\epsilon]} = o_P(1)$$

as $n \to \infty$, followed by $\epsilon \to 0$. Both relations are fulfilled because of the Kolmogorov-type inequalities. The result now follows. □

2.1.2 Change-Point Estimation

This sub-subsection presents a change-point estimate under the epidemic change-point model and we prove some rates for the speed of convergence. Since ρ_i is the expectation of $Z_i^{(0)}$, a change-point estimate in a correlation model is similar to a change-point estimate in a "change in the mean" model, for which many different results are known, cf. Bai (1994), Lombard and Hart (1994), Bai (1997), Bai and Perron (1998), Lavielle and Moulines (2000), Qu and Perron (2007), and for a survey we refer to Aue and Horváth (2013).

In an AMOC model and under the assumptions of Theorem 2.1.1 we already know that $B_n^{0,0,0}(\cdot) = B_n(\cdot) + R_{n,\rho}(\cdot)$, where B_n converges towards a Brownian bridge and $R_{n,\rho}$ satisfies that

$$\hat{D}_0^{1/2}\frac{|R_n(\cdot)|}{\sqrt{n}|\Delta_n|} \to |(\cdot - \theta_\rho)^+ - \cdot(1-\theta_\rho)|$$

uniformly on $[0,1]$. Here, $\theta_\rho \in (0,1)$ is the change-point. Since the function on the right–hand side is continuous and its maximum is at θ_ρ, it remains to show that $B_n = o_P(\hat{D}^{1/2}(\sqrt{n}|\Delta_n|)^{-1})$ to get with the help of Kim and Pollard (1990, Th. 2.7) that

$$\arg\max |B_n^{0,0,0}| \xrightarrow{P} \theta. \tag{2.1.20}$$

In the following, we are interested in the convergence rates of the estimates, i.e., we look for sequences $a_n \to \infty$ such that

$$a_n|\hat{\theta} - \theta| = O_P(1).$$

Epidemic Change-Points Firstly, we consider the slightly more general model, the epidemic change-point setting, cf. Yao (1993), Hušková (1995), or Antoch and Hušková (1996). For the sake of completeness we prove the following result, which is based on the proof of Bucchia and Heuser (2015, Th. 4.1), which is stated under the more general setting of multidimensional panel data.

Theorem 2.1.15. *Under Assumption* $\mathbf{H_A}$ *set* $(k_1^*,k_2^*) = ([n\theta_1^*],[n\theta_2^*])$ *with* $0 < \theta_1^* < \theta_2^* < 1$. *Additionally, let* $\{Z_n^{(0)} - \rho_n\}$ *fulfill* $(\mathcal{K}_\mathbf{r}^{(\mathbf{2})})$ *and* $(\mathcal{K}_\mathbf{r}^{(\mathbf{3})})$ *for* $r_z > 1$. *Then, it holds that*

$$n\|\tilde{\theta} - \theta^*\| = O_P(1), \tag{2.1.21}$$

where $\tilde{\theta} = (\tilde{\theta}_1,\tilde{\theta}_2) \in \arg\max\{\tilde{Q}_n(s,t) \ : \ 0 \leq s < t \leq 1\}$ *with*

$$\tilde{Q}_n(s,t) = \left(\sum_{i=1+[ns]}^{[nt]} (Z_i^{(0)} - \overline{Z^{(0)}}_n) \right) \left(\sum_{i=1+[ns]}^{[nt]} (Z_i^{(0)} - \overline{Z^{(0)}}_n) \right). \tag{2.1.22}$$

Proof. See appendix (p. 177). □

If we additionally allow that the change-points are local, we obtain the following result.

Theorem 2.1.16. *Under the assumption of Theorem 2.1.15 with* $\Delta_{\rho,n}^{-1} = o(n^{-(r_z-1)/r_z})$*, it holds that*

$$n\Delta_{\rho,n}\|\tilde{\theta} - \theta^*\| = O_P(1). \tag{2.1.23}$$

Proof. See appendix (p. 179). □

Remark 2.1.17. *The preceding theorem shows that the local epidemic change-points can be estimated with some rate if the change size does not vanish "too fast", which is the case under Assumption* $\mathbf{H_{LA}}$ *(*$\Delta_{\rho,n}^{-1} = O(n^{1/2})$*). In the case of* $r_z = 2$ *and a local change vanishing slightly more "slowly", we will still get some estimation rate.*
Furthermore, it is possibly to extend these results by allow large or small change sets, cf. Hušková (1995). Thereby, rate conditions of a combination of $\Delta_{\rho,n}$, k_1^*, $n - k_2^*$, $k_2^* - k_1^*$, *and* $n - (k_2^* - k_1^*)$ *are necessary.*

Multiple Change-Points In the following we are interested in the more general multiple change-point setting, Assumption $\mathbf{H_A^{(M)}}$.

Firstly, we assume that the number of abrupt changes $R^* = R \in \mathbb{N}_{>0}$ is known and we want to estimate the change-points $\theta_1,\ldots,\theta_R$. To estimate these we use the well-known least square estimate and define

$$(\hat{k}_1^{(\psi)},\ldots,\hat{k}_R^{(\psi)}) \in \arg\min\left\{Q_n^{(\psi)}(k_1,\ldots,k_R) \ : \ 1 = k_0 < k_1 < \ldots < k_{R+1} = n\right\} \tag{2.1.24}$$

with

$$Q_n^{(\psi)}(k_1,\ldots,k_R) = \sum_{r=1}^{R+1} \mathbb{1}_{\{k_r > k_{r-1}\}} \sum_{i=k_{r-1}+1}^{k_r} (Z_i^{(\psi)} - \overline{Z^{(\psi)}}_{k_{r-1}}^{k_r})^2, \quad \psi = 0,1,\ldots \tag{2.1.25}$$

which has already been considered with $\psi = 0$, for instance by Lavielle and Moulines (2000) in a modified form. These authors use the assumption that the distances between two change-points have a rate of n to construct their estimates:

$$(k_1',\ldots,k_R') \in \arg\max\left\{Q_n^{(0)}(k_1,\ldots,k_R) \ : \ k_r - k_{r-1} \geq n\Delta_n\right\},$$

where $\Delta_n \to 0$ as $n \to \infty$. Note that if we define $\Delta_n = 2/n$, we will get the same estimates as for $\hat{k}_1,\ldots,\hat{k}_R$. However, in this case the results of Lavielle and Moulines (2000) do not provide the possible estimation rate of $|\hat{k}_r - k_r^*| = O_P(1)$.

Before we consider the general case we want to keep an eye on the special case of estimating one change-point which has already been considered by Bai (1994). Here, we additionally investigate the influence of the positions and the sizes of the structural breaks.

Theorem 2.1.18. *Under* $\mathbf{H_A^{(M)}}$ *set* $R = 1$, *define* $k^* = k_{1,n}^*$, $\Delta_{\rho,n} = \Delta_{\rho,1,n}$, *and assume*

$$|\Delta_{\rho,n}|^{-\frac{2r_z}{r_z-1}} = O\left(k^* \vee (n - k^*)\right). \tag{2.1.26}$$

If $\{Z_n^{(0)} - \rho_n\}$ *fulfills* $(\mathcal{K}_\mathbf{r}^{(1)})$ *and* $(\mathcal{K}_\mathbf{r}^{(2)})$ *for some* $r_z > 1$, *then it holds that*

$$|\hat{k}_1^{(0)} - k^*| = O_P\left(|\Delta_{\rho,n}|^{-\frac{2r_z}{r_z-1}}\right). \tag{2.1.27}$$

Proof. See appendix (p. 179). □

Remark 2.1.19. *Under* H_0 *we obtain that*

$$\begin{aligned}\arg\min Q_n^{(0)}(k) &= \arg\max\left[k^{-1}\left(\sum_{i=1}^{k}(Z_i^{(0)}-\rho)\right)^2+(n-k)^{-1}\left(\sum_{i=k+1}^{n}(Z_i^{(0)}-\rho)\right)^2\right]\\ &= \arg\max\frac{n}{k(n-k)}\left(\sum_{i=1}^{k}(Z_i^{(0)}-\rho)-\frac{k}{n}\sum_{i=1}^{n}(Z_i^{(0)}-\rho)\right)^2,\end{aligned}$$

where the right-hand side has already been investigated by several authors, cf. Lombard and Hart (1994), Ferger (1994), or Ferger (2001). Under suitable assumptions the estimate $\hat{k}/n$ *converges in distribution towards a Bernoulli distributed random variable* Z *with* $P(Z=0)=P(Z=1)=\frac{1}{2}$, *cf. Ferger (2001, Th. 2.7).*

In the following theorem we consider the behavior of the estimate sequentially calculated. More precisely, in a sequential procedure we obtain the data one by one. After a stopping time exceeds the critical value and asserts that there is a break in structure, we want to estimate the location of this change. Without stopping our observations and assuming that no second change–point will occur, we use each new observation to estimate the change-point again and again. In the next theorem, we will see that the maximal estimation error of all calculations after a certain period of time $b_{n,m}$ after the change-point k^* is of the same rate as in the above one-time estimation.

Theorem 2.1.20. *Set* $b_{n,m}=m|\Delta_{\rho,n}|^{-\frac{2r_z}{r_z-1}}$. *Then, under the assumptions of Theorem 2.1.18 it holds that*

$$\max_{k^*_{1,n}+b_{n,m}\le N\le n}|\hat{k}_{1,N}-k^*_{1,n}|=O_P\left(|\Delta_{\rho,n}|^{-\frac{2r_z}{r_z-1}}\right),\quad \textit{as } n\to\infty, \textit{ followed by } m\to\infty.$$

Proof. Firstly, for $a_{n,M}=b_{n,M}$ we obtain that

$$\begin{aligned}&P(\max_{k^*_{1,n}+b_{n,m}\le N\le n}|\hat{k}_{1,N}-k^*_{1,n}|\ge a_{n,M}+1)\\ &\le P\left(\bigcup_{k^*_{1,n}+b_{n,m}\le N\le n}\left\{\min_{k;|k-k^*|\ge a_{n,M}}Q_N^{(0)}(k)\le Q_N^{(0)}(k^*)\right\}\right)\end{aligned}$$

and with $\tilde{Z}_i=Z_i^{(0)}-\rho_i$ that

$$\begin{aligned}&Q_N^{(0)}(k)-Q_N^{(0)}(k^*)\\ &=\Delta^2_{\rho,n}(k^*-k)\frac{N-k^*}{N-k}\left(1+\frac{\Delta^{-2}_{\rho,n}}{k^*-k}\sum_{i=k+1}^{k^*}\tilde{Z}_i\rho_i+\frac{\Delta^{-2}_{\rho,n}}{N-k^*}\sum_{i=k^*+1}^{N}\tilde{Z}_i\rho_i\right)\mathbb{1}_{\{k<k^*\}}\\ &\quad+\Delta^2_{\rho,n}(k-k^*)\frac{k^*}{k}\left(1+\frac{\Delta^{-2}_{\rho,n}}{k^*}\sum_{i=1}^{k^*}\tilde{Z}_i\rho_i+\frac{\Delta^{-2}_{\rho,n}}{k-k^*}\sum_{i=k^*+1}^{k}\tilde{Z}_i\rho_i\right)\mathbb{1}_{\{k\ge k^*\}}.\end{aligned}$$

Since the Hájek–Rényi–type inequality implies that

$$\max_{k^*_{1,n}+b_{n,m}\le N\le n}\left|\frac{1}{N-k^*}\sum_{i=k^*+1}^{N}\tilde{Z}_i\rho_i\right|=O_P\left(\left(\sum_{k=b_{n,m}}^{n-k*}k^{-r_z}\right)^{1/r_z}\right)=O_P(b_{n,m}^{-(r_z-1)/r_z})$$

as $n\to\infty$, followed by $m\to\infty$, we obtain with the arguments used in the proof of Theorem 2.1.18 that the content of the previous brackets is asymptotically positive, as $n\to\infty$, $m\to\infty$, and $M\to\infty$. Hence, the claim follows. □

Now, we focus on the multiple change-point estimation. Therefore, we need the following technical Lemma.

Lemma 2.1.21. *Define a class of step-functions with* m *jumps on* $[1,n]$, $n>1$, *by*

$$D_{m,n}=\Big\{g:[1,n]\to\mathbb{R}\ :\ g(x)=\rho_0+\sum_{j=1}^{m}\Delta_j^{(m)}\mathbb{1}_{\{k_j^{(m)}\le[x]\}},\ \text{with some}\ k_i^{(m)}\in\mathbb{N}$$
$$\text{such that}\ 1=k_0^{(m)}<k_1^{(m)},\ldots<k_m^{(m)}<k_{m+1}^{(m)}=n,\ \inf_{1<j}|\Delta_j^{(m)}|>0\Big\}.$$

Then, it holds for any $m\in\mathbb{N}$, $m<n$ *that for all* $g_1\in D_{m,n}$ *and* $g_2\in\bigcup_{i=1}^{m-1}D_{i,n}$

$$\int_1^n(g_1(x)-g_2(x))^2dx\ge\frac{1}{2}\min_{1\le i\le m}(\Delta_i^{(m)})^2\min_{1\le i\le m+1}(k_i^{(m)}-k_{i-1}^{(m)}).$$

Proof. See appendix (p. 180). □

Theorem 2.1.22. *Under Assumption* $\mathbf{H_A^{(M)}}$ *let* $\{Z_n^{(0)}-\rho_n\}$ *fulfill* $(\mathcal{K}_\mathbf{r}^{(1)})$ *and* $(\mathcal{K}_\mathbf{r}^{(2)})$ *for* $r_z>1$ *and let their* r*th* $(r>2)$ *moments be uniformly bounded. In addition, let*

$$n^{1/r_z+1/r}=o(\min_i\Delta_{\rho,i,n}^2\underline{\Delta}_{k^*,n}) \tag{2.1.28}$$

as $n\to\infty$. *Then, it holds that*

$$\max_{1\le j\le R}a_n|\theta_{\rho,j}-\hat{\theta}_{\rho,j}^{(0)}|=O_P(1),$$

where $\theta_{\rho,j}=\lim_{n\to\infty}k_j^*/n$, $\hat{\theta}_j^{(0)}=\hat{k}_j^{(0)}/n$, *and*

$$a_n=n\left[\left(\min_{1\le i\le R}\Delta_{\rho,i}^2/\max_{1\le i\le R}|\Delta_{\rho,i}|\right)^{r_z/(r_z-1)}\wedge\left(\min_i\Delta_{\rho,i,n}^2/\max_r|\Delta_{k^*,r,n}|^{2/r_z-1}\right)\right]. \tag{2.1.29}$$

Remark 2.1.23. *1. The above theorem contains some scenarios where, on the one hand, the change sizes may asymptotically vanish and, on the other hand, the asymptotic change-points* $\theta_1,\ldots,\theta_R$ *do not have to be different.*

2. *If the number of change-points* R *is misspecified and is expected to be too high, then there are* R *change-point estimates which satisfy* $a_n|\theta_{\rho,j}-\hat{\theta}_{\rho,j}|=O_P(1)$. *Otherwise, if the number of change-points is expected to be too low, the change-point estimates approximate a subset of the change-points depending on* $k_j^*-k_{j-1}^*$ *and* $\Delta_{\rho,j}$ *such that* $a_n|\theta_{\rho,j}-\hat{\theta}_{\rho,j}|=O_P(1)$.

3. *The second condition of the preceding theorem can be replaced by*

$$\max_{1\le k_1<k_2\le n}\frac{1}{k_2-k_1}\left(\sum_{i=k_1+1}^{k_2}(Z_i^{(0)}-\rho_i)\right)^2=o_P(\min_i\Delta_{\rho,i,n}^2\underline{\Delta}_{k^*,n}).$$

4. *If* $\min_{i,n}|\Delta_{\rho,i,n}|\ge\epsilon>0$, $\underline{\Delta}_{k^*,n}\sim n$, $r>2$, *and* $r_z=2$, *then we can choose* $a_n=n$.

Proof of Theorem 2.1.22. Define $a_{1,n,N}=Nn/a_n$.
Firstly, we obtain with $k_0=0$, $k_{R+1}=n$, and with some $r_0\in\{1,\ldots,R\}$ that

$$P(a_n|\hat{\theta}_{r_0}-\theta_{r_0}|\ge N+1)\le P\left(\min_{1<k_1<\ldots<k_R<n;|k_{r_0}-k_{r_0}^*|\ge a_{1,n,N}}Q_n^{(0)}(k)\le Q_n^{(0)}(k^*)\right).$$

Set $\tilde{Z}_i^{(0)} = Z_i^{(0)} - \rho_i$. Then, we get

$$Q_n^{(0)}(k) - Q^{(0)}(k^*) = \sum_{r=1}^{R+1} \sum_{i=k_{r-1}+1}^{k_r} \left[2\left(\rho_i - \overline{\rho}(k_{r-1},k_r)\right) \tilde{Z}_i^{(0)} + \left(\rho_i - \overline{\rho}(k_{r-1},k_r)\right)^2 \right]$$
$$+ \sum_{r=1}^{R+1} \left[(k_r^* - k_{r-1}^*) \left(\overline{\tilde{Z}^{(0)}}(k_{r-1}^*,k_r^*) \right)^2 - (k_r - k_{r-1}) \left(\overline{\tilde{Z}^{(0)}}(k_{r-1},k_r) \right)^2 \right],$$

where we will show that $\sum_{r=1}^{R+1} \sum_{i=k_{r-1}+1}^{k_r} (\rho_i - \overline{\rho}(k_{r-1},k_r))^2$ is the dominating summand. Furthermore, we obtain that $\sum_{r=1}^{R+1} (k_r^* - k_{r-1}^*) \left(\overline{\tilde{Z}^{(0)}}(k_{r-1}^*,k_r^*) \right)^2$ is non-negative.
Secondly, we define

$$g_\rho : [1,n] \to [-1,1],\ x \mapsto \rho_{[x]} \quad \text{and} \quad g_{\overline{\rho},k} : [1,n] \to [-1,1],\ x \mapsto \sum_{r=1}^{R+1} \mathbb{1}_{[x] \in (k_{r-1},k_r]} \overline{\rho}(k_{r-1},k_r)$$

and obtain that

$$\sum_{r=1}^{R+1} \sum_{i=k_{r-1}+1}^{k_r} (\rho_i - \overline{\rho}(k_{r-1},k_r))^2 = \int_1^n (g_\rho(x) - g_{\overline{\rho},k}(x))^2 dx.$$

Furthermore, it holds for each fixed $r^* \in \{1,\dots,R\}$ with $k_{r^*} \neq k_{r^*}^*$ that:

1. If $k_{r^*} \in [1,k_{r*}^* - 1]$, then g_ρ and $g_{\overline{\rho},k}$ are step-functions with at least $R - r^* + 1$ and exactly $R - r^*$ many steps on $[k_{r^*},n]$, respectively.

2. If $k_{r^*} \in [k_{r*}^* + 1,n]$, then g_ρ and $g_{\overline{\rho},k}$ are step-functions with at least r^* and exactly $r^* - 1$ many steps on $[1,k_{r^*}]$, respectively.

In the first case it holds with Lemma 2.1.21 that

$$\int_1^n (g_\rho(x) - g_{\overline{\rho},k}(x))^2 dx \geq \int_{k_{r^*}}^n (g_\rho(x) - g_{\overline{\rho},k}(x))^2 dx$$
$$\geq \frac{1}{2} \left((k_{r^*}^* - k_{r^*}) \wedge \min_{r^* \leq i \leq R} (k_{i+1}^* - k_i^*) \right) \min_{r^* \leq i \leq R} \Delta_{\rho,i,n}^2,$$

and in the second case that

$$\int_1^n (g_\rho(x) - g_{\overline{\rho},k}(x))^2 dx \geq \int_1^{k_{r^*}} (g_\rho(x) - g_{\overline{\rho},k}(x))^2 dx$$
$$\geq \frac{1}{2} \left((k_{r^*}^* - k_{r^*}) \wedge \min_{1 \leq i \leq r^*} (k_i^* - k_{i-1}^*) \right) \min_{1 \leq i \leq r^*} \Delta_{\rho,i,n}^2.$$

Hence, we get for $\Delta_{k^*,i,n} = k_i^* - k_{i-1}^*$ that

$$\int_1^n (g_\rho(x) - g_{\overline{\rho},k}(x))^2 dx \geq \frac{1}{2} \left(\|k^* - k\| \wedge \underline{\Delta}_{k^*,n} \right) \min_{1 \leq i \leq R} \Delta_{\rho,i,n}^2.$$

Thus if there is a $r^* \in \{1,\dots,R\}$ and an $\epsilon > 0$ so that $|k_{r^*} - k_{r^*}^*| \geq \epsilon \underline{\Delta}_{k^*,n} \vee a_{1,n,N}$, then it is quite clear that

$$\max_k \left| \sum_{r=1}^{R+1} \sum_{i=k_{r-1}+1}^{k_r} (\rho_i - \overline{\rho}(k_{r-1},k_r)) \tilde{Z}_i^{(0)} \right| \leq c(R+1) \max_{1 \leq i \leq R} |\Delta_{\rho,i,n}| \max_{1 \leq k_1 < k_2 \leq n} \left| \sum_{i=k_1+1}^{k_2} \tilde{Z}_i^{(0)} \right|$$
$$= O_P(n^{1/r_z} \max_{1 \leq i \leq R} |\Delta_{\rho,i,n}|).$$

Hence, the above yields

$$\max_k \sum_{r=1}^{R+1} (k_r - k_{r-1}) \left(\widetilde{Z^{(0)}}(k_{r-1},k_r)\right)^2 \leq (R+1) \left(\max_{1\leq k_1<k_2\leq n} \frac{1}{\sqrt{k_2-k_1}} \left| \sum_{i=k_1+1}^{k_2} \tilde{Z}_i^{(0)} \right| \right)^2,$$

where we can split the index set over $k_1 < k_2$ into parts where the difference between k_1 and k_2 is either bigger or smaller than $b_n = n^{1/r_z - 1/r}$. This implies that

$$\begin{aligned} \max_{1\leq k_1<k_2\leq n} \frac{1}{\sqrt{k_2-k_1}} \left| \sum_{i=k_1+1}^{k_2} \tilde{Z}_i^{(0)} \right| &\leq b_n^{1/2} \max_{1\leq k\leq n} |\tilde{Z}_k^{(0)}| + b_n^{-1/2} \max_{k_2-k_1\geq b_n} \left| \sum_{i=k_1+1}^{k_2} \tilde{Z}_i^{(0)} \right| \\ &= O_P(b_n^{1/2} n^{1/r} + b_n^{-1/2} n^{1/r_z}) \end{aligned}$$

and, in addition, that

$$\max_k \sum_{r=1}^{R+1} (k_r - k_{r-1}) \left(\widetilde{Z^{(0)}}(k_{r-1},k_r)\right)^2 = O_P(b_n n^{2/r} + b_n^{-1} n^{2/r_z}) = O_P(n^{1/r_z + 1/r})$$

as $n \to \infty$. Hence, using the rate as displayed in (2.1.28) we get that

$$\begin{aligned} &P\left(\min_{1<k_1<\ldots<k_R<n; \|k-k^*\|\geq \epsilon\underline{\Delta}_{k^*,n}} Q_n^{(0)}(k) - Q_n^{(0)}(k^*) \leq 0 \right) \\ &= P\Bigg(1 - O_P\left(\frac{n^{1/r_z}}{\epsilon\underline{\Delta}_{k^*,n} \vee a_{1,n,N}} \frac{\max_{1\leq i\leq R} |\Delta_{\rho,i,n}|}{\min_{1\leq i\leq R} |\Delta_{\rho,i,n}|} \right) \\ &\quad - O_P\left(\frac{n^{1/r_z+1/r}}{\epsilon\underline{\Delta}_{k^*,n} \vee a_{1,n,N} \min_{1\leq i\leq R} |\Delta_{\rho,i,n}|} \right) \leq 0 \Bigg) \to 0 \end{aligned}$$

as $n \to \infty$, followed by $N \to \infty$. Finally, it remains to consider the case in which we minimize over each k_r, which is in an ϵ-neighborhood of k_r^* with a radius of $\epsilon\underline{\Delta}_{k^*,n}$. Then, we get

$$\begin{aligned} &Q_n^{(0)}(k) - Q_n^{(0)}(k^*) \\ &\geq c\|k^* - k\| \min_{1\leq i\leq R} \Delta_{\rho,i}^2 + \sum_{r=1}^{R+1} \left[(k_r^* - k_{r-1}^*) \left(\widetilde{Z^{(0)}}(k_{r-1}^*,k_r^*)\right)^2 - (k_r - k_{r-1}) \left(\widetilde{Z^{(0)}}(k_{r-1},k_r)\right)^2 \right] \\ &\quad - 2 \left| \sum_{r=0}^{R} \left[c_{k_r,k_{r+1}}^{(1)} \sum_{i=k_r\vee k_r^*+1}^{k_{r+1}\wedge k_{r+1}^*} \tilde{Z}_i^{(0)} + c_{k_r,k_{r+1}}^{(2)} \sum_{i=k_r\wedge k_r^*+1}^{k_r\vee k_r^*} \tilde{Z}_i^{(0)} + c_{k_r,k_{r+1}}^{(3)} \sum_{i=k_{r+1}\wedge k_{r+1}^*+1}^{k_{r+1}\vee k_{r+1}^*} \tilde{Z}_i^{(0)} \right] \right|, \end{aligned}$$

where the second summand of the first line is in this case equal to $O_P(\max_{1\leq r\leq R+1} |\Delta_{k^*,r,n}|^{2/r_z - 1})$ due to the Kolmogorov-type inequalities. Here, c_{k_{r-1},k_r} fulfills the following properties as $n \to \infty$

$$\max_{|k_l - k_l^*| \leq \epsilon\underline{\Delta}_{k^*,n}, l=r,r+1} |c_{k_{r-1},k_r}^{(1)}| \|k - k^*\|^{-1} = O((k_{r+1}^* - k_r^*)^{-1} \max_r |\Delta_{\rho,r,n}|)$$

and

$$\max_{l=2,3} \|c_{k_{r-1},k_r}^{(l)}\| = O(\max_{1\leq i\leq R} |\Delta_{\rho,i,n}|).$$

Hence, using the Kolmogorov-type inequalities we observe that

$$\begin{aligned} &P(a_n|\hat{\theta}_{r_0} - \theta_{r_0}| \geq N+1) \leq o(1) \\ &\quad + P\Bigg(c \min_{1\leq i\leq R} \Delta_{\rho,i}^2 - O_P\left(\frac{\max_r |\Delta_{k^*,r,n}|^{2/r_z-1}}{Nn/a_n} \right) - O_P\left(\max_r \Delta_{\rho,r,n} \underline{\Delta}_{k^*,n}^{1/r_z - 1} \right) \\ &\quad - C(R+1) \max_{1\leq i\leq R} |\Delta_{\rho,i,n}| \max_{1\leq r\leq R} \max_{1\leq |k_r - k_r^*| \leq \epsilon n} \frac{|\sum_{i=k_r\wedge k_r^*+1}^{k_r\vee k_r^*} \tilde{Z}_i^{(0)}|}{|k_r - k_r^*| \vee (Nn/a_n)} \geq 0 \Bigg) \end{aligned}$$

as $n \to \infty$ followed by $N \to \infty$. Now, we use the Hájek-Rényi-type inequalities to obtain

$$\max_{1 \leq |k_r - k_r^*| \leq \epsilon n} \frac{|\sum_{i=k_r \wedge k_r^*+1}^{k_r \vee k_r^*} \tilde{Z}_i^{(0)}|}{|k_r - k_r^*| \vee (Nn/a_n)} = O_P\left((Nn/a_n)^{-(r_z-1)/r_z}\right).$$

Due to the assumed displays (2.1.28) and (2.1.29), each of the three $O_P(\cdot)$ terms is of order $o_P(\underline{\Delta}_\rho^2)$ so that the claim follows. □

Remark 2.1.24. *Suppose* $\underline{\Delta}_{k^*,n} \sim n$, $k_i^* \sim n$, $r_z = 2$, *and* $r > 2$. *Then, we could set* ρ_i *equal to a* $g_\rho(i/n)$, *where* g_ρ *is an integrable step function, and obtain that* $n^{-1}Q_n^{(0)}([n\cdot]) \in D([0,1]^R)$. *Furthermore, we obtain with the above proof that* $n^{-1}Q_n^{(0)}([n\cdot])$ *converges in probability towards a deterministic function, i.e., it holds that*

$$\sup_{\substack{x_1,\ldots,x_R \in [0,1] \\ x_1 < \ldots < x_R}} \left| n^{-1}Q_n^{(0)}([nx_1],\ldots,[nx_R]) - \sum_{r=1}^{R+1} \int_{x_{r-1}}^{x_r} \left(g_\rho(z) - \frac{\int_{x_{r-1}}^{x_r} g_\rho(y)dy}{x_r - x_{r-1}} \right)^2 dz \right| = o_P(1).$$

Now, we consider the assumption that the number of change-points $R = R^*$, $R^* \in \mathbb{N}$, is unknown. Let the estimates be given by

$$(\hat{k}^{(\psi)},\hat{R}^{(\psi)}) \in \arg\min \Big\{ Q_n^{(\psi)}(k_1,\ldots,k_R) + \beta_n^{(\psi)} R : 1 = k_0 < k_1 < \ldots < k_R < k_{R+1} = n, R \leq C^* \Big\}, \tag{2.1.30}$$

where $C^* \in \mathbb{N}$ is some upper bound for the number of change-points and $\psi = 0,1\ldots$ is a design index. Here, β_n is a sequence of non-negative numbers so that $\beta_n R$ is a penalty term which is applied to avoid over-fitting. This concept is not new and was already considered by Lavielle and Moulines (2000) in a slightly modified way.

Theorem 2.1.25. *Under the assumptions of Theorem 2.1.22 let be given that*

$$d_n^{(0)} = n^{1/r+1/r_z} \ll \beta_n^{(0)} \leq \frac{1}{4C^*} \min_{1 \leq i \leq m} \Delta_{\rho,i,n}^2 \underline{\Delta}_{k^*,n}. \tag{2.1.31}$$

Then, the estimate $\hat{R}^{(0)}$ *is consistent for the number of change-points* R^*.

Remark 2.1.26. 1. *The lower (upper) bound for* $\beta_n^{(0)}$ *is required to prevent overestimation (underestimation) of the numbers of change-points. Heuristically, we get the best asymptotic behavior if we choose* $\beta_n^{(0)}$ *equal to the upper bound since the error terms of the change-point estimation and of the change-point number estimation have the same upper bound in this case.*

2. *The problem is still that we have to postulate that we already know the minimal step size and the distance between the closest change-points to choose an optimal* β_n. *However, if* $r_z = 2$, *which is not unusual, and* r *is not bounded, i.e., each moment of* $Z_n^{(0)}$ *exists, it is sufficient that* $\beta_n^{(0)}$ *tends to infinity as* $n \to \infty$.

Proof of Theorem 2.1.25. Set $\beta_n = \beta_n^{(0)}$. It is obviously sufficient to show that the sets $\{\hat{R} < R^*\}$ and $\{\hat{R} > R^*\}$ are asymptotically empty. Firstly, we consider the set $\{\hat{R} < R^*\}$. It holds that

$$\begin{aligned} P(\hat{R} < R^*) &= P\left(\min_{1 \leq k \leq n; R < R^*} Q_n^{(0)}(k) + \beta_n R < \min_{1 \leq k \leq n; R \geq R^*} Q_n^{(0)}(k) + \beta_n R \right) \\ &\leq P\left(\min_{1 \leq k \leq n; R < R^*} Q_n^{(0)}(k) - Q_n^{(0)}(k^*) - \beta_n(R^*-1) < 0 \right) \\ &= P\left(o_P(\min_{1 \leq i \leq m} \Delta_{\rho,i,n}^2 \underline{\Delta}_{k^*,n}) + \frac{1}{2} \min_{1 \leq i \leq m} \Delta_{\rho,i,n}^2 \underline{\Delta}_{k^*,n} - \beta_n(R^*-1) < 0 \right) = o(1), \end{aligned}$$

where we use the same arguments as in the proof of Theorem 2.1.22 together with the assumption that $\beta_n \le \frac{1}{4C^*} \min_{1\le i\le m} \Delta^2_{\rho,i,n}\Delta_{k^*,n}$.

Now, we consider the set $\{\hat{R} > R^*\}$ and obtain

$$P(\hat{R} > R^*) = P\left(\min_{1\le k\le n; R>R^*} Q_n^{(0)}(k) + \beta_n R < \min_{1\le k\le n; R\ge R^*} Q_n^{(0)}(k) + \beta_n R\right)$$
$$\le P\left(\min_{1\le k\le n; R>R^*} Q_n^{(0)}(k) - Q_n^{(0)}(k^*) + \beta_n < 0\right).$$

Furthermore, it holds that

$$Q_n^{(0)}(k) - Q_n^{(0)}(k^*) = \sum_{r=1}^{R+1} \sum_{i=k_{r-1}+1}^{k_r} \left[2\left(\rho_i - \overline{\rho}(k_{r-1},k_r)\right)\tilde{Z}_i^{(0)} + \left(\rho_i - \overline{\rho}(k_{r-1},k_r)\right)^2\right]$$
$$+ \sum_{r=1}^{R^*+1} (k_r^* - k_{r-1}^*)\left(\overline{\tilde{Z}^{(0)}}(k_{r-1}^*,k_r^*)\right)^2 - \sum_{r=1}^{R+1} (k_r - k_{r-1})\left(\overline{\tilde{Z}^{(0)}}(k_{r-1},k_r)\right)^2.$$

We already know from the proof of Theorem 2.1.22 that the latter terms are of the order $O_P(n^{1/r+1/r_z})$.

Next, we consider the double sum on the right-hand side of the above equation: For the first term, for all $r \in \{1,\dots,R+1\}$, and each $l \in \{2,\dots,R^*+1\}$ with $k_{l-1}^* < k_{r-1} < k_r \le k_l^*$ we obtain that $\rho_i - \overline{\rho}(k_{r-1},k_r) = 0$ for all $i \in (k_{r-1},k_r]$. Hence, it remains to consider the cases where $k_{l-2}^* < k_{r-1} \le k_{l-1}^* < k_r \le k_l^*$ and $k_{r-1} \le k_{l-1}^* < k_l^* < k_r$. In the last case the inner sum is uniformly bounded in k with rate

$$(k_r - k_{r-1})^{1/r_z} a_n(k_{r-1},k_r)\left(O_P(1) + (k_r - k_{r-1})^{(r_z-1)/r_z} a_n(k_{r-1},k_r)\right),$$

where we use the Kolmogorov-type inequalities and $a_n(k_{r-1},k_r)$ equal to $|\rho_i - \overline{\rho}(k_{r-1},k_r)| \le 2$ with i in $(k_{r-1},k_{l-1}^*]$, $(k_{l-1}^*,k_l^*]$, or $(k_l^*,k_r]$. In the case where $a_n(k_{r-1},k_r) \gg (k_r - k_{r-1})^{-(r_z-1)/r_z}$ the above rate is asymptotically non-negative and it remains to show that $\beta_n \to \infty$ to obtain that the inner sum is dominated by β_n. In the case of $a_n(k_{r-1},k_r) = O((k_r - k_{r-1})^{-(r_z-1)/r_z}) = O((k_l^* - k_{l-1}^*)^{-(r_z-1)/r_z})$, the above rate has a lower bound which has a rate of $-(k_l^* - k_{l-1}^*)^{(2-r_z)/r_z}$.

In the case where $k_{l-2}^* < k_{r-1} \le k_{l-1}^* < k_r \le k_l^*$, we obtain rates for the inner sum of the forms

$$|k_{l-1}^* - k_v|^{1/r_z} a_n(k)\left(O_P(1) + |k_{l-1}^* - k_v|^{(r_z-1)/r_z} a_n(k)\right)$$

and

$$|k_w^* - k_r|^{1/r_z} a_n(k)\left(O_P(1) + |k_w^* - k_r|^{(r_z-1)/r_z} a_n(k)\right)$$

with $v \in \{r-1,r\}$ and $w \in \{l-1,l\}$, where, again, $a_n(k)$ stands for $|\rho_i - \overline{\rho}(k_{r-1},k_r)|$. The latter expression has again three different values depending on i. The same arguments yield that $Q_n^{(0)}(k) - Q_n^{(0)}(k^*)$ is dominated by β_n if $\beta_n \gg n^{1/r+1/r_z}$. Hence, we get $|\hat{R} - R^*| = o_P(1)$ if (2.1.31) holds and the proof is complete.

□

Remark 2.1.27. *If the sample size n is sufficiently large, the numerical complexity is of order n^{R^*} or n^{R^*+1}, depending on the number of changes being known or not. This order is quite high compared to a binary segmentation algorithm, cf. Eckley et al. (2011). Wied and Galeano (2014) presented an algorithm which could only detect some special kind of change-points. However, it is also possible to combine the presented detector with a binary segmentation idea which yields a faster estimation. To this goal, we have to calculate $Q_n = \sum_{i=1}^n (Z_i^{(0)} - \overline{Z^{(0)}})^2$ and $\min_{1\le k\le n} Q_n^{(0)}(k) + \beta_n$ and compare both. If the first term is smaller, it will be assumed that there is no change and we stop the procedure. Otherwise, we can reason that there is at least one change, so we can argue on subsets such as $\{1,\dots,[n/2]\}$ and $\{[n/2]+1,\dots,n\}$. Then, the change-point estimates, $\hat{k}_1,\dots,\hat{k}_{\hat{R}}$, are the collection of the arguments which minimize $Q_n^{(0)}(\cdot)$ on the disjoint sets $\{1,\dots,n_1\}$, $\dots\{n_{\hat{R}}+1,\dots,n\}$, where*

on subsets $\{n_i+1,\dots,n_i+(n_{i+1}-n_i)/2\}$ *and* $\{n_i+(n_{i+1}-n_i)/2,\dots,n_{i+1}\}$, $i=1,\dots \hat{R}$, *the terms* Q_n *are smaller than the corresponding* $\min Q_n^{(0)}(k)+\beta_n$.
The difference between this algorithm and the one of Wied and Galeano (2014) is the detector and the bounds of the subsets. Here they are non-random whereas in Wied and Galeano (2014) they depend on the estimates calculated by the iteration step before. If we assume that the change-points and change size are such that for all $\theta_1\in\{0,\theta_1^*,\dots,\theta_{R-1}^*\}$ *and* $\theta_2\in\{\theta_1,\dots,\theta_R^*,1\}\setminus\theta_1$

$$\arg\min\left\{\int_{\theta_1}^{x}\left(g_\rho(z)-\frac{\int_{\theta_1}^{x}g_\rho(y)dy}{x-\theta_1}\right)^2dz\right.$$
$$\left.+\int_{x}^{\theta_2}\left(g_\rho(z)-\frac{\int_{x}^{\theta_2}g_\rho(y)dy}{\theta_2-x}\right)^2dz\ \middle|\ x\in[\theta_1,\theta_2]\right\}\in\{\theta_1^*,\dots,\theta_R^*\}\cap(\theta_1,\theta_2),$$

it is possible to use the algorithm presented in Wied and Galeano (2014) with our detection rule. This technical assumption is suitable for the situation that one change-point between two others is dominated in some sense.

In the preceding remark we make use of the fact that the statistic $Q_n^{(0)}$ is and is not smaller than $\min_{1\le k\le n}Q_n^{(0)}(k)+\beta_n$ under H_0 and Assumption $\mathbf{H_A}$, respectively. Thereby, we can construct a quite simple test in the following theorem.

Theorem 2.1.28. *Let* $\{Z_i^{(0)}-\rho_i\}$ *fulfill* $(\mathcal{K}_\mathbf{r}^{(\mathbf{1})})$ *and* $(\mathcal{K}_\mathbf{r}^{(\mathbf{2})})$ *for* $r_z>1$. *Furthermore, we set*

$$D_n^{(0)}(X,Y)=\begin{cases}1, & T_n^{(0)}>1,\\ 0, & T^{(0)}\le 1\end{cases}\qquad \textit{with}\qquad T_n^{(0)}=\frac{\sum_{i=1}^n(Z_i^{(0)}-\overline{Z^{(0)}})^2}{\min_{1\le k\le n}Q_n^{(0)}(k)+\beta_n}. \tag{2.1.32}$$

Then, it holds under H_0 *with* $\log(n)^2\vee n^{2/r_z-1}=o(\beta_n)$ *that*

$$D_n^{(0)}(X,Y)\xrightarrow{P}0$$

and under Assumption $\mathbf{H_A}$ *with* $\beta_n\vee n^{1/r_z}=o(\lambda(R_{k^*}^c)\lambda(R_{k^*})^2\Delta_{\rho,n}^2n^{-2})$ *that*

$$D_n^{(0)}(X,Y)\xrightarrow{P}1.$$

Proof. Under H_0 we obtain by using $\tilde{Z}_i^{(0)}=Z_i^{(0)}-\rho_0$ that

$$\begin{aligned}\left\{T_n^{(0)}>1\right\}&=\left\{\sum_{i=1}^n(Z_i^{(0)}-\overline{Z^{(0)}})^2-\min_{1\le k\le n}Q_n^{(0)}(k)>\beta_n\right\}\\&=\left\{\max_{1\le k\le n-1}\frac{n}{k(n-k)}\left(\sum_{i=1}^k\tilde{Z}_i^{(0)}-\frac{k}{n}\sum_{i=1}^n\tilde{Z}_i^{(0)}\right)^2>\beta_n\right\}\\&=\left\{O_P\left(n^{2/r_z-1}\mathbb{1}_{\{r_z\ne2\}}+\log(n)^2\mathbb{1}_{\{r_z=2\}}\right)>\beta_n\right\},\end{aligned}$$

in which we use the Hájek-Rényi-type inequalities in the last line. Hence, it remains to show that $\log(n)^2\vee n^{2/r_z-1}=o(\beta_n)$ as $n\to\infty$ to obtain the claim. Under Assumption $\mathbf{H_A}$ we obtain

$$\begin{aligned}\left\{T_n^{(0)}>1\right\}&=\left\{\sum_{i=1}^n(Z_i^{(0)}-\overline{Z^{(0)}})^2-\min_{1\le k\le n}Q_n^{(0)}(k)>\beta_n\right\}\\&=\left\{\max_{1\le k\le n-1}\frac{k(n-k)}{n}(\overline{\rho}_1^k-\overline{\rho}_{1+k}^n)^2+\frac{n}{k(n-k)}\left(\sum_{i=1}^k\tilde{Z}_i^{(0)}-\frac{k}{n}\sum_{i=1}^n\tilde{Z}_i^{(0)}\right)^2\right.\\&\qquad\left.+2\left[\sum_{i=1}^k\rho_i\overline{\tilde{Z}}_1^k+\sum_{i=1+k}^n\rho_i\overline{\tilde{Z}}_{1+k}^n-\sum_{i=1}^n\rho_i\overline{\tilde{Z}}_1^n\right]>\beta_n\right\}\end{aligned}$$

$$
\begin{aligned}
&= \left\{ \max_{1\le k\le n-1} \frac{k(n-k)}{n} (\overline{\rho}_1^k - \overline{\rho}_{1+k}^n)^2 + O_P(n^{1/r_z}) > \beta_n \right\} \\
&\supset \left\{ \frac{\lambda(R_{k^*}^c)\lambda(R_{k^*})^2}{2n^2} \Delta_{\rho,n}^2 + O_P(n^{1/r_z}) > \beta_n \right\}.
\end{aligned}
$$

Hence, using $\beta_n \vee n^{1/r_z} = o(\lambda(R_{k^*}^c)\lambda(R_{k^*})^2\Delta_{\rho,n}^2 n^{-2})$ completes the proof. □

Remark 2.1.29. *1. It remains unclear how to choose* β_n *such that the test is in some sense optimal. However, the test is more conservative if a larger value of* β_n *is chosen.*

2. *Assuming* $\beta_n = \log(n)^2$ *and a single change-point is existent at* $k_1^* = n - \sqrt{n}$, *we obtain that in the preceding proof the positive deterministic function does not necessarily dominate the random term. However, if* $k_1^* = n - n^{1/2+\epsilon}$ *for an arbitrarily small* $\epsilon > 0$, *this cannot happen.*

3. *Under some additional assumptions we can apply Darling and Erdös (1956) to obtain that* $\log(\log(n)) \ll \beta_n$ *still holds under* H_0.

The following corollary will be used in Section 4. There, we sequentially estimate a change in the means which has to be detected before the change in the mean influences the stopping time controlling the change in the correlation.

Corollary 2.1.30. *Assume a change-point* $k_1^* = [n\theta]$, $\theta \in (0,1)$, *in an AMOC model. Assume the assumptions of Theorem 2.1.28 to be fulfilled with* $r_z = 2$. *Moreover, set* $\beta_n = \sqrt{n}$ *and*

$$
\hat{k}_n = \begin{cases} n, & \text{if } T_n^{(0)} \le 1, \\ \arg\min Q_n^{(0)}(k), & \text{if } T_n^{(0)} > 1, \end{cases} \tag{2.1.33}
$$

where $T_n^{(0)}$ *is defined in* (2.1.32). *Then, for any sequence* $c_n \to \infty$ *as* $n \to \infty$ *we get*

$$
P\left(\max_{1\le k\le n} \hat{k}_k \le k_1^* + c_n\sqrt{n} \right) \to 1, \quad \text{as } n \to \infty.
$$

Proof. Firstly, we obtain with $a_n = c_n\sqrt{n}$ that

$$
\begin{aligned}
\max_{1\le N\le n} \hat{k}_N &= \max_{1\le N\le n} \left(N * \mathbb{1}_{\{T_N^{(0)}\le 1\}} + \mathbb{1}_{\{T_N^{(0)}>1\}} \arg\min_{1\le k\le N} Q_N^{(0)}(k) \right) \\
&\le \Big(n * \mathbb{1}_{\{\min_{k_1^*+a_n\le N\le n} T_N^{(0)}\le 1\}} \\
&\qquad + \mathbb{1}_{\{\max_{k_1^*+a_n\le N\le n} T_N^{(0)}>1\}} \max_{k_1^*+a_n\le N\le n} \arg\min_{1\le k\le N} Q_N^{(0)}(k) \Big).
\end{aligned}
$$

Thus, it holds that

$$
\begin{aligned}
&P\left(\min_{k_1^*+a_n\le N\le n} T_N^{(0)} \le 1 \right) = 1 - P\left(\min_{k_1^*+a_n\le N\le n} T_N^{(0)} > 1 \right) \\
&\le 1 - P\left(\min_{k_1^*+a_n\le N\le n} \max_{1\le k\le N-1} \frac{k(N-k)}{N} (\overline{\rho}_1^k - \overline{\rho}_{1+k}^N)^2 - O_P(n^{1/2}) > \beta_{k_1^*+a_n} \right) \\
&\le 1 - P\left(\frac{k_1^* a_n}{k_1^* + a_n} (\Delta_{\rho,1})^2 - O_P(n^{1/2}) > \beta_{k_1^*+a_n} \right) = o(1)
\end{aligned}
$$

if $n^{1/2} = o\left(\frac{k_1^* a_n}{k_1^*+a_n}(\Delta_{\rho,1})^2\right)$. Theorem 2.1.20 yields

$$
P\left(\max_{k_1^*+a_n\le N\le n} \arg\min_{1\le k\le N} Q_N^{(0)}(k) \le k_1^* + a_n \right) \longrightarrow 1,
$$

which finally implies the claim. □

2.1.3 Long-run Variance Estimation

This paragraph investigates the convergence rates of some LRV estimates. These estimates are needed to normalize the limit process and have been investigated i.a. by Newey and West (1986) and Andrews (1991) for multivariate time series. Juhl and Xio (2009) and Hušková and Kirch (2010) focus on LRV estimates under changes in the location.

In the following, we assume that we already have a consistent kernel estimate for LRV using the exact correlation coefficient ρ_i. Since these coefficients are usually unknown, we replace them by some estimates, i.e., we are interested in

$$\hat{D}_0 = \hat{D}_{0,n} = \frac{1}{n}\sum_{i=1}^{n}\sum_{j=1}^{n} \mathfrak{f}\left(\frac{i-j}{q_n}\right)(Z_i^{(0)} - \tilde{\rho}_n(i))(Z_j^{(0)} - \tilde{\rho}_n(i)), \tag{2.1.34}$$

where q_n is the bandwidth which tends towards infinite as $n \to \infty$. Instead of assuming a specific type of correlation estimates, we specify general conditions on the estimation errors of $\tilde{\rho}_n(i) - \rho_i$.

Definition 2.1. *A* ***kernel*** *is a non-negative, real-valued, integrable function* $\mathfrak{f}$ *which fulfills* $\mathfrak{f}(0) = 1$,

$$\int_{-\infty}^{\infty} \mathfrak{f}(x)\,dx = 1 \qquad \text{and} \qquad \mathfrak{f}(-x) = \mathfrak{f}(x) \quad \text{for all } x \in \mathbb{R}.$$

Remark 2.1.31. *For many results the symmetry of* $\mathfrak{f}$ *is not necessary but simplifies the proof.*

In the following theorem we consider eight different types, (A)-(H), of estimation errors. Case (A) is motivated by the situation where the correlations ρ_i and their estimates $\hat{\rho}_i$ are constant for all i. Case (B) represents the scenario where the constant correlation is disturbed by a deterministic error (e.g. in the situation of misspecified parameters). Case (C) includes the setting of a local change-point assumption. In the case of (D) we consider the situation where the estimation error is just bounded as in the case of $\mathbf{H_A}$. Cases (E)-(H) represent situations (A)-(D) respectively, where the estimation errors are now split on random intervals. This becomes necessary in the case of piecewise constant correlation estimates.

Remark 2.1.32. *Let* $\{X_{i,n}\}_{i=1,\dots,n,n\in\mathbb{N}}$ *be a random array and* d_{in} *a deterministic array. We use the notation* $X_{i,n} = O_P(1) + d_{in}$ *to denote:* $X_{i,n} - d_{in}$ *is independent of the index* i *and its absolute value is equal to* $O_P(1)$.

The following result is well-known and has already been investigated in many different contexts. We prove it for the sake of completeness.

Theorem 2.1.33. *Assume the following conditions:*

1. $\mathfrak{f}$ *is a kernel and* q_n *is a bandwidth with* $q_n \to \infty$ *and* $q_n = o(n)$ *as* $n \to \infty$,
2. $\{Z_n^{(0)} - \rho_n\}$ *fulfills* $(\mathcal{K}_{\mathbf{r}}^{(1)})$ *for some* $r_z > 1$,
3. *it holds that* $n^{-1}\mathrm{Var}\left[\sum_{i=1}^{n} Z_i^{(0)}\right] \to D > 0$ *and*

$$\hat{D}_n = \frac{1}{n}\sum_{i=1}^{n}\sum_{j=1}^{n} \mathfrak{f}\left(\frac{i-j}{q_n}\right)(Z_i^{(0)} - \rho_i)(Z_j^{(0)} - \rho_j) \xrightarrow{P} D, \tag{2.1.35}$$

4. *let the estimation error* $\rho_i - \tilde{\rho}_n(i)$ *fulfill one of the following cases:*

$$\rho_i - \tilde{\rho}_n(i) = \hat{R}_{\rho,n}(i) = \begin{cases} O_P(n^{-\delta_1}), & \text{case (A)},\\ O_P(n^{-\delta_1}) + d_{in}^{(1)}, & \text{case (B)},\\ O_P(n^{-\delta_1}) + d_{in}^{(2)}, & \text{case (C)},\\ O_P(n^{-\delta_1}) + d_{in}^{(3)}, & \text{case (D)},\\ \sum_{j=1}^{m} \mathbb{1}_{\hat{C}_j}(i) O_P(n^{-\delta_j}), & \text{case (E)},\\ \sum_{j=1}^{m} \mathbb{1}_{\hat{C}_j}(i) O_P(n^{-\delta_j}) + d_{in}^{(1)}, & \text{case (F)},\\ \sum_{j=1}^{m} \mathbb{1}_{\hat{C}_j}(i) O_P(n^{-\delta_j}) + d_{in}^{(2)}, & \text{case (G)},\\ \sum_{j=1}^{m} \mathbb{1}_{\hat{C}_j}(i) O_P(n^{-\delta_j}) + d_{in}^{(3)}, & \text{case (H)}, \end{cases}$$

where $\delta_j \geq 0$ *for all* $j \in \{1,\ldots,m\}$, $m \in \mathbb{N}$ *and where*

- $\sum_{i=1}^n |d_{in}^{(1)}| = o(\sqrt{n})$, $\sum_{i=1}^n |d_{in}^{(2)}| = O(\sqrt{n})$, $\sup_{1\leq i\leq n} |d_{i,n}^{(2)}| = O(n^{-1/2})$, $\sup_{n,i\in\mathbb{N}} |d_{i,n}^{(3)}| \leq C < \infty$;
- $\frac{1}{n}\left|\sum_{i,j=1}^n \mathfrak{f}\left(\frac{i-j}{q_n}\right) d_{jn}^{(k)}(Z_i^{(0)} - \rho_i)\right| = O_P(a_n^{(k)})$ *with* $a_n^{(k)} \to 0$, *for* $k = 1,2,3$;
- *in cases of* (E)-(G) *we assume that* $\bigcup_{j\in\{1,\ldots,m\}} \hat{C}_j = \{1,\ldots,n\}$ *and that there are sets* $C_1,\ldots,C_n$ *so that* $P\left(\bigcap_{j\in\{1,\ldots,m\}}\left\{\hat{C}_j \subset C_j\right\}\right) \to 1$.

Then, it holds that $\hat{D}_{0,n} = \hat{D}_n + \hat{R}_n^{(0)}$ *with*

$$\hat{R}_n^{(0)} = \begin{cases} \tilde{R}_n^{(A)} = O_P(q_n n^{-2\delta_1}) + O_P(q_n n^{-(1-1/r_z+\delta_1)}), & \text{under } (A), \\ \tilde{R}_n^{(A)} + O_P(q_n n^{-(1/2+\delta_1)}) + O(q_n n^{-1}) + O_P(a_n^{(1)}), & \text{under } (B), \\ \tilde{R}_n^{(A)} + O_P(q_n n^{-(1/2+\delta_1)}) + O(q_n n^{-1}) + O_P(a_n^{(2)}), & \text{under } (C), \\ \tilde{R}_n^{(A)} + O_P(q_n n^{-\delta_1}) + O(q_n \|d_{in}^{(3)}\|^2) + O_P(a_n^{(3)}), & \text{under } (D) \\ O_P\left(q_n n^{-1}(n^{1/r-\min_k \delta_k} + \max_{k_1,k_2} n^{-\delta_{k_1}-\delta_{k_2}} \#(C_{k_1}) \wedge \#(C_{k_1}))\right), & \text{under } (E), \\ \tilde{R}_n^{(E)} + O_P(a_n^{(1)})) + o_P(1) + o_P(q_n n^{-1/2-\min_k \delta_k}), & \text{under } (F), \\ \tilde{R}_n^{(E)} + O_P(a_n^{(2)}) + O(q_n n^{-1}) + O_P(q_n n^{-1-1/2} \max_k \#C_k n^{-\delta_k}), & \text{under } (G), \\ \tilde{R}_n^{(E)} + O(q_n) + O_P(a_n^{(3)}) + O_P(q_n n^{-1} \max_k \#C_k n^{-\delta_k}), & \text{under } (H). \end{cases} \tag{2.1.36}$$

Proof. See appendix (p. 180). □

Remark 2.1.34. *1. Using Markov's inequality it is obvious that* $a_n^{(1)} = o(q_n n^{-1/2})$, $a_n^{(2)} = O(q_n n^{-1/2})$, *and* $a_n^{(3)} = O(q_n)$ *hold true. The latter is a rough estimate.*

2. Later on, we will use the preceding theorem also in the case where $\{Z_n^{(0)} - \rho_n\}$ *fulfills a FCLT, the* $(\mathcal{K}_r^{(1)})$ *for* $r = 2$, *and the* r'*th* ($r' > 2$) *moments are uniformly bounded. Note that this implies that the sample mean of* $Z_n^{(0)}$ *has an estimation error of* $O_P(n^{-1/2})$ *if the correlations* ρ_i *are constant. Thus, in this case* $\delta_1 = 1/2$.

In this subsection we have considered the asymptotic behavior of test statistic under known means and variances. Therefore, we have presented some suitable LRV estimates. Moreover, in this section we have presented change-point estimates and have proven their convergence rates.

2.2 Sequential Analysis under General Dependency Framework

This subsection presents an open-end and a closed-end procedure.

2.2.1 Closed-end Procedure

In the following theorem we consider the asymptotic behavior of the stopping times. It uses the known parameters $\mu_{1,n}$, $\mu_{2,n}$, $\sigma^2_{1,n}$, and $\sigma^2_{2,n}$. In the context of sequential analysis for changes in the mean this result is well-known. We prove it for the sake of completeness.

Theorem 2.2.1. *Given that*

$$\frac{1}{\sqrt{n}}\sum_{i=1}^{[n\cdot]}(Z_i^{(0)}-\rho_i) \overset{\mathcal{D}[0,1+m]}{\longrightarrow} D^{1/2}W(\cdot), \tag{2.2.1}$$

where $W(\cdot)$ is a Brownian motion and $D, m>0$, it holds
(i) under H_0 and $|\hat{D}_0-D|=o_P(1)$ that

$$P(\tau^{(c)}_{n,\iota,0,0}<\infty)\to P\left(\sup_{z\in[0,m]} u_\iota(G)(z)>c_\alpha\right), \tag{2.2.2}$$

(ii) under Assumption $\mathbf{H}^{(\mathbf{c})}_{\mathbf{LA}}$ and $|\hat{D}_0-D|=o_P(1)$ with $\tilde{g}(z)=D^{-1/2}\frac{1}{1+z}\int_1^{1+z}g_\rho(x)dx$ that

$$P(\tau^{(c)}_{n,\iota,0,0}<\infty)\to P\left(\sup_{z\in[0,m]} u_\iota(G+\tilde{g})(z)>c_\alpha\right), \tag{2.2.3}$$

(iii) and under Assumption $\mathbf{H}^{(\mathbf{c})}_{\mathbf{A}}$, $\Delta_\rho^{-2}|\hat{D}_0|=o_P(n)$ and $\lim_{\|x\|\to\infty}\sup_{z\in\mathbb{R}^+}u_\iota(x)(z)=\infty$ that

$$P\left(\tau^{(c)}_{n,\iota,0,0}<\infty\right)\to 1,$$

where G is a centered Gaussian process with covariance structure $\mathbb{E}[G(s)G(t)]=\frac{s}{1+s}\wedge\frac{t}{1+t}$.

Proof. Firstly, under $H_0^{(2)}$ we obtain that

$$\frac{n}{n+[n\cdot]}\left(\frac{1}{\sqrt{n}}\sum_{i=n+1}^{n+[n\cdot]}(Z_i^{(0)}-\rho_0)-\frac{[n\cdot]}{n\sqrt{n}}\sum_{i=1}^{n}(Z_i^{(0)}-\rho_0)\right)$$
$$=f\left(\frac{1}{\sqrt{n}}\sum_{i=1}^{[n\cdot]}(Z_i^{(0)}-\rho_i),\frac{[n\cdot]}{n},\frac{n}{n+[n\cdot]}\right),$$

where $f:\mathcal{D}[0,1+m]^3\to\mathcal{D}[0,m]$ with $f(x,y,z)=z(\cdot)(x(1+\cdot)-x(1)-y(\cdot)x(1))$. Since f is continuous it follows with the CMT, that under $H_0^{(2)}$

$$\tilde{B}_n^{1,0,0}([n\cdot])=\hat{D}^{-1/2}f\left(\frac{1}{\sqrt{n}}\sum_{i=1}^{[nz]}(Z_i^{(0)}-\rho_0),\frac{[nz]}{n},\frac{n}{n+[nz]}\right) \tag{2.2.4}$$

$$\overset{\mathcal{D}[0,m]}{\longrightarrow}\frac{1}{1+\cdot}(W(1+\cdot)-(1+\cdot)W(1))=G(\cdot). \tag{2.2.5}$$

Under the assumption $\mathbf{H}^{(\mathbf{c})}_{\mathbf{LA}}$ we get

$$\tilde{B}_n^{1,0,0}([n\cdot])=\hat{D}^{-1/2}f\left(\frac{1}{\sqrt{n}}\sum_{i=1}^{[nz]}(Z_i^{(0)}-\rho_i),\frac{[nz]}{n},\frac{n}{n+[nz]}\right) \tag{2.2.6}$$

$$+\hat{D}^{-1/2}\frac{n}{n+[n\cdot]}\frac{1}{n}\sum_{i=n+1}^{n+[n\cdot]}g(i/n) \tag{2.2.7}$$

$$\overset{\mathcal{D}[0,m]}{\longrightarrow}G(\cdot)+D^{-1/2}\frac{1}{1+\cdot}\int_1^{1+\cdot}g(x)dx=G(\cdot)+\tilde{g}(\cdot). \tag{2.2.8}$$

Now, we weight the statistic by $\mathbb{1}_{\{\cdot\leq m\}}$ and apply the CMT with the continuous functions $u_\iota : \mathcal{D}[0,\infty) \to \mathcal{D}[0,\infty)$ and $f_1 : \mathcal{D}[0,\infty) \to \mathbb{R}$ with $f_1(h) = \sup_{z\in[0,\infty)} h(z)$ so that it implies

$$
\begin{aligned}
P\left(\tau^{(c)}_{n,\iota,1,0} < \infty\right) &= P\left(f_1(u_\iota(\tilde{B}^{1,0,0}_n([n\cdot]))) > c_\alpha\right) \\
&\to \begin{cases} P\left(\sup_{z\in[0,\infty)} u_\iota(\mathbb{1}_{\{\cdot\leq m\}}G(\cdot))(z) > c_\alpha\right), & \text{under } H_0^{(2)}, \\ P\left(\sup_{z\in[0,\infty)} u_\iota(\mathbb{1}_{\{\cdot\leq m\}}(G(\cdot)+\tilde{g}(\cdot)))(z) > c_\alpha\right), & \text{under } \mathbf{H}^{(\mathbf{c})}_{\mathbf{LA}}, \end{cases} \\
&= \begin{cases} P\left(\sup_{z\in[0,m]} u_\iota(G)(z) > c_\alpha\right), & \text{under } H_0^{(2)}, \\ P\left(\sup_{z\in[0,m]} u_\iota(G+\tilde{g})(z) > c_\alpha\right), & \text{under } \mathbf{H}^{(\mathbf{c})}_{\mathbf{LA}}, \end{cases}
\end{aligned}
$$

where we use in the last step the main property 2 of u_ι. Under the assumption $\mathbf{H}^{(\mathbf{c})}_{\mathbf{A}}$ we just have to replace the summand in (2.2.7) by

$$
R_n(\cdot) = \hat{D}^{-1/2}\frac{n}{n+[n\cdot]}\frac{1}{\sqrt{n}}\sum_{i=n+1}^{n+[n\cdot]}\Delta_\rho 1_{R_{k^*}}(i) = \hat{D}^{-1/2}\frac{n}{n+[n\cdot]}\frac{\lambda(R_{k^*}\cap(n,n+[n\cdot]])}{\sqrt{n}}\Delta_\rho,
$$

where the maximum increases as $|\Delta_\rho|\hat{D}^{-1/2}n^{1/2}$. Using the above displays and the property of u_ι we get that

$$
\|u_\iota(\tilde{B}^{1,0,0}_n)([n\cdot])\| \xrightarrow{P} \infty. \tag{2.2.9}
$$

Hence, this implies that even under the assumption that $\Delta_\rho^{-2}|\hat{D}| = o_P(n)$ the stopping times are asymptotically finite with probability one. □

Remark 2.2.2. *1. In (iii), the condition* $\lim_{\|x\|\to\infty}\sup_{z\in\mathbb{R}^+} u_\iota(x)(z) = \infty$ *can be replaced by the condition:* u_ι *satisfies the triangle inequality and*

$$
\left\| u_\iota\left(\hat{D}^{-1/2}\frac{n}{n+[n\cdot]}\frac{\lambda(R_{k^*}\cap(n,n+[n\cdot]])}{\sqrt{n}}\Delta_\rho\right)\right\| \xrightarrow{P} \infty.
$$

2. Considering $R_n(\cdot)$ *under Assumption* $\mathbf{H}^{(\mathbf{c})}_{\mathbf{A}}$ *makes it clear that we detect a change-point with probability one within* $c_n n^{1/2}$ *time-points after the change appears. Therefore, we assume* $c_n \to \infty$ *and* $\hat{D}^{-1/2} = O_P(1)$.

In the following, we employ the set of weighting functions for the closed-end sequential analysis

$$
\mathcal{WF}^{(S_c)} = \left\{ w_\gamma : (0,m) \to \mathbb{R}_{>0} \ : w_\gamma \text{ continuous}, \ w_\gamma(t) = O(t^{-\gamma}), \right. \\
\left. \text{as } t \to 0, \text{for some } \gamma \in \left[0,\frac{1}{2}\right) \right\}. \tag{2.2.10}
$$

Theorem 2.2.3. *Let* $\{Z^{(0)}_n - \rho_n\}$ *satisfy* $(\mathcal{K}^{(\mathbf{3})}_{\mathbf{r}})$ *for* $r = 2$, *let* $w_\gamma \in \mathcal{WF}^{(S_c)}$ *and*

$$
\frac{1}{\sqrt{n}}\sum_{i=1}^{[n\cdot]}(Z^{(0)}_i - \rho_i) \overset{\mathcal{D}[0,1+m]}{\longrightarrow} D^{1/2}W(\cdot), \tag{2.2.11}
$$

where $W(\cdot)$ *is a Brownian motion and* $D > 0$. *Then, it holds*

(i) under $H_0^{(2)}$ *and* $|\hat{D}_0 - D| = o_P(1)$ *that*

$$
P(\tau^{(c)}_{n,\iota,0,\gamma} < \infty) \to P\left(\sup_{z\in[0,m]} u_\iota(w_\gamma G)(z) > c_\alpha\right), \tag{2.2.12}
$$

(ii) underAssumption $\mathbf{H}_{\mathbf{LA}}^{(\mathbf{c})}$ *and* $|\hat{D}_0 - D| = o_P(1)$ *with* $\tilde{g}(z) = D^{-1/2}\frac{1}{1+z}\int_1^{1+z} g_\rho(x)dx$ *that*

$$P(\tau_{n,\iota,0,\gamma}^{(c)} < \infty) \to P\left(\sup_{z\in[0,m]} u_\iota(w_\gamma(G+\tilde{g}))(z) > c_\alpha\right), \tag{2.2.13}$$

(iii) and under Assumption $\mathbf{H}_{\mathbf{A}}^{(\mathbf{c})}$, $|\hat{D}_0| = o_P(\sqrt{n})$, *and* $\lim_{\|x\|\to\infty}\sup_{z\in\mathbb{R}^+} u_\iota(x)(z) = \infty$ *that*

$$P\left(\tau_{n,\iota,0,\gamma}^{(c)} < \infty\right) \to 1,$$

where G *is a Gaussian process with covariance structure* $I\!E[G(s)G(t)] = \frac{s}{1+s} \wedge \frac{t}{1+t}$.

Proof. Due to arguments quite analogous to those in the proof of Theorem 2.1.4, it is sufficient to prove

$$\sup_{z\in[1/n,[\epsilon n]/n]} \left|w_\gamma\left(\frac{[nz]}{n}\right)\tilde{B}_n^{1,0,\gamma}([nz])\right| \xrightarrow{P} 0,$$

as $n \to \infty$ followed by $\epsilon \to 0$. Using the triangle inequality and the FCLT yields

$$\sup_{z\in\left[\frac{1}{n},\frac{[\epsilon n]}{n}\right]} \left|w_\gamma\left(\frac{[nz]}{n}\right)\tilde{B}_n^{1,0,\gamma}([nz])\right| = O(1) \sup_{z\in\left[\frac{1}{n},\frac{[\epsilon n]}{n}\right]} \left|\left(\frac{n}{[nz]}\right)^\gamma \frac{(S_{n+[nz]} - S_n)}{\sqrt{n}}\right| + o_P(1),$$

as $n \to \infty$, followed by $\epsilon \to 0$ and with $S_k = \sum_{i=1}^k (Z_i^{(0)} - \rho_i)$. Now we apply the Hájek-Rényi-type inequality, which is equivalent to the Kolmogorov-type inequality and obtain that the right–hand side above is equal to $o_P(1)$. The rest of the proof follows quite analogously to a combination of the proofs of Theorem 2.1.4 and Theorem 2.2.1. □

Remark 2.2.4. *1. In the third result, the condition* $\lim_{\|x\|\to\infty}\sup_{z\in\mathbb{R}^+} u_\iota(x)(z) = \infty$ *can be replaced by the condition* u_ι *fulfills the triangle inequality and*

$$\left\|u_\iota\left(w_\gamma\left(\frac{[n\cdot]}{n}\right)\hat{D}^{-1/2}\frac{n}{n+[n\cdot]}\frac{\lambda(R_{k^*}\cap(n,n+[n\cdot]])}{\sqrt{n}}\Delta_\rho\right)\right\| \xrightarrow{P} \infty.$$

2. If we consider an AMOC model with an early change $k_1^* = n^\epsilon$, $\epsilon > 0$, *we use* $u_\iota(\cdot) = |\cdot|$ *and* $w_\gamma(z) = \left(\frac{1+z}{z}\right)^\gamma$, $\gamma \in [0,\frac{1}{2})$, *with a* γ *near* $\frac{1}{2}$ *to detect asymptotically the change within* $n^\delta\Delta_\rho^{-1}$ *time-points after the change occurs with probability one. Here,* $\delta > \frac{1-2\gamma}{2(1-\gamma)}$ *and we assumed* $\hat{D}_n = O_P(1)$.

2.2.2 Open-end Procedure

In this sub-subsection, we augment the (un-)weighted closed-end procedure presented before to an open-end one. In order to do so, we define the weighting function in this sub-subsection as follows:

$$w_\gamma : (0,\infty) \to \mathbb{R}, \quad w_\gamma(z) = \left(\frac{1+z}{z}\right)^\gamma, \quad \gamma \in \left[0,\frac{1}{2}\right). \tag{2.2.14}$$

Theorem 2.2.5. *Let* $\{Z_i^{(0)} - \rho_i\}$ *fulfill* $(\mathcal{K}_{\mathbf{r}}^{(\mathbf{3})})$ *for* $r = 2$ *and let*

$$\frac{1}{\sqrt{n}}\sum_{i=1}^{[n\cdot]}(Z_i^{(0)} - \rho_i) \overset{\mathcal{D}[0,1+m]}{\longrightarrow} D^{1/2}W(\cdot), \tag{2.2.15}$$

where $W(\cdot)$ *is a Brownian motion and* $D, m > 0$. *Then, it holds*

1. *under* $H_0^{(2)}$ *and* $|\hat{D}_0 - D| = o_P(1)$ *that*

$$P(\tau^{(o)}_{n,\iota,0,0,\gamma} < \infty) \to P\left(\sup_{z\in(0,\infty)} u_\iota(w_\gamma G)(z) > c_\alpha\right) \tag{2.2.16}$$

if for each $\lambda > 0$ *and as* $m \to \infty$

$$P\left(\sup_{x>m} u_\iota\left(w_\gamma G\mathbb{1}_{[0,m]}(\cdot) - [w_\gamma(\cdot)\frac{\cdot}{\cdot+1}W(1) + \delta N m^{-\lambda}]\mathbb{1}_{\{\cdot\geq m\}}\right)(x) - \sup_{0\leq x\leq m} u_\iota(w_\gamma G)(x) \geq \epsilon\right) \to 0; \tag{2.2.17}$$

2. *under Assumption* $\mathbf{H_{LA}^{(o)}}$, *under the absolute integrability of* g_ρ *on* $\mathbb{R}_+$ *and for* $|\hat{D}_0 - D| = o_P(1)$ *that*

$$P(\tau^{(o)}_{n,\iota,0,0,\gamma} < \infty) \to P\left(\sup_{z\in(0,\infty)} u_\iota\left(w_\gamma(G+\tilde{g})\right)(z) > c_\alpha\right) \tag{2.2.18}$$

with $\tilde{g}(z) = D^{-1/2}\frac{1}{1+z}\int_1^{1+z} g_\rho(x)dx$ *if for each* $\lambda > 0$ *and as* $m \to \infty$

$$P\left(\sup_{x>m} u_\iota\left(w_\gamma(G+\tilde{g})\mathbb{1}_{[0,m]}(\cdot) - [w_\gamma(\cdot)(\frac{\cdot}{\cdot+1}W(1) + \tilde{g}) + \delta N m^{-\lambda}]\mathbb{1}_{\{\cdot\geq m\}}\right)(x) - \sup_{0\leq x\leq m} u_\iota(w_\gamma(G+\tilde{g}))(x) \geq \epsilon\right) \to 0; \tag{2.2.19}$$

3. *and under Assumption* $\mathbf{H_A^{(o)}}$, $|\hat{D}_0| = o_P(\sqrt{n})$ *and* $\lim_{\|x\|\to\infty}\sup_{z\in\mathbb{R}^+} u_\iota(x)(z) = \infty$ *that*

$$P\left(\tau^{(c)}_{n,\iota,0,0,\gamma} < \infty\right) \to 1,$$

where $G(\cdot) = \frac{1}{1+\cdot}(W(1+\cdot) - (1+\cdot)W(1))$ *and* $\delta = sign(W(1))$.

Proof. The convergence under Assumption $\mathbf{H_A^{(o)}}$ obviously holds true by Theorems 2.2.1 and 2.2.3. Under $H_0^{(2)}$ and $\mathbf{H_{LA}^{(c)}}$ we obtain

$$\begin{aligned} P(\tau^{(o)}_{n,\iota,0,\gamma} < \infty) &= P\left(\sup_{0\leq x} u_\iota(\tilde{B}_n^{1,0,\gamma})(x/n) \geq c_\alpha\right) \\ &= P\left(\sup_{0\leq x\leq nm} u_\iota(\tilde{B}_n^{1,0,\gamma})(x/n) \geq c_\alpha\right) \\ &\quad + P\left(\sup_{0\leq x\leq mn} u_\iota(\tilde{B}_n^{1,0,\gamma})(x/n) < c_\alpha,\ \sup_{x>nm} u_\iota(\tilde{B}_n^{1,0,\gamma})(x/n) \geq c_\alpha\right). \end{aligned}$$

Now, it is sufficient to show that the first summand fulfills the claimed convergence and the second vanishes as $n \to \infty$, followed by $m \to \infty$. We start with the first summand: From Theorems 2.2.1 and 2.2.3 follow that $\tilde{B}_n^{1,0,\gamma}([n\cdot])$ converges in distribution towards $w_\gamma(\cdot)G(\cdot)$ and towards $w_\gamma(\cdot)(G(\cdot) + \tilde{g}(\cdot))$ under $H_0^{(2)}$ and under $\mathbf{H_{LA}^{(o)}}$, respectively, where G is a Gaussian process with covariance structure $\mathbb{E}\left[G(s)G(t)\right] = \frac{s}{1+s} \wedge \frac{t}{1+t}$.

Thus, the convergence of a monotone sequence of real numbers yields

$$P\left(\sup_{0\leq x\leq nm} u_\iota(\tilde{B}_n^{1,0,\gamma})(x/n) \geq c_\alpha\right) \to P\left(\|u_\iota(w_\gamma G)\|_{[0,m]} \geq c_\alpha\right) \to P\left(\|u_\iota(w_\gamma G)\|_{[0,\infty)} \geq c_\alpha\right).$$

Now, it remains to show that the second term of the right–hand side in the first display vanishes as $n \to \infty$, followed by $m \to \infty$. We can estimate this summand by

$$P\left(\sup_{x>nm} u_\iota(\tilde{B}_n^{1,0,\gamma})(x/n) - \sup_{0\leq x\leq mn} u_\iota(\tilde{B}_n^{1,0,\gamma})(x/n) \geq \epsilon\right)$$

for an $\epsilon > 0$. Under $H_0^{(2)}$ we obtain that

$$\tilde{B}_n^{1,0,\gamma}(z) = \hat{D}^{-1/2} w_\gamma\left(\frac{[zn]}{n}\right)\frac{n[zn]n^{1/2}}{(n+[zn])n}(\hat{\rho}_{n+1,0}^{n+[zn]} - \rho) - \hat{D}^{-1/2} w_\gamma\left(\frac{[zn]}{n}\right)\frac{n[zn]n^{1/2}}{(n+[zn])n}(\hat{\rho}_{1,0}^{n} - \rho)$$

and $\sup_{z\geq m}\left|w_\gamma([nz]/n)\frac{n[nz]}{(n+[nz])n} - w_\gamma(z)\frac{z}{z+1}\right| = o(1)$ as $n \to \infty$. Thus, the second summand converges in distribution uniformly towards $-w_\gamma(z)\frac{z}{z+1}W(1)$. For the first summand, we obtain

$$P\left(\max_{k>nm}|n^{1/2}(\hat{\rho}_{n+1,0}^{n+k} - \rho)| \geq \eta\right) \leq \sum_{j=0}^{\infty} P\left(\max_{2^j nm \leq k \leq 2^{j+1}nm}\frac{\sqrt{n}}{k}|S_{n+k} - S_n| \geq \eta\right)$$

$$\leq \sum_{j=0}^{\infty} P\left(\frac{1}{2^j m\sqrt{n}}\max_{1\leq k\leq 2^{j+1}nm}|S_{n+k} - S_n| \geq \eta\right) \leq \frac{1}{(c\eta)^2}\sum_{j=0}^{\infty}\frac{1}{(2^j m\sqrt{n})^2}\sum_{i=1}^{2^{j+1}nm}\alpha_i = O(m^{-1})$$

as $n \to \infty$, followed by $m \to \infty$. Here, we use the assumed Kolmogorov-type inequality. Thus, it holds under $H_0^{(2)}$ that

$$P\left(\sup_{x>nm} u_\iota(\tilde{B}_n^{1,0,\gamma})(x/n) - \sup_{0\leq x\leq mn} u_\iota(\tilde{B}_n^{1,0,\gamma})(x/n) \geq \epsilon\right)$$

$$= P\left(\sup_{x>m} u_\iota\left(\tilde{B}_n^{1,0,\gamma}(\cdot)\mathbb{1}_{\{\cdot\leq m\}} + \tilde{B}_n^{1,0,\gamma}(\cdot)\mathbb{1}_{\{\cdot>m\}}\right)(x) - \sup_{0\leq x\leq m} u_\iota(\tilde{B}_n^{1,0,\gamma})(x) \geq \epsilon\right)$$

$$\leq P\Bigg(\sup_{x>m} u_\iota\Bigg(N\eta^{1/2-\gamma}\mathbb{1}_{\{\cdot\leq\eta\}} + \tilde{B}_n^{1,0,\gamma}(\cdot)\mathbb{1}_{\{\eta<\cdot\leq m\}}$$

$$-[\hat{D}^{-1/2}n^{-1/2}w_\gamma([nz]/n)\frac{n[nz]}{(n+[nz])n}(\hat{\rho}_1^n - \rho) + \hat{\delta}_n N m^{-1/2}]\mathbb{1}_{\{\cdot>m\}}\Bigg)(x)$$

$$- \sup_{0\leq x\leq m} u_\iota(\tilde{B}_n^{1,0,\gamma}\mathbb{1}_{\{\cdot\geq\eta\}} - \mathbb{1}_{\{\cdot\leq\eta\}}N\eta^{1/2-\gamma})(x) \geq \epsilon\Bigg)$$

$$+ P\left(\max_{k>nm} w_\gamma(k/n)\frac{nk}{(n+k)n}|n^{1/2}(\hat{\rho}_{n+1,0}^{n+k} - \rho)| \geq m^{-1/2}N\right)$$

$$+ P\left(\max_{k<n\eta}|\tilde{B}_n^{1,0,\gamma}(k)| \geq \eta^{1/2-\gamma}N\right)$$

$$\to P\Bigg(\sup_{x>m} u_\iota\left(\eta^{1/2-\gamma}N\mathbb{1}_{\{\cdot\leq\eta\}} + w_\gamma G\mathbb{1}_{[\eta,m]}(\cdot) - [w_\gamma(\cdot)\frac{\cdot}{\cdot+1}W(1) + \delta N m^{-1/2}]\mathbb{1}_{\{\cdot\geq m\}}\right)(x)$$

$$- \sup_{0\leq x\leq m} u_\iota(w_\gamma G\mathbb{1}_{\{\cdot\geq\eta\}} - \mathbb{1}_{\{\cdot\leq\eta\}}N\eta^{1/2-\gamma})(x) \geq \epsilon\Bigg) + \frac{C}{N^2}$$

$$\to P\Bigg(\sup_{x>m} u_\iota\left(w_\gamma G\mathbb{1}_{[0,m]}(\cdot) - [w_\gamma(\cdot)\frac{\cdot}{\cdot+1}W(1) + \delta N m^{-1/2}]\mathbb{1}_{\{\cdot\geq m\}}\right)(x)$$

$$- \sup_{0\leq x\leq m} u_\iota(w_\gamma G)(x) \geq \epsilon\Bigg) + \frac{C}{N^2} \to 0$$

as $n \to \infty$, followed by $\eta \to 0$, $m \to \infty$, and $N \to \infty$, where $G(\cdot) = \frac{1}{1+\cdot}(W(1+\cdot) - (1+\cdot)W(1))$, $\hat{\delta} = \mathrm{sign}(\hat{\rho}_1^n - \rho)$, and $\delta = \mathrm{sign}(W(1))$. Here, we use the monotony of u_ι, for the first and second convergence the CMT, as well as for the last one the assumed convergence. Both applications of the CMT use that $u_\iota : \mathcal{D}[0,\infty) \to \mathcal{D}[0,\infty)$ is continuous with respect to $\|\cdot\|_{[0,\infty)}$ and for the second we additionally use $Gw_\gamma\mathbb{1}_{\{\cdot\geq\eta\}} \xrightarrow{P} Gw_\gamma$ as $\eta \to 0$ on $\mathcal{D}[0,m]$.

Under Assumption $\mathbf{H_{LA}^{(c)}}$ we similarly obtain that

$$\tilde{B}_n^{1,0,\gamma}(k) = w_\gamma(k/n)\frac{nk}{(n+k)n}\frac{n^{1/2}}{k}\sum_{i=n+1}^{n+k}(Z_i^{(0)} - \rho_{i,n}) - w_\gamma(k/n)\frac{nk}{(n+k)n}n^{-1/2}\sum_{i=1}^{n}(Z_i^{(0)} - \rho_{i,n})$$

$$+ w_\gamma(k/n)\frac{nk}{(n+k)n}\frac{1}{k}\sum_{i=n+1}^{n+k} g_\rho(i/n) - w_\gamma(k/n)\frac{nk}{(n+k)n}\frac{1}{n}\sum_{i=1}^{n} g_\rho(i/n),$$

and that the first and second summand have the same behavior as before. The third converges as $n \to \infty$ uniformly on $[m,\infty)$ towards $(\cdot + 1)^{-1} \int_1^{1+\cdot} g_\rho(x)dx$, where we use the integrability of g_ρ on $\mathbb{R}$. The last summand is zero by definition. Thus, we obtain with the same arguments

$$P\left(\sup_{x>nm} u_\iota(\tilde{B}_n^{1,0,\gamma})(x/n) - \sup_{0\le x\le mn} u_\iota(\tilde{B}_n^{1,0,\gamma})(x/n) \ge \epsilon\right)$$
$$\to P\left(\sup_{x>m} u_\iota\left(w_\gamma(G+\tilde{g})\mathbb{1}_{[0,m]}(\cdot) - [w_\gamma(\cdot)(\frac{\cdot}{\cdot+1}W(1)+\tilde{g}) + \delta N m^{-1/2}]\mathbb{1}_{\{\cdot\ge m\}}\right)(x)\right.$$
$$\left. - \sup_{0\le x\le m} u_\iota(w_\gamma(G+\tilde{g}))(x) \ge \epsilon\right) \to 0$$

as $n \to \infty$, followed by $m \to \infty$. Hence, under Assumption $\mathbf{H}_{\mathbf{LA}}^{(\mathbf{o})}$ we get the convergence in display (2.2.18). □

Remark 2.2.6. *In the third result, the condition* $\lim_{\|x\|\to\infty} \sup_{z\in\mathbb{R}^+} u_\iota(x)(z) = \infty$ *can be replaced by the condition* u_ι *fulfills the triangle inequality and*

$$\left\| u_\iota\left(w_\gamma\left(\frac{[n\cdot]}{n}\right)\hat{D}^{-1/2}\frac{n}{n+[n\cdot]}\frac{\lambda(R_{k^*}\cap(n,n+[n\cdot]])}{\sqrt{n}}\Delta_\rho\right)\right\| \xrightarrow{P} \infty.$$

Furthermore, the convergences as displayed in (2.2.17) *and* (2.2.19) *do not have to hold for each* $\lambda > 0$ *but rather for certain* λ*s. In many cases this difference does not matter.*

Proposition 2.2.7. *The functions* $u_1(g)(x) = |g(x)|$ *and* $u_2(g)(x) = \int_0^x (1+z)^{-2}|g(z)|dz$ *fulfill the rate displayed in* (2.2.17).

Proof. u_1 is obvioulsy continous. For u_2 we obtain: Let $g_1, g_2 \in \mathcal{D}[0,\infty)$ with $\|g_1 - g_2\| < \epsilon$ then it holds that

$$\|u_2(g_1) - u_2(g_2)\| \le \|g_1 - g_2\| \int_0^\infty (1+z)^{-2}dz < \epsilon.$$

Thus, u_2 is also continuous. Obviously, u_1 and u_2 fulfill the two main assumptions. Now, we show that the rate displayed in (2.2.17) holds:

$$P\left(\sup_{x>m} |w_\gamma(x)\frac{x}{x+1}W(1) + \delta N m^{-\lambda}| - \sup_{0\le x\le m} |w_\gamma(x)G(x)| \ge \epsilon\right)$$
$$\le P\left(|W(1)| - \sup_{m/2\le x\le m} |w_\gamma(x)G(x)| \ge \epsilon - Nm^{-1/2}\right)$$
$$\le P\left(2\sup_{m/2\le x\le m}\left|\frac{W(z+1)}{(1+z)}\right| \ge \epsilon - Nm^{-\lambda}\right) = o(1)$$

due to the law of the iterated logarithm. For u_2 we obtain the following upper bound

$$P\left(\int_m^\infty w_\gamma(z)\frac{z}{(z+1)^3}|W(1)|dz + Nm^{-\lambda}\int_m^\infty (z+1)^{-2}dz \ge \epsilon\right) = o(1).$$

□

Proposition 2.2.8. *Suppose* G *is a centered Gaussian process with covariance structure* $\mathbb{E}[G(s)G(t)] = \frac{s}{1+s} \wedge \frac{t}{1+t}$ *and* W *a standard Brownian motion. Then, it holds that*

1. *if* $u_1(g(\cdot))(z) = |g(z)|$ *then*

$$\sup_{z\in(0,\infty)} u_\iota(w_\gamma G)(z) \stackrel{D}{=} \sup_{z\in(0,1)} |z^{-\gamma}W(z)|$$

and

$$\sup_{z\in(0,\infty)} u_\iota(w_\gamma(G+\tilde{g}))(z) \stackrel{D}{=} \sup_{z\in(0,1)} |z^{-\gamma}(W(z) + \tilde{g}(z/(1-z)))|,$$

2. *if* $u_2(g(\cdot))(z) = \int_0^z \frac{(1+x)^2}{(1+2x)^4}|g(x)|dx$ *then*

$$\sup_{z\in(0,\infty)} u_2(G)(z) \overset{D}{=} \int_0^{1/2} B(t)^2\,dt$$

Proof. 1.: Using

$$\{G(z)\ :\ z \in [0,\infty)\} \overset{D}{=} \{W(z/(1+z))\ :\ z \in [0,\infty)\}$$

implies

$$\sup_{z\in(0,\infty)} u_\iota(w_\gamma G)(z) \overset{D}{=} \sup_{z\in(0,\infty)} \left| \left(\frac{z}{1+z}\right)^{-\gamma} W\left(\frac{z}{1+z}\right) \right| \overset{D}{=} \sup_{z\in(0,1)} |z^{-\gamma}W(z)|.$$

Analogously, we can show the second equality.
2.: Using the first equality implies

$$\begin{aligned}
&\sup_{z\in(0,\infty)} \int_0^z \frac{(1+s)^2}{(1+2s)^4} G(s)^2 ds \overset{D}{=} \sup_{z\in(0,\infty)} \int_0^z \frac{(1+s)^2}{(1+2s)^4} W(s/(s+1))^2 ds \\
&= \sup_{z\in(0,\infty)} \int_0^{z/(1+2z)} \frac{(1+\frac{t}{1-2t})^2}{(1+2\frac{t}{1-2t})^4} W\left(\frac{t}{1-2t}/(\frac{t}{1-2t}+1)\right)^2 \frac{1}{(1-2t)^2} dt \\
&= \sup_{z\in(0,1)} \int_0^{1/2} (1-t)^2 W\left(\frac{t}{1-t}\right)^2 dt \\
&\overset{D}{=} \int_0^{1/2} B(t)^2\,dt,
\end{aligned}$$

where we substitute s by $t/(1-2t)$ and use $\left\{(1-t)W\left(\frac{t}{1-t}\right)\right\} \overset{D}{=} \{B(t)\}$, where B is a Brownian bridge. □

2.3 Examples

This subsection highlights three examples for the main model (1.2.1). For each example we prove that the assumptions on the main theorems of this section are fulfilled. Additionally, we show under which cases which ones of the LRV estimates are useful.
For each example we assume that the following assumption holds:

Assumption (LRV). *Let for each fixed* $t > 0$

$$D_{[nt]} = \frac{1}{n} I\!E\left[\left(\sum_{i=1}^{[nt]} Z_i^{(0)} - I\!E\left[Z_i^{(0)}\right]\right)^2\right] \to tD > 0, \quad as \quad n \to \infty.$$

2.3.1 Identical Independent Distribution

In the first example we consider the special case of i.i.d. innovations.

Assumption (IID). *In the main model let* $\{(\tilde{\epsilon}_{1,n},\tilde{\epsilon}_{2,n})^T\}$ *be i.i.d. Additionally, let* $\|\tilde{\epsilon}_{l,1}\|_{r'_l} < \infty$ *for* $l = 1,2$ *with* $r'_1, r'_2 > 2$ *and* $\frac{r'_1 r'_2}{r'_1 + r'_2} > 2$.

Remark 2.3.1. 1. *The assumption that the vectors* $\{(\tilde{\epsilon}_{1,n},\tilde{\epsilon}_{2,n})^T\}$ *are identically distributed can be weakened without much effort.*

2. *The differentiation in* r'_1 *and* r'_2 *makes sense, since it could be possible that*

$$B_t = \begin{pmatrix} \sigma_{1,t} & 0 \\ 0 & \sigma_{2t} \end{pmatrix} \cdot \begin{pmatrix} 1 & 0 \\ \rho_t & \sqrt{1-\rho_t^2} \end{pmatrix}$$

and $\|\tilde{\epsilon}_{1n}\|_{r'_1} < \infty$ *and* $\|\tilde{\epsilon}_{2n}\|_{r'_2} < \infty$ *for some* $r'_1 > r'_2$, *yielding* $\|Y_n\|_{r'_2} < \infty$ *but not necessarily* $\|Y_n\|_{r'_1} < \infty$.

3. *We already know that* B_t *is not unique, which implies that* $\{Z_n^{(0)}\}$ *is a sequence of independent but not necessarily identically distributed random variables. In particular,* B_t *has direct influence on* $\operatorname{Var}\left[Z_t^{(0)}\right]$. *However, Assumption (LRV) limits the fluctuation of* $\operatorname{Var}\left[Z_t^{(0)}\right]$ *and thereby that of the matrices* $B_1, B_2, \ldots$.

1. Example for the WFC. *Let Assumptions (IID) and (LRV) be fulfilled. Then,* $\{Z_i^{(0)} - \rho_i\}$ *satisfies the WFC so that Theorem 2.1.4 holds. This essentially follows from the following steps:*

1. *Since* $r = \frac{r'_1 r'_2}{r'_1 + r'_2} > 2$, $Z_i^{(0)}$ *is* L_r*-bounded, where*

$$Z_i^{(0)} = \frac{\begin{pmatrix} 1 & 0 \end{pmatrix} B_i \begin{pmatrix} \tilde{\epsilon}_{1,i} \\ \tilde{\epsilon}_{2,i} \end{pmatrix} \begin{pmatrix} \tilde{\epsilon}_{1,i} & \tilde{\epsilon}_{2,i} \end{pmatrix} B_i^T \begin{pmatrix} 0 \\ 1 \end{pmatrix}}{\sigma_{1,i}\sigma_{2,i}}.$$

Furthermore, $\{Z_i^{(0)} - \rho_i\}$ *is centered and independent.*

2. *With item* 1. *and Assumption (LRV) it is easy to show that Lyapounov's condition is fulfilled and hence, the sequence* $\{Z_n^{(0)} - \rho_n\}$ *satisfies a CLT.*

3. *In analogy to the proof of Theorem 10.1 in Billingsley (1968) the finite-dimensional distributions converge.*

4. *Now, we apply Theorem 15.6 of Billingsley (1968), where the second condition of this theorem follows similarly to the second item. Hence,* $\{Z_i^{(0)} - \rho_i\}$ *fulfills a FCLT with an asymptotic LRV* $D = \lim_{n\to\infty} \frac{1}{n} \sum_{i=1}^n I\!E\left[(Z_i^{(0)} - \rho_i)^2\right]$.

5. *Due to the independence of* $\{Z_i^{(0)} - \rho_i\}$ *and the boundary of the second moments, we get the Kolmogorov-type inequalities* $(\mathcal{K}_{\mathbf{r}}^{(\mathbf{1})})$, $(\mathcal{K}_{\mathbf{r}}^{(\mathbf{2})})$ *for* $r_z = 2$. *This implies the WFC.*

1. Example of Theorem 2.1.4. *Let Assumption (IID) and Assumption (LRV) be fulfilled and set*

$$w_\gamma : (0,1) \to \mathbb{R}_+, \qquad w_\gamma(z) = \left(\frac{1}{z(1-z)}\right)^\gamma.$$

Then, Theorem 2.1.4 holds with

$$f_0(g(\cdot)) = \sup_{z\in[0,1]} |g(z)| \quad \text{or} \quad f_0(g(\cdot) = \int_0^1 |g(z)|dz.$$

2. Example of Theorem 2.1.4. *Under the assumption of 1. Example of Theorem 2.1.4 and* $k_1^* = n^\epsilon$, $k_2^* = n - n^\epsilon$ *for some fixed* $\epsilon > 0$ *and* $|\Delta_\rho| > 0$, *there exists a* $\gamma > 0$ *so that*

$$a_n|\Delta_\rho|n^{-(1-\gamma)(1-\epsilon)} \to \infty \quad \text{with} \quad a_n = \sqrt{n}.$$

This implies that the test $\phi_{\iota,0,0}^\gamma$ *is consistent if* $\hat{D}_n^{-1} = O_P(1)$.

The preceding example shows that the test detects early and late changes if we choose γ high and the inverse of the LRV is bounded.

1. Example of Theorem 2.1.15. *Under Assumption (IID) and* $\mathbf{H_A}$ *let* $[n\theta_1^*]$ *and* $[n\theta_2^*]$ *be the unknown change-points where* $0 < \theta_1^* < \theta_2^* < 1$ *and* $\Delta_\rho \neq 0$ *are independent of* n. *Since* $\{Z_n^{(0)} - \rho_n\}$ *satisfies the Kolmogorov-type inequalities for* $r_z = 2$ *and have uniformly bounded second moments, we get from Theorem 2.1.15 an estimation rate of* $n\|\hat{\theta} - \theta^*\| = O_P(1)$.

1. Example of Theorem 2.1.16. *If we replace the condition* $\Delta_\rho \neq 0$ *in the 1. Example of Theorem 2.1.15 by* $\Delta_{\rho,n}^{-1} = o(n^{1/2})$, *we get* $n\Delta_{\rho,n}\|\hat{\theta} - \theta^*\| = O_P(1)$.

Thus, the estimates approximate the change-points by Theorem 2.1.16 if they do not vanish too fast. The possibility of successful estimation with this estimate is only given (in some sense) if $\Delta_{\rho,n}^{-1} = O(n^{1/2})$. Additionally, the change size influences the estimation rate.

1. Example of Theorem 2.1.22. *Under Assumption (IID) and* $\mathbf{H_A^{(M)}}$ *we have* $r_z = 2$, $r = r_1'r_2'/(r_1' + r_2') > 2$, *and assume that the condition*

$$n^{1/2+1/r} = o(\min_i \Delta_{\rho,i,n}^2 \underline{\Delta}_{k^*,n})$$

of Theorem 2.1.22 is fulfilled. Then, we can choose

$$a_n = n\left[\left(\min_i \Delta_{\rho,i}^2 / \max_{1\le i\le R} |\Delta_{\rho,i}|\right)^2 \wedge \min_i \Delta_{\rho,i,n}^2\right]$$

and get $a_n n^{-1}\|k^* - \hat{k}\| = O_P(1)$. *In particular, we obtain* $a_n = n$ *if* $\min_{i,n}|\Delta_{\rho,i,n}| > \epsilon > 0$.

1. Example of Theorem 2.1.25. *Under the assumption of 1. Example of Theorem 2.1.22 the condition*

$$n^{1/2+1/r} \ll \beta_n \le \frac{1}{4C^*} \min_{1\le i\le R^*} \Delta_{\rho,i,n}^2 \underline{\Delta}_{k^*,n}$$

is sufficient for $|\hat{R} - R^*| = o_P(1)$ *as* $n \to \infty$.

1. Example of Theorem 2.1.28. *If we modify Assumption (IID) by just assuming that* $r_z = 2 \wedge r_1'r_2'/(r_1' + r_2') > 1$, *the Kolmogorov-type inequalities hold for* $r_z > 1$. *Moreover, we can choose* β_n *with*

$$\log(n)^2 \vee n^{2/r_z - 1} \ll \beta_n \quad \text{and} \quad \beta_n \vee n^{1/r_z} \ll \lambda(R_{k^*}^c)\lambda(R_{k^*})^2|\Delta_{\rho,n}^2 n^{-2}|$$

to get an asymptotic test by Theorem 2.1.28.

The above example demonstrates the possibility of a test with fewer assumptions than under 1. Example of Theorem 2.1.4. In particular, a FCLT is not necessary and even the second moments of $Z_i^{0)}$ do not have to be bounded.

1. Example of Theorem 2.1.33. *Under H_0 and Assumptions (IID), (LRV) let $\mathfrak{f}$ be an absolute integrable kernel and $q_n = o(n)$. Then, we see that conditions 1., 2., and 3. of Theorem 2.1.33 are fulfilled, since*

$$|\hat{D} - D| \leq |D_n - D| + |\hat{D} - D_n| = o(1) + \left| \frac{1}{n} \sum_{i=1}^{n} \left((Z_i^{(0)} - \rho_i)^2 - \operatorname{Var}\left[Z_i^{(0)}\right] \right) \right| + \left| \frac{1}{n} \sum_{i,j=1, i \neq j}^{n} \mathfrak{f}\left(\frac{i-j}{q_n}\right) (Z_i^{(0)} - \rho_i)(Z_j^{(0)} - \rho_j) \right| = o_P(1).$$

Here, the first order follows from Assumption (LRV), the second from Kolmogorov's inequality, and the third from Markov's inequality, the absolute boundary of $\mathfrak{f}$ as well as $q_n = o(n)$. Under H_0 and with $\hat{\rho}_n(i) \equiv \overline{Z^{(0)}}$, case (A) can be applied with $\delta_1 = 1/2$. Hence, from Theorem 2.1.33 we get that $|D - \hat{D}_{0,n}| = o_P(1)$.

2. Example of Theorem 2.1.33. *If we just replace the assumption of H_0 by $\mathbf{H_{LA}}$ in 1. Example of Theorem 2.1.33, we obtain the same result by case (B).*

3. Example of Theorem 2.1.33. *If we just replace the assumption of H_0 by $\mathbf{H_A}$ in 1. Example of Theorem 2.1.33, we obtain for $\hat{\rho}_n(i) \equiv \overline{Z^{(0)}}$ that case (D) could be applied with $\delta_1 = 0$. Hence, we get $|D - \hat{D}_{0,n}| = O_P(q_n)$ which implies a consistent test if $q_n = o(n^{1/2})$.*

4. Example of Theorem 2.1.33. *In the model of multiple (finite) change-points which do not vanish asymptotically and where the distances between the locations diverge with rate n, i.e., $n \sim \underline{\Delta}_{k^*,n}$, let Assumption (IID) and Assumption (LRV) be fulfilled. Then, we use a correlation estimate based on a change-point estimate:*

$$\hat{\rho}_n(i) = \sum_{j=1}^{m+1} \mathbb{1}_{i \in \hat{R}_{j,n}} \overline{Z^{(0)}}_{\hat{k}_{j-1}+1}^{\hat{k}_j},$$

where the $\hat{k}_j$'s are the change-point estimates. Furthermore, we define

$$\hat{C}_1 = [1,\hat{k}_1] \cap [1,k_1^*] \quad \hat{C}_2 = [\hat{k}_1 + 1,\hat{k}_2] \cap [1,k_1^*], \ldots, \quad \hat{C}_{(m+1)^2} = [\hat{k}_m + 1,n] \cap [k_m^* + 1,n],$$

where we get $\|k^* - \hat{k}\| = O_P(1)$ by the 1. Example of Theorem 2.1.22. This implies that case (E) of Theorem 2.1.33 is fulfilled with $\delta_j = 1/2$ for all $j = (k-1) \cdot m + k$, $k = 1, \ldots ,m+1$, and $\delta_j = 0$ for all $j \neq (k-1) \cdot m + k$, $k = 1, \ldots ,m+1$. Additionally, we define for instance

$$C_1 = [1,k_1^* + \log(n)], \quad C_2 = [k_1^* - \log(n),k_1^* + \log(n)], \ldots, C_{(m+1)^2} = [k_m^* - \log(n),n]$$

which implies $P(\bigcap_j \{\hat{C}_j \subset C_j\}) \to 1$. Hence, we get $|D - \hat{D}_{0,n}| = O_P(q_n n^{-1/2})$ from Theorem 2.1.33, which only implies a consistent estimate if $q_n = o(n^{1/2})$. For a consistent test, it is just required that $q_n = o(n)$, cf. item 3. of Theorem 2.1.4.

Remark 2.3.2. *In each of the preceding examples of Theorem 2.1.33 we can clearly replace the (piecewise) sample means by some generally weighted estimate types. In this case, we just have to prove the necessary asymptotic property. Explicitly, we have the weighted sample mean in mind:*

$$\hat{\rho}_n(i) = \sum_{j \in M \cap (0,n]} Z_j^{(0)} k\left(\frac{|j-i|}{\#(M \cap (0,n])}\right) \Big/ \sum_{j \in M \cap (0,n]} k\left(\frac{|j-i|}{\#(M \cap (0,n])}\right) \quad \text{for all } i \in M,$$

where $M \in \{[1,n],\hat{C}_1,\ldots,\hat{C}_{m+1}\}$ and $k(\cdot)$ is a kernel function such as the Uniform, Triangular, Epanechnikov, or Gaussian Kernel. Due to the independence of $\{Z_n^{(0)}\}$ and the change-point estimation rate $\|k^* - \hat{k}\| = O_P(1)$, it is quite easy to obtain the same results in the case of (multiple) change-point(s).

1. Example of Theorem 2.2.1, 2.2.3 and 2.2.5. *Under Assumptions (IID), (LRV)* $\{Z_i^{(0)} - \rho_i\}$ *satisfies the WFC so that Theorem 2.2.1 holds. As in 1. Example for the WFC we prove the FCLT on* $\mathcal{D}[0,1]$ *and that* $\{Z_i^{(0)} - \rho_i\}$ *fulfills the Kolmogorov-type inequalities. Since a Brownian motion is scale invariant we can extend the convergence on* $\mathcal{D}[0,1+m]$*. Possible (weighting) functions are, with* $\gamma \in [0,1/2)$,

$$w_\gamma(z) = \left(\frac{1+z}{z}\right)^\gamma \quad \text{and} \quad u_1(g)(\cdot) = |g(\cdot)|, \quad \text{or} \quad u_2(g)(\cdot) = \int_0^{\cdot} (1+x)^{-2}|g(x)|dx.$$

2.3.2 Mixing

Definition 2.2. *A process* $\{Z_n\}_{n\in\mathbb{Z}}$ *is called* α*-mixing and respectively* ϕ*-mixing of size* $-\xi$ *if the coefficient holds that*

$$\alpha(n) := \sup_{k\in\mathbb{Z}} \alpha(\mathcal{F}^k_{-\infty},\mathcal{F}^\infty_{k+n}) := \sup_{k\in\mathbb{Z}} \sup_{A\in\mathcal{F}^k_{-\infty}, B\in\mathcal{F}^\infty_{k+n}} |P(A\cap B) - P(A)P(B)| = o(n^{-\xi})$$

and respectively that

$$\phi(n) := \sup_{k\in\mathbb{Z}} \phi(\mathcal{F}^k_{-\infty},\mathcal{F}^\infty_{k+n}) := \sup_{k\in\mathbb{Z}} \sup_{A\in\mathcal{F}^k_{-\infty}, B\in\mathcal{F}^\infty_{k+n}} |P(A|B) - P(B)| = o(n^{-\xi}).$$

Throughout this sub-subsection, we assume that the following assumption is fulfilled:

Assumption (MIX). *Let*

$$\begin{pmatrix} X_n \\ Y_n \end{pmatrix} = \begin{pmatrix} \mu_{1n} \\ \mu_{2n} \end{pmatrix} + \begin{pmatrix} \sigma_{1,n} & 0 \\ 0 & \sigma_{2,n} \end{pmatrix} \begin{pmatrix} f_{1,n}(\epsilon_{1,n},\epsilon_{2,n},\ldots) \\ f_{2,n}(\epsilon_{1,n},\epsilon_{2,n},\ldots) \end{pmatrix},$$

where the processes $\{\epsilon_{m,n}, n \in \mathbb{Z}\}$ *are* α*–mixing (*ϕ*–mixing) with coefficients* $\alpha_n(m)$ $(\phi_n(m))$ *for each fixed* $n \in \mathbb{N}$ *and* $\{\epsilon_{1,n}, n \in \mathbb{Z}\}, \{\epsilon_{2,n}, n \in \mathbb{Z}\}, \ldots$ *are totally independent of each other. Let* $f_{1,n}(\cdot)$ *and* $f_{2,n}(\cdot)$ *be Borel-measurable functions for each* n*. Additionally, we assume that* X_n *and* Y_n *possess the correlation* ρ_n *and*

1. $\tilde\epsilon_{1,n} := f_{1,n}(\epsilon_{1,n},\epsilon_{2,n},\ldots)$ *and* $\tilde\epsilon_{2,n} := f_{2,n}(\epsilon_{1,n},\epsilon_{2,n},\ldots)$ *are centered and normalized;*
2. $\|\tilde\epsilon_{1,n}\|_{r'_{1,n}} < \infty$ *and* $\|\tilde\epsilon_{1,n}\|_{r'_{2,n}} < \infty$*, where* $r'_{1,n}, r'_{2,n} \geq 2$ *and* $\inf_t \frac{r'_{1,n}r'_{2,n}}{r'_{1,n}+r'_{2,n}} > 2$*;*
3. *for* $r' \in (2,r]$ *with* $r = \inf_n \frac{r'_{1,n}r'_{2,n}}{r'_{1,n}+r'_{2,n}}$ *let* $\tilde\alpha(m) = \sum_{n=1}^\infty \alpha_n(m)$ *fulfill*

$$\sum_{m=0}^\infty (\tilde\alpha(m))^{2(1/r'-1/r)} < \infty \tag{2.3.1}$$

and

$$\sum_{m=0}^\infty \left(\sum_{k=0}^m (\tilde\alpha(k))^{-(1-\frac{2}{r})}\right)^{-1/2} < \infty. \tag{2.3.2}$$

Remark 2.3.3. *1. The random variable* $\epsilon_{n,t}$ *can be vector-valued.*

2. *If* (2.3.1) *is only fulfilled for* $r' = 2$*, we can only prove the FCLT but not the WFC.*
3. (2.3.2) *is only used to show that the Kolmogorov-type inequalities hold.*
4. *If* $(\tilde\epsilon_{1,t},\tilde\epsilon_{2,t})$ *is* α*-mixing of size* $-r/(r-2)$, (2.3.1) *and* (2.3.2) *hold true.*

Additionally, for the LRV estimation we need the following conditions for the kernel:

Assumption (K2). *For all* $x \in \mathbb{R}$ *let* $|\mathfrak{f}(x)| \leq 1$, $\mathfrak{f}(x) = \mathfrak{f}(-x)$*, and* $\mathfrak{f}(0) = 1$*. Additionally, let* $\mathfrak{f}$ *be continuous at zero and for almost all* $x \in \mathbb{R}$, $\int_{\mathbb{R}} |\mathfrak{f}(x)|dx < \infty$*, and there is a non-increasing function* $l(x) \geq |\mathfrak{f}(x)|$ *so that* $\int_{\mathbb{R}} |x|l(x)dx < \infty$.

Remark 2.3.4. *The above kernel assumption includes Assumptions 1.1. and 1.4. of De Jong (2000).*

2. Example for the WFC. *Let Assumptions (LRV) and (MIX) be fulfilled. Then,* $\{Z_i^{(0)} - \rho_i\}$ *satisfies the WFC and Theorem 2.1.4 holds. This follows from the following:*

1. *We apply Herrndorf (1984, Corollary 1) to get the FCLT, for which we need Assumption (LRV), the centering as well as the upper bound of the second moments of* $\{Z_n^{(0)} - \rho_n\}$*, and that* $\{Z_n^{(0)} - \rho_n\}$ *is* α*-mixing with coefficient* $\tilde{\alpha}(m)$ *with* $\sum_{m=1}^{\infty} \tilde{\alpha}(m)^{1-\frac{2}{r}} \le C \sum_{m=1}^{\infty} \tilde{\alpha}(m)^{2(1/r'-1/r)} < \infty$*: Since* $\sigma(X_n)$*,* $\sigma(Y_n)$*, and* $\sigma(X_nY_n)$ *are sub–*σ*-fields of* $\sigma(X_n,Y_n)$ *(cf. Davidson (1994, Th. 10.4)) it is sufficient to prove the mixing size of* $(X_n,Y_n)^T$*:*

$$\begin{aligned}&\alpha\left(\sigma\left((X_{-\infty},Y_{-\infty}),\ldots,(X_n,Y_n)\right),\sigma\left((X_{n+m},Y_{n+m}),\ldots,(X_{\infty},Y_{\infty})\right)\right)\\ &\le \alpha\left(\bigvee_{k=1}^{\infty}\left\{\bigvee_{i=-\infty}^{n}\sigma(\epsilon_{k,i})\right\},\bigvee_{k=1}^{\infty}\left\{\bigvee_{i=n+m}^{\infty}\sigma(\epsilon_{k,i})\right\}\right)\\ &\le \sum_{k=1}^{\infty}\alpha\left(\bigvee_{i=-\infty}^{n}\sigma(\epsilon_{k,i}),\ \bigvee_{i=n+m}^{\infty}\sigma(\epsilon_{k,i})\right)=\sum_{k=1}^{\infty}\alpha_k(m)=\tilde{\alpha}(m),\end{aligned}$$

   where the first inequality follows from the monotonicity of α *and the second from Bradley (2009, Th. 6.2). Hence, by the third condition of Assumption (MIX),* $\{Z_n^{(1)} - \rho_n\}$ *is an* α*-mixing process with coefficients* $\tilde{\alpha}(m)$*.*

2. *On the one hand, the Kolmogorov's type inequalities are fulfilled, since* $\{(Z_n^{(1)} - \rho_n),\mathcal{F}_n\}$ *is an* L_2*-mixingale with* $\mathcal{F}_n := \bigvee_{i=1}^{n} \bigvee_{j=1}^{\infty} \sigma(\epsilon_{j,i})$ *and sequence* $\xi_n = \tilde{\alpha}(n)^{1/2-1/r}$*, which follows directly from Davidson (1994, Th. 14.2). This implies a maximal moment inequality (cf. Davidson (1994, p. 255)), which goes back to McLeish (1975) and directly implies* $(\mathcal{K}_{\mathbf{r}}^{(1)})$*, since*

$$\sum_{m=0}^{\infty}\left(\sum_{k=0}^{m}\xi_k^{-2}\right)^{-1/2} < \infty.$$

   On the other hand, we know that an α*-mixing process is time–reversible so that* $\{Z_{m-n}^{(1)} - \rho_{m-n}\}_{1\le n\le m}$ *is an* α*-mixing process with sequence* $\tilde{\alpha}(n)$ *for each fixed, but arbitrary* $m \in \mathbb{N}$*. Hence, with the same arguments as before* $(\mathcal{K}_{\mathbf{r}}^{(2)})$ *and* $(\mathcal{K}_{\mathbf{r}}^{(3)})$ *are fulfilled. Thus, the WFC is confirmed.*

2. Example of Theorem 2.1.4, 2.1.15, 2.1.16, 2.1.22, 2.1.25, 2.1.28, and 2.2.1, 2.2.3, 2.2.5. *Replace Assumption (IID) by Assumption (MIX) in each of the first examples. Then they hold true.*

5. – 8. Example of Theorem 2.1.33. *Under the assumptions on the 2. Example for the WFC let* $\mathfrak{f}$ *additionally fulfill Assumption (K2) and let* $q_n = o(n^{1/2-1/r'})$ *with* r' *from the third condition of Assumption (MIX). Then, we apply De Jong (2000, Th. 2) to obtain* $|\hat{D}_n - D| = o_P(1)$*, where we use Assumption (LRV), Assumption (K2), and the condition* $\sum_{m=1}^{\infty} \tilde{\alpha}(m)^{1-\frac{2}{r}-\epsilon} < \infty$ *for an arbitrarily small* $\epsilon > 0$*. Additionally, we know by 2. Example for the WFC that the Kolmogorov-type inequalities are fulfilled. Hence, we get the same estimation rates as* $|\hat{D}_{0,n} - D| = o_P(1)$ *as in the settings of Example 1 to 4 of Theorem 2.1.33.*

2.3.3 Near Epoch Dependent

In this sub-subsection, we assume a near epoch dependent structure.

Definition 2.3. *Davidson (1994, Def. 17.2) For a stochastic array* $\{\{V_{nt}\}_{t=-\infty}^{\infty}\}_{n=1}^{\infty}$*, possibly vector-valued, on a probability space* $(\Omega,\mathcal{F},P)$*, let* $\mathcal{F}_{n,t,t-m}^{t+m} = \sigma/V_{n,t-m},\ldots,V_{n,t+m})$*. If an integrable array* $\{\{X_{nt}\}_{t=-\infty}^{\infty}\}_{n=1}^{\infty}$*, satisfies*

$$\|X_{nt} - I\!E\left[X_{nt}|\mathcal{F}_{n,t,t-m}^{t+m}\right]\|_p \le d_{nt}\nu_m,$$

where $\nu_m \to 0$*, and* $\{d_{nt}\}$ *is an array of positive constants, it is said to be* L_p*-NED on* $\{V_{nt}\}$*.*

First, we want to reduce the dependency of $Z_t^{(0)}$ to the dependency of X_t and Y_t.

Assumption (NED). *Let X_n be L_{p_1}-NED on $\{V_n\}$ of size $-a_1$ and $\sup_n \|X_n\|_{r_1'} < \infty$. Let Y_n be L_{p_2}-NED on $\{V_n\}$ of size $-a_2$ and $\sup_n \|Y_n\|_{r_2'} < \infty$, where V_n is either α-mixing of size $-a_V \leq -r/(r-2)$ or ϕ-mixing of size $-r/(2r-2)$ with $r = r_1' r_2'/(r_1'+r_2')$. Furthermore, let* $\mathrm{Corr}\,(X_n,Y_n) = \rho_n$,

$$(a_1,a_2,p_1,p_2,r_1',r_2') \in \mathcal{M}_2^{\frac{1}{2}} = M_2^{\frac{1}{2}} \bigcap \mathbb{R}_+^4 \times \left\{ (r_1',r_2') \in \mathbb{R}_+^2 \ : \frac{r_1' r_2'}{r_1'+r_2'} > 2 \right\}, \tag{2.3.3}$$

$$M_k^\alpha = \left\{ (a_1,a_2,p_1,p_2,r_1',r_2') \in \mathbb{R}_+^6 \ : \ -a_{a_1,p_1,r_1'}^{a_2,p_2,r_2'}(p) \leq -\alpha \text{ with } p = k \right\}, \tag{2.3.4}$$

and

$$-a = -a_{a_1,p_1,r_1'}^{a_2,p_2,r_2'}(p) = \begin{cases} -\min\{a_1,a_2\}, & \text{if } U_1 \leq p_2 \text{ and } U_2 \leq p_1, \\ -\min\{a_1 C^1_{p_1,r_1',r_2',p}, a_2 C^2_{p_1,r_1',r_2',p}\}, & \text{if } 2r_2' \geq U_1 > p_2 \text{ and } 2r_1' \geq U_2 > p_1, \\ -\min\{a_1, a_2 C^2_{p_1,r_1',r_2',p}\}, & \text{if } 2r_2' \geq U_1 > p_2 \text{ and } U_2 \leq p_1 , \\ -\min\{a_1 C^1_{p_1,r_1',r_2',p}, a_2\}, & \text{if } U_1 \leq p_2 \text{ and } 2r_1' \geq U_2 > p_1, \\ 0, & \text{else}, \end{cases} \tag{2.3.5}$$

where $U_i = \frac{p r_i}{r_i - p}$ *for* $i = 1,2$, $C^1_{p_1,r_1',r_2',p} = \frac{p_1(r_1'r_2' - r_1'p - r_2'p)}{r_2'p(r_1'-p_1)}$ *and* $C^2_{p_1,r_1',r_2',p} = \frac{p_2(r_1'r_2' - r_1'p - r_2'p)}{r_1'p(r_2'-p_2)}$.

Remark 2.3.5. *Later, we will see that the condition $\frac{r_1'r_2'}{r_1'+r_2'} > 2$ in (2.3.3) yields that $\|Z_n^{(1)} - \rho_n\|_r < \infty$ as $r > 2$ and implies that $r_1',r_2' > 2$. Hence, the second moments of X_n and Y_n exist. The parameter set M_k^α ensures with $p = k$ that $\{Z_n^{(n)} - \rho_n\}$ is L_p-NED of size $-\alpha$. Thus, $\mathcal{M}_2^{\frac{1}{2}}$ contains parameters so that $\{Z_n^{(n)} - \rho_n\}$ is L^r-bounded, $r > 2$, and L_2-NED of size $-\frac{1}{2}$ on $\{V_n\}$, where V_n satisfies a certain mixing condition.*

Assumption (K3). *Davidson and De Jong (2000, Assumption 1) Suppose $\mathfrak{f} \in \mathcal{K}$, where*

$$\mathcal{K} = \Big\{ \mathfrak{f} : \mathbb{R} \to [-1,1] \ : \ \mathfrak{f}(0) = 1, \mathfrak{f}(x) = \mathfrak{f}(-x)\, \forall x \in \mathbb{R}, \int_{-\infty}^{\infty} |\mathfrak{f}(x)| dx < \infty, \int_{-\infty}^{\infty} |\psi(x)| dx < \infty,$$
$$\mathfrak{f} \text{ is continuous at } 0 \text{ and at all points except for a finite number} \Big\}$$

with $\psi(x) = (2\pi)^{-1} \int_{-\infty}^{\infty} \mathfrak{f}(z) e^{ixz} dz$.

Remark 2.3.6. *The Bartlett, Parzen, Quadratic Spectral, and Tukey-Hanning kernels are all elements of $\mathcal{K}$ (Davidson and De Jong (2000, p. 409)).*

Before we consider the examples under the NED dependency, we first analyze the NED–size of a product of two NED time series. On the one hand, it holds if X_n and Y_n are L_p-NED on $\{V_n\}$ of respective sizes $-\phi_X$ and $-\phi_Y$, that $X_n Y_n$ is $L_{p/2}$-NED of sizes $-\min\{\phi_X,\phi_Y\}$ (Davidson (1994, Th. 17.9)). On the other hand, if X_n and Y_n are L_2-NED on $\{V_n\}$ of size $-a$ and $\|X_n\|_{2r} < \infty$ and $\|Y_n\|_{2r} < \infty$ for $r > 2$, it holds that $X_n Y_n$ is L_2-NED of size $-a(r-2)/2(r-1)$ (Davidson (1994, Th. 17.17)). In the following lemma we combine both conclusions.

Lemma 2.3.7. *Let X_n be L_{p_1}-NED on $\{V_n\}$ of size $-a_1$ and $\|X_n\|_{r_1'} < \infty$. Moreover, let Y_n be L_{p_2}-NED on $\{V_n\}$ of size $-a_2$ and $\|Y_n\|_{r_2'} < \infty$. Then, $X_n Y_n$ is L_p-NED on $\{V_n\}$, where $1 \leq p < \min\{r_1',r_2'\}$ and $p_i \leq r_i$ for $i = 1,2$, of size $-a = -a_{a_1,p_1,r_1'}^{a_2,p_2,r_2'}(p)$ defined in (2.3.5).*

Proof. Since

$$\begin{aligned} &\|X_n Y_n - \mathbb{E}_{n-m}^{n+m}[X_n Y_n]\|_p \leq \|X_n (Y_n - \mathbb{E}_{n-m}^{n+m}[Y_n])\|_p \\ &\quad + \|(X_n - \mathbb{E}_{n-m}^{n+m}[X_n])\mathbb{E}_{n-m}^{n+m}[Y_n]\|_p + \|\mathbb{E}_{n-m}^{n+m}[(X_n - \mathbb{E}_{n-m}^{n+m}[X_n])(Y_n - \mathbb{E}_{n-m}^{n+m}[Y_n])]\|_p \\ &\leq \|X_n\|_{r_1'} \|Y_n - \mathbb{E}_{n-m}^{n+m}[Y_n]\|_{k_1} \\ &\quad + \|X_n - \mathbb{E}_{n-m}^{n+m}[X_n]\|_{k_2} \|\mathbb{E}_{n-m}^{n+m}[Y_n]\|_{r_2'} + \|X_n - \mathbb{E}_{n-m}^{n+m}[X_n]\|_{r_1'} \|Y_t - \mathbb{E}_{n-m}^{n+m}[Y_n]\|_{k_1}, \end{aligned} \tag{2.3.6}$$

by use of the Hölder inequality with selected k_i so that $\frac{1}{r_1'} + \frac{1}{k_1} = \frac{1}{r_2'} + \frac{1}{k_2} = \frac{1}{p}$. Now, we consider $\|Y_n - \mathbb{E}_{n-m}^{n+m}[Y_n]\|_{k_1}$ and $\|X_n - \mathbb{E}_{n-m}^{n+m}[X_n]\|_{k_2}$.
If $\frac{pr_1'}{r_1'-p} = k_1 \le p_2$ and $\frac{pr_2'}{r_2'-p} = k_2 \le p_1$, X_nY_n is L_p–NED of size $-\min\{a_1,a_2\}$.
In the other case if $p_2 < \frac{pr_1'}{r_1'-p} = k_1$ and $p_1 < \frac{pr_2'}{r_2'-p} = k_2$, where p_1 and p_2 are smaller than r_2' and r_1', respectively, we can apply the L_p–interpolation inequality with selected $\theta_1 \in (0,1)$ so that $\frac{1}{k_1} = \frac{1-\theta_1}{p_2} + \frac{\theta_1}{r_2'}$:

$$\|Y_n - \mathbb{E}_{n-m}^{n+m}[Y_n]\|_{k_1} \le \|Y_n - \mathbb{E}_{n-m}^{n+m}[Y_n]\|_{p_2}^{1-\theta_1} \|Y_n - \mathbb{E}_{n-m}^{n+m}[Y_n]\|_{r_2'}^{\theta_1} \le c d_{n,1}^{1-\theta_1} v_{m,2}^{1-\theta_1},$$

where $1-\theta_1 = \frac{p_2(r_1'r_2' - r_1'p - r_2'p)}{r_1'p(r_2'-p_2)}$. In the same way we get

$$\|X_n - \mathbb{E}_{n-m}^{n+m}[X_n]\|_{k_2} \le c d_{n,1}^{1-\theta_2} v_{m,1}^{1-\theta_2},$$

where $1-\theta_2 = \frac{p_1(r_1'r_2' - r_1'p - r_2'p)}{r_2'p(r_1'-p_1)}$. Hence, the three above summands can be estimated to $\tilde{d}_n\tilde{v}_m$, where $\tilde{d}_n = c\max\{d_{1,n},d_{n,2}\}$ with a suitable constant $c < \infty$, and $\tilde{v}_m = \max\{v_{m,1}^{1-\theta_2}, v_{m,2}^{1-\theta_1}\}$. Thus, X_nY_n is L_p–NED of size

$$-\min\left\{a_1\frac{p_1(r_1'r_2' - r_1'p - r_2'p)}{r_2'p(r_1'-p_1)}, a_2\frac{p_2(r_1'r_2' - r_1'p - r_2'p)}{r_1'p(r_2'-p_2)}\right\},$$

which implies the claim. □

Remark 2.3.8. *1. The proof shows that the constants $C^i_{p_1,r_1',r_2',p}$, $i = 1,2$, lie in $(0,1)$ if they are in use.*

2. If we insert the parameters of the two NED-series given by the two aforesaid results of Davidson (1994, Th. 17.9 and 17.17), we get the same size by Lemma 2.3.7.

3. If we assume $p = p_1 = p_2 = 2$, the NED-size is reduced to

$$-\min\left\{a_1\frac{(r_1'r_2' - 2r_1' - 2r_2')}{r_2'(r_1'-2)}, a_2\frac{(r_1'r_2' - 2r_1' - 2r_2')}{r_1'(r_2'-2)}\right\}.$$

To obtain a NED-size of $-\frac{1}{2}$ for $\{Z_n^{(1)} - \rho_n\}$, we have to postulate that the two time series X_t and Y_t fulfill a NED-size of $-\frac{1}{2}$ since $\frac{(r_1'r_2' - 2r_1' - 2r_2')}{r_{3-i}(r_i-2)} = C^i_{2,r_1',r_2',2} \in (0,1)$ for $i = 1,2$.

Proposition 2.3.9. *Lemma 2.3.7 holds true, even if the first, second, and third $\{V_n\}$ is replaced by $\{V_n^{(1)}\}$, $\{V_n^{(2)}\}$, and $\{(V_n^{(1)},V_n^{(2)})\}$, respectively.*

Proof. Set $\mathbb{E}_n^s[\cdot] = \mathbb{E}[\cdot|V_n^{(1)},V_n^{(2)},\dots,V_s^{(1)},V_s^{(2)}]$, $\mathbb{E}_{X,n}^s[\cdot] = \mathbb{E}[\cdot|V_n^{(1)},\dots,V_s^{(1)}]$, and $\mathbb{E}_{Y,n}^s[\cdot] = \mathbb{E}[\cdot|V_n^{(2)},\dots,V_s^{(2)}]$. Then, it is sufficient to show that X_n and Y_n are L_p–NED on $\{(V_n^{(1)},V_n^{(2)})\}$ for $p = r_1'$ and $p = r_2'$, respectively:

$$\|X_n - \mathbb{E}_{n-m}^{n+m}[X_n]\|_{r_1'} \le \|X_n - \mathbb{E}_{X,n-m}^{n+m}[X_n]\|_{r_1'} + \|\mathbb{E}_{n-m}^{n+m}[X_n - \mathbb{E}_{X,n-m}^{n+m}[X_n]]\|_{r_1'} \le 2c_n\xi_m$$

and for Y_n analogously. □

Remark 2.3.10. *If $V_n^{(1)}$ and $V_n^{(2)}$ are α-mixing, this does not imply that $\{V_n\} = \{(V_n^{(1)},V_n^{(2)})\}$ is α–mixing, too.*

Corollary 2.3.11. *Under Assumption (NED) $\{Z_n^{(n)} - \rho_n\}$ is L^r bounded with $r = \frac{r_1'r_2'}{r_1'+r_2'}$ as well as it is L_2-NED of size $-\frac{1}{2}$ on $\{V_n\}$.*

Proof. Firstly, we obtain that $\frac{X_t-\mu_{1,t}}{\sigma_{1,t}}$ and $\frac{Y_t-\mu_{2,t}}{\sigma_{2,t}}$ fulfill the same assumptions as X_t and Y_t, respectively. Due to the definition of $M_2^{\frac{1}{2}}$, $\{\frac{X_t-\mu_{1,t}}{\sigma_{1,t}}\frac{Y_t-\mu_{2,t}}{\sigma_{2,t}} - \rho_t\}$ is L_2-NED of size $-\frac{1}{2}$ since $U_i \le r_{3-i}$ by the assumption of $\frac{r_1' r_2'}{r_1'+r_2'} > p = 2$ and the definition of U_i from Assumption (NED). Let $\tilde{r} = \max\{r_1',r_2'\}$ and w.l.o.g. be equal to r_1'. Moreover, let k be chosen so that $\frac{1}{r} = \frac{1}{\tilde{r}} + \frac{1}{k}$, then, the Hölder inequality yields

$$\|X_i Y_i\|_r \le \|X_i\|_{\tilde{r}} \|Y_i\|_k < \infty.$$

So, for $k = r_2' = \min\{r_1',r_2'\}$, the maximal possible r is equal to $\frac{r_1' r_2'}{r_1'+r_2'} > 2$. □

Remark 2.3.12. $(\frac{1+\delta}{2\delta},\frac{1+\delta}{2\delta},2,2,2+\delta,2+\delta) \in \mathcal{M}_2^{\frac{1}{2}}$ *holds for an arbitrary* $\delta > 0$.

3. Example for the WFC. *Let Assumption (LRV) and Assumption (NED) be fulfilled. Then,* $\{Z_n^{(0)} - \rho_i\}$ *satisfies the WFC and Theorem 2.1.4 holds true. This essentially follows from the following:*

1. *The FCLT follows by combining Corollary 2.3.11 and Assumption (LRV) to apply Davidson (2002, Th. 1.2).*
2. *We know from Davidson (1994, Th. 17.5(i)) that* $\{Z_n,\mathcal{F}_{-\infty}^n\}$ *is a* L_2*-mixingale of size* $-\frac{1}{2}$ *with uniformly bounded constants* $c_n \le \max\{\|Z_n\|_r,d_n\}$*, where we define* $Z_n = Z_n^{(0)} - \rho_n$*.* *On the other hand,* $\{Z_{nt}\}$ *is also a* L_2*-NED on* $\{V_{nt}\}$ *of size* $-\frac{1}{2}$ *with uniformly bounded constants* d_{nt}*, where* $Z_{nt} = Z_{n-t}$*,* $V_{nt} = V_{n-t}$ *and* $d_{nt} = d_{n-t}$*. Hence, the Kolmogorov-type inequalities* $(\mathcal{K}_{\mathbf{r}}^{(1)})$ *and* $(\mathcal{K}_{\mathbf{r}}^{(2)})$ *are fulfilled.*

Similarly, it can be proven that $\{Z_n^{(0)} - \rho_n\}$ satisfies $(\mathcal{K}_{\mathbf{r}}^{(3)})$.

3. Example of Theorems 2.1.4, 2.1.15, 2.1.16, 2.1.22, 2.1.25, 2.1.28, and 2.2.1, 2.2.3, 2.2.5. *Replace Assumption (IID) by Assumption (NED) in each of the first examples. Then each example holds true.*

9. – 12. Example of Theorem 2.1.33. *Under the assumptions of 3. Example for the WFC let* $\mathfrak{f}$ *additionally fulfill Assumption (K3) and let* $q_n = o(n)$*. Then, we can apply Davidson and De Jong (2000, Th. 2.1) with* $X_{nt} = \frac{1}{\sqrt{n}}(Z_t^{(0)} - \rho_t)$*,* $d_{nt} = \frac{1}{\sqrt{n}}d_t$*, and* $c_{nt} = n^{-1/2}$*, which implies that* $|\hat{D}_n - D| = o_P(1)$*. Hence, we get the same estimation rate for* $|\hat{D}_{0,n} - D| = o_P(1)$ *as in the settings of the 1. to 4. Examples of Theorem 2.1.33 so that these examples are also fulfilled under the assumptions here.*

2.3.4 Time Series

This sub-subsection presents different time series, which satisfy one of the different examples presented before. Essentially they are applications of different literature works. Firstly, we consider the well-known moving-average model (MA):

MA(∞) Let

$$X_t = \sum_{k=0}^{\infty} \theta_k^{(1)} u_{t-k}^{(1)} \quad \text{and} \quad Y_t = \sum_{k=0}^{\infty} \theta_k^{(2)} u_{t-k}^{(2)},$$

where $\{(u_n^{(1)},u_n^{(2)})\}_{n\in\mathbb{Z}}$ has zero mean and is a bivariate sequence of i.i.d. random variables with $\operatorname{Var}\left[u_1^{(i)}\right] = 1$, $i = 1,2$ as well as $\operatorname{Var}\left[u_1^{(1)}u_1^{(2)}\right] > 0$. Additionally, let $\sup_t \|u_t^{(1)}\|_{r_1'} < \infty$ and $\sup_t \|u_t^{(2)}\|_{r_2'} < \infty$, where $r_1',r_2' > 2$ and $\frac{r_1' r_2'}{r_1'+r_2'} > 2$. Furthermore, let $\theta_n^{(l)} = O(n^{-1-a_l-\epsilon}))$ for

$l = 1,2$ and $\epsilon > 0$, $a_1 = \frac{r_2'(r_1'-2)}{r_1'r_2'-2r_1'-2r_2'} \vee \frac{1}{2}$, and $a_2 = \frac{r_1'(r_2'-2)}{r_1'r_2'-2r_1'-2r_2'} \vee \frac{1}{2}$.

Due to Davidson (1994, Example 17.3), $\{X_t\}$ and $\{Y_t\}$ are L_2-NED of size $-a_1$ and $-a_2$ on $\{V_n\} = \{(u_t^{(1)},u_t^{(2)})\}$ with uniformly bounded r_1'th and r_2'th moments, respectively. Hence, Assumption (NED) is fulfilled. Direct calculations yield that Assumption (LRV) is fulfilled, too.

Remark 2.3.13. *1. The i.i.d assumption on $\{(u_n^{(1)},u_n^{(2)})\}$ can be replaced by an α-mixing condition. In doing so, it has to be ensured that Assumption (LRV) holds true. Additionally, the constant mean and variance can be replaced since we assume that they are known.*

2. Let $\{\epsilon_t\}$ be i.i.d. Gaussian with zero mean and variance σ^2, then $u_t = \sum_{j=0}^{\infty} \theta_j \epsilon_{t-j}$ is stationary and we have a spectral density function

$$f(x) = \frac{1}{2\pi} \left| \sum_{j=0}^{\infty} \theta_j^{(1)} e^{ixj} \right|^2$$

if $\sum_{i=0}^{\infty} |\theta_i| < \infty$ and $\theta_0 > 0$, cf. Davidson (1994, p. 215). Furthermore, we know from Ibragimov and Linnik (1971, Th. 17.3.3.) that $\{u_t\}$ satisfies the strong-mixing condition. A closer look at the proof of this theorem shows that we even get the size of

$$a_m \leq c \sum_{j=[m/2]+1}^{\infty} |\theta_j| = O\left(m^{-\frac{r_1'r_2'}{r_1'r-r_1'-r_2'}-\tilde{\epsilon}}\right) = O\left(m^{-\frac{r}{r-2}-\tilde{\epsilon}}\right)$$

for some $\tilde{\epsilon} > 0$ and $r = \frac{2r_1'r_2'}{r_1'+r_2'}$ if $|\theta_m| = O\left(m^{-1-\frac{r_1'r_2'}{r_1'r-r_1'-r_2'}}\right)$. Hence, under the conditions described, u_t is α-mixing of size $-\frac{r}{r-2}$ so that we can replace $u_t^{(1)}$ and $u_t^{(2)}$ by u_t if Assumption (LRV) holds true.

ARMA(p,q) Let

$$X_t = \sum_{j=1}^{p_1} \lambda_j^{(1)} X_{t-j} + u_t^{(1)} + \sum_{j=1}^{q_1} \theta_j^{(1)} u_{t-j}^{(1)} \qquad \text{and} \qquad Y_t = \sum_{j=1}^{p_2} \lambda_j^{(2)} Y_{t-j} + u_t^{(2)} + \sum_{j=1}^{q_2} \theta_j^{(2)} u_{t-j}^{(2)},$$

where $u_n^{(1)}$ and $u_n^{(2)}$ fulfill the same conditions as in the above MA(∞) example. Additionally, we assume that the characteristic roots of

$$z^{p_1} - \lambda_1^{(1)} z^{p_1-1} - \ldots - \lambda_{p_1}^{(1)} = 0 \quad \text{and} \quad z^{p_2} - \lambda_1^{(2)} z^{p_2-1} - \ldots - \lambda_{p_2}^{(2)} = 0$$

lie inside the unit circle.

Due to Qiu and Lin (2011, Lemma 3.1), $\{X_t\}$ and $\{Y_t\}$ can be expressed as $X_t = \sum_{j=0}^{\infty} \theta_j^{(1)} u_{t-j}^{(1)}$ and $Y_t = \sum_{j=0}^{\infty} \theta_j^{(2)} u_{t-j}^{(2)}$, respectively, where $|\theta_m^{(i)}| = O(\rho_i^m)$ as $m \to \infty$ for some $0 < \rho_i < 1$, $i = 1,2$. Now, we can apply the considerations of the MA(∞) model.

Remark 2.3.14. *As noted in Remark 2.3.13, we can weaken the assumption in the same way here.*

Actually, there are many other different time series such as the Bilinear models, the GARCH(p,q) models, cf. Davidson (2002), the IGARCH and the FIGARCH models as well as the ARCH(∞) models, cf. Davidson (2004), which are NED under some suitable assumptions so that they can satisfy Assumptions (LRV) and (NED).

At the end of this sub-subsection, let us note that obviously $\{X_n\}$ and $\{Y_n\}$ do not have to possess the same series type. However, since we assume a linear dependency between both time series, we do expect related time series.

3 Assumptions on the Unknown Means and Variances

This section presents the assumptions on the unknown parameters and their estimates. Hereby, we do not go into detail about the estimate design but rather about the behavior of the estimation errors. In particular, we want to get a general overview of two different approaches: On the one hand, we want to optimally estimate the unknown parameters $\{\mu_{l,i}\}$ and $\{\sigma_{l,i}\}$. On the other hand, we want to optimally estimate the correlation on the basis of the observations on $1,\dots,k$ and on $1,\dots,n$, which is motivated by our introduction. Moreover, we divide both approaches into the cases where the parameters are nearly constant and where the parameters might have, in the limit, infinitely many structural breaks.

3.1 Assumptions in the a Posteriori Analysis

In what follows, we present four different assumptions for a parameter $x \in \{\mu_1, \mu_2, \sigma_1, \sigma_2, \sigma_1^2, \sigma_2^2\}$, which will be used in the next sections. We use the design index $\psi = 1,2,3,4$ to distinguish the different estimates $\{\hat{x}_{i,k,n}^{(\psi)}\}_{i=1,\dots,k,\,k=1,\dots,n}$ and their properties. Subsequently, we will say "the parameter x fulfills the assumption" instead of the extensive phrase "the sequences $\{x_i\}_{i=1,\dots,n}$ and their estimates $\{\hat{x}_{i,k,n}^{(\psi)}\}$ fulfill the assumption".

Firstly, we look at the cases where we assume that the considered parameters are nearly constant.

Assumption (PEE1). *For the considered parameter x with sequence $\{x_i\}$ and estimates $\{\hat{x}_{i,k,n}^{(1)}\}$ there are sequences $\{d_{x,i}\}$ and $\{e_{x,n}\}$, an estimate $\hat{x}_n$, and some constant $\delta_x \geq 0$ so that $\sum_{i=1}^{n} |d_{x,i}| = o(\sqrt{n})$ as $n \to \infty$, $\hat{x}_{i,k,n}^{(1)} \equiv \hat{x}_n$, and*

$$e_{x,n} = x_i - \hat{x}_n - d_{x,i} \qquad \textit{with} \qquad |e_{x,n}| = O_P(n^{-\delta_x}).$$

Remark 3.1.1. *1. We call the sequences $\{d_{x,i}\}_{i\in\mathbb{N}}$ parameter error sequences (p.e.s.).*

2. The above assumption implies that the considered parameter, e.g. μ_1 with $\mu_{1,i}$, is nearly constant, i.e., $\mu_{1,i} = \mu_{1,0} + d_{\mu,1,i}$ for some $\mu_{1,0} \in \mathbb{R}$ and all i.

3. As estimates we have the sample means in mind.

4. The p.e.s. can be interpreted as the error of some incorrect information about the exact parameters which have already been treated in Lemma 2.1.12.

In contrast to the previous assumption, where the estimates just depend on the sample size n and which could result in an optimal estimate for the considered parameter, we now look at estimates depending on $k = 1,\dots,n$, which is motivated by our introduction, p. 7.

Assumption (PEE2). *For the considered parameter x with sequence $\{x_i\}$ and estimates $\{\hat{x}_{i,k,n}^{(2)}\}$ there exist sequences $\{d_{x,i}\}, \{a_{x,N}\}$, and $\{e_{x,n}\}$, an estimate $\hat{x}_n$, and a constant $\delta_x \geq 0$ so that $\{d_{x,i}\}$ satisfies the condition of Assumption (PEE1), $\hat{x}_{i,k,n}^{(2)} \equiv \hat{x}_k$, and*

$$\max_{1\leq k\leq N} |e_{x,k}| = \max_{1\leq k\leq N} |x_i - \hat{x}_k - d_{x,i}| = O_P(1), \text{ as } n \to \infty,$$

$$\max_{N\leq k\leq n} |e_{x,k}| = \max_{N\leq k\leq n} |x_i - \hat{x}_k - d_{x,i}| = O_P(n^{-\delta_x} a_{x,N}) \text{ as } n \to \infty, \text{ followed by } N \to \infty,$$

where $a_{x,N} = o(1)$ as $N \to \infty$.

Remark 3.1.2. *1. We will see, that N can be replaced by a sequence $N_n \to \infty$ with $N_n = o(n^{1/2})$.*

2. For example, we have in mind to use the sample means and sample variances as estimates based on the observations $X_1,\dots,X_k$ and $Y_1,\dots,Y_k$. In this special case the correlation estimate is equal to the empirical correlation coefficient, which is under special assumptions the maximum likelihood estimate, cf. Anderson (1984, p. 65), and which leads us to the design of the test statistic considered in Wied et al. (2012).

The two previous assumptions imply that the considered parameters are nearly constant. Now, we implicitly allow that the parameters could have significant structural breaks. We distinguish the estimates depending on the index either i or i and k. Again, we present sufficient properties of the estimation error so that both main results of Subsection 2.1 hold true. In doing so, we assume that the estimation error is a random step-function. Furthermore, we simplify the following theorems of this paragraph in notation by assuming that the random intervals are subsets of well-chosen deterministic intervals. To illustrate what we have in mind and where the motivation originates from, we consider the following example:

Example 3.1.3. *Let the mean parameters be given as*

$$\mu_l(i) = \sum_{j=1}^{m_{l,n}+1} \mu_{l,j} \mathbb{1}_{\{i \in ([n\theta_{l,j-1}]\ [n\theta_{l,j}]]\}}$$

for $l = 1,2$, some fixed unknowns $0 = \theta_{l,0} < \ldots < \theta_{l,m_n+1} = 1$ and $m_{l,j}$. Let $\hat{\theta}_{l,i,n}$ be some estimates for $\theta_{l,i}$ with $0 = \hat{\theta}_{l,0} < \ldots < \hat{\theta}_{l,m_n+1} = 1$. We define the index sets $J_{l,i} = \{1 + [n\theta_{l,i-1}], \ldots, [n\theta_{l,i}]\}$ with their estimates

$$\tilde{J}_{l,i,n} = \begin{cases} \{1 + [n\hat{\theta}_{l,i-1,n}], \ldots, [n\hat{\theta}_{l,i,n}]\}, & \text{if } \hat{\theta}_{l,i-1,n} < \hat{\theta}_{l,i,n}, \\ \emptyset, & \text{if } \hat{\theta}_{l,i-1,n} = \hat{\theta}_{l,i,n}. \end{cases}$$

Moreover, we define the mean estimates for each $i \in \{1, \ldots, n\}$ in the following way

$$\tilde{\mu}_{l,i} = \sum_{j=1}^{m_{l,n}} \hat{\mu}_{l,j} \mathbb{1}_{i \in \tilde{J}_{l,j,n}}, \quad l = 1,2,$$

where $\hat{\mu}_{l,j}$ is an estimate for $\mu_{l,j}$. Then, we obtain with

$$\operatorname{vec}(\hat{\mathbf{I}}) = \operatorname{vec}\begin{pmatrix} \hat{I}_{l,1,n} & \cdots & \hat{I}_{l,m_{l,n},n} \\ \hat{I}_{l,m_{l,n}+1,n} & \cdots & \vdots \\ \vdots & \cdots & \vdots \\ \hat{I}_{l,m_{l,n}^2-m_{l,n},n} & \cdots & \hat{I}_{l,m_{l,n}^2,n} \end{pmatrix} = \operatorname{vec}\begin{pmatrix} \tilde{J}_{l,1} \cap J_{l,1} & \cdots & \tilde{J}_{l,1} \cap J_{l,m_{l,n}} \\ \tilde{J}_{l,2} \cap J_{l,1} & \cdots & \vdots \\ \vdots & \cdots & \vdots \\ \tilde{J}_{l,m_{l,n}} \cap J_{l,1} & \cdots & \tilde{J}_{l,m_{l,n}} \cap J_{l,m_{l,n}} \end{pmatrix}$$

that the mean estimation error is a random step-function with at most $2(m_{l,n}+1)$ unknown steps

$$\mu_l(i) - \tilde{\mu}_{l,i} = \sum_{j_1=1}^{m_{l,n}} \sum_{j_2=1}^{m_{l,n}} \mathbb{1}_{i \in J_{l,j_1} \cap \tilde{J}_{l,j_2}} (\mu_{l,j_1} - \hat{\mu}_{l,j_2}).$$

Furthermore, the error $|\mu_{l,j_1} - \hat{\mu}_{l,j_2}|$ should be small on $\hat{\mathbf{I}}_{i,i}$ for each $i = 1, \ldots, m_{l,n}$ and at least bounded on the remaining sets $\hat{\mathbf{I}}_{i,j}$, $i \neq j$, which should possess at least a small cardinality. In addition, we obtain that each $\hat{I}_{l,j}$ could have another asymptotic property which can be categorized as follows:

1. *$\hat{I}_{l,j}$ converges in probability towards a non-empty set $I'_{l,j} \subset (0,n]$, i.e., $\#\hat{I}_{l,j} \Delta I'_{l,i} = o_P(n)$;*
2. *$\hat{I}_{l,j}$ converges in probability towards the empty set, i.e., $\#\hat{I}_{l,i} = o_P(1)$;*
3. *$\hat{I}_{l,j}$ does not converge in probability.*

At least in the third case, for each $j \in \{1, \ldots, m_{l,n}^2\}$ exists a deterministic set $I''_{l,j} \subset (0,n]$ so that $\hat{I}_{l,j} \subset I''_{l,j}$. Hence, we can split the index set $\{1, \ldots, n\}$ in the following way

$$\{1, \ldots, n\} = \bigcup_{j=1}^{m_{l,n}^2} I_{l,j} \quad \text{with } P\left(\bigcap_{j=1}^{m_{l,n}^2} \{\hat{I}_{l,j} \subseteq I_{l,j}\}\right) \to 1,$$

where $I_{l,i} \in \{\emptyset, I'_{l,j}, I''_{l,j}\}$. Furthermore, we note that we do not assume that $I_{l,1}, \ldots, I_{l,m_{l,n}}$ are disjoint.

At first, we are just interested in the second half of the preceding example, i.e., in the behavior of the mean estimation errors which are random step-functions. Later, when we will present some examples, the first part of the preceding example, i.e., the construction, will be interesting, too.

Assumption (PEE3). *For the considered parameter* x *with sequence* $\{x_i\}$ *and estimate* $\hat{x}^{(3)}_{i,k,n}$ *there exist a sequence* $\{d_{x,i}\}$*, a subsequence* $m_{x,n}$ *of* n*, an array* $\{e_{x,j}\}_{j=1,\dots,m_{x,n}}$*, estimates* $\{\hat{x}_{j,n}\}_{j=1,\dots,m_{x,n}}$*, and an array of random intervals* $\{\hat{I}_{j,n}\}_{j=1,\dots,m_{x,n}}$ *so that* $\sum_{i=1}^{n}|d_{x,i}| = o(\sqrt{n})$ *as* $n\to\infty$*,* $\bigcup_{j=1}^{m_n}\hat{I}_{x,j,n} = \{1,\dots,n\}$*, and* $\hat{x}^{(3)}_{i,k,n}\equiv\hat{x}_{j,n}$ *for all* $i\in\hat{I}_{x,j}$ *and* k*. Furthermore, for all*

$$e_{x,j} = \begin{cases} x_i - \hat{x}_{j,n} - d_{x,i}, & \forall i\in\hat{I}_{x,j,n}, \\ 0, & \hat{I}_{x,j,n} = \emptyset \end{cases}$$

with $j = 1,\dots,m_{x,n}$ *there exists an array* $b_{x,j,n}$ $(j = 1,\dots,m_{x,n}, n = 1,\dots)$ *so that*

$$\max_{1\le j\le m_{l,n}} \frac{|e_{l,j}|}{b_{x,j,n}} = O_P(1). \tag{3.1.1}$$

Additionally, there exists a sequence of intervals $\{I_{x,j}\}_{j=1,\dots,m_{x,n}}$ *so that*

$$P\left(\bigcap_{j=1}^{m_{x,n}}\{\omega\ :\ \hat{I}_{x,j}(\omega)\subseteq I_{x,j}\}\right)\to 1. \tag{3.1.2}$$

Remark 3.1.4. *1. For example, the above assumption is fulfilled if the parameters are nearly constant on a partition* $\{C_j\}_{j=1,\dots,v_n}$ *of the set* $\{1,\dots,n\}$*. Assuming we have estimates* $\hat{C}_j$ *for* $j=1,\dots,v_n$ *and the parameter estimate is constant on them, then the estimation error is nearly constant on* $\hat{I}_{x,j_1+j_2\cdot v_n} = C_{j_1}\cap\hat{C}_{j_2}$ *for* $j_1,j_2 = 1,\dots v_n$*, in which case* m_n *is qual to* v_n^2*.*

2. The rate assumption displayed in (3.1.2) *can be weakened if we additionally assume that the sequence of the deterministic sets* $\{I_{x,j}\}_{1\le j\le m_{x,n}}$ *depends on another control variable* N' *which tends towards infinity after* n *does. In particular, each of the following theorems, which use Assumption (PEE3), would hold true.*

3. Assumption (PEE3) is a generalization of Assumption (PEE1).

The next assumption is motivated by a generalization of Assumption (PEE2). To that purpose, we would like to take up a modification of Example 3.1.3.

Example 3.1.5. *Under the settings of Example 3.1.3 we define for all* $i\in\{1,\dots,k\}$ *and* $k\in\{1,\dots,n\}$

$$\tilde{\mu}_{l,i,k} = \sum_{j=1}^{m_{l,n}} \mathbb{1}_{i\in\tilde{J}_{l,j,n}\cap(0,k]}\hat{\mu}_{l,j,k},\quad l = 1,2,$$

where

$$\hat{\mu}_{l,j,k} = \left(\#\tilde{J}_{l,j,n}\cap(0,k]\right)^{-1}\sum_{v\in\tilde{J}_{l,j,n}\cap(0,k]} Z_{l,v},\quad \text{for } \tilde{J}_{l,j,n}\cap(0,k]\neq\emptyset.$$

Then, we obtain with

$$\mu_l(i) - \tilde{\mu}_{l,i,k} = \sum_{j=1}^{m_{l,n}^2}\mathbb{1}_{\{i\in\hat{I}_{l,j}\cap(0,k]\}}(\mu_{l,j} - \hat{\mu}_{l,[j/m_{l,n}],k})$$

that the mean estimation error is just depending on k *if* i *and* k *lie in* $\tilde{J}_{l,j}$*. As in Example 3.1.3 we expect that the estimation error should be small on* $\mathbf{I}_{i,i}\cap(0,k]$*,* $i=1,\dots,m_{l,n}$*, especially for large* k*. Hence, we introduce a second sequence of random sets* $\{\hat{I}_{2,l,j}\}$ *to categorize the estimation rate depending on the size of* k*.*

Assumption (PEE4). *For the considered parameter* x *with sequence* $\{x_i\}$ *and estimate* $\hat{x}^{(4)}_{i,k,n}$ *there exist a sequence* $\{d_{x,i}\}$, *subsequences* $m_{x,1,n}$ *and* $m_{x,2,n}$ *of* n, *an array* $\{e_{x,k,j_1,j_2}\}$, *and arrays of random intervals* $\{\hat{I}_{x,1,j,n}\}_{j=1,\ldots,m_{x,1,n}}$ *and* $\{\hat{I}_{x,2,j,n}\}_{j=1,\ldots,m_{x,2,n}}$ *so that* $\sum_{i=1}^{n}|d_{x,i}| = o(\sqrt{n})$ *as* $n \to \infty$, $\cup_{j=1}^{m_n}\hat{I}_{x,l,j,n} = \{1,\ldots,n\}$, $l = 1,2$. *Furthermore, for all*

$$e_{x,k,j_1,j_2} = \begin{cases} x_i - \hat{x}^{(4)}_{i,k,n} - d_{x,i}, & \text{if } i \in \hat{I}_{x,1,j_1,n} \cap (0,k],\ k \in \hat{I}_{x,2,j_2,n} \\ 0, & \text{else,} \end{cases} \tag{3.1.3}$$

with $j = 1,\ldots,m_{x,n}$ *there exists an array* $\{b_{x,j_1,j_2,n}\}$ *so that*

$$\max_{1\le j_1\le m_{x,1,n},1\le j_2\le m_{x,2,n}} \frac{\max_{k\in\hat{I}_{x,2,j_2,n}}|e_{x,k,j_1,j_2}|}{b_{x,j_1,j_2,n}} = O_P(1), \quad \text{as } n \to \infty. \tag{3.1.4}$$

Furthermore, we assume that there exists a sequence of deterministic sets $\{I_{x,l,i,n}\}$, $l = 1,2$, *so that they satisfy* (3.1.2).

Remark 3.1.6.

1. *It is possible that some sets* $\hat{I}_{x,2,j,n}$ *are independent of* n *and depend on another control parameter* N *which tends towards infinity after* n *does. Thereby, Assumption (PEE4) could be interpreted as a generalization of Assumption (PEE2).*
2. *Under each of the four presented assumptions we will suppress the indices* x *and* n *of the sequences* $m.,I.,\hat{I}.,\ldots$ *if it is clear which sequence is mentioned.*

3.2 Assumptions in the Sequential Analysis

This subsection presents the assumptions on the unknown parameters which will be used in the sequential testing procedure. We distinguish between the estimates sequentially using the whole observation, i.e., from 1 until $n+k=n+1,\dots$ to estimate the unknown parameters and the estimates using the observations from 1 until n as well as the observations from $n+1$ until $k=n+1,\dots$

Similarly to the assumptions on the a posteriori setting, we first assume nearly constant parameters.

Assumption (PEE5). *For the considered parameter x with sequence $\{x_i\}$ and estimate $\hat{x}^{(\psi)}_{i,k,n}$, $\psi=1$, there are sequences $\{d_{x,i}\}$ and $\{e_{x,n}\}$, an estimate $\hat{x}_n$, and a constant $\delta_x \geq 0$ so that $\sum_{i=1}^{n}|d_{x,i}|=o(\sqrt{n})$ as $n\to\infty$, $\hat{x}^{(1)}_{i,n+k,n}=\hat{x}_{k,n}$ for all $i=1,\dots,n+k$, $k=1,\dots$, and*

$$e_{x,k,n}=x_i-\hat{x}_{k,n}-d_{x,i} \quad \textit{with} \quad \max_{k\in M_{n,m}}|e_{x,k,n}|=O_P(n^{-\delta_x}),$$

where $M_{n,m}=\{1,\dots,[nm]\}$, $m>0$, and $M_{n,m}=\{1,\dots\}$ in the closed- and open-end setting, respectively.

Assumption (PEE6). *For the considered parameter x with sequence $\{x_i\}$ and estimate $\hat{x}^{(\psi)}_{i,k,n}$, $\psi=2$, there are sequences $\{d_{x,i}\},\{a_{x,N}\}$ and $\{e_{x,n}\}$, an estimate $\hat{x}_n$ and some constant $\delta_{2,x}\geq 0$ so that $\sum_{i=1}^{n}|d_{x,i}|=o(\sqrt{n})$ as $n\to\infty$, and $\hat{x}^{(2)}_{i,k,n}=\mathbb{1}_{i\leq n}\hat{x}^n_1+\mathbb{1}_{i>n}\hat{x}^{n+k}_{n+1}$ $(i=1,\dots,n+k;\ k=1,\dots)$. Furthermore, we assume that*

1. *$\{x_i\}_{i\leq n}$ fulfills Assumption (PEE1) with the estimate $\hat{x}^n_1$,*
2. *$\max_{1\leq k\leq N}|e_{x,k}|=\max_{1\leq k\leq N}|x_i-\hat{x}^{n+k}_{n+1}-d_{x,i}|=O_P(1)$, and*
3. *$\max_{N\leq k\leq [nm]}|e_{x,k}|=\max_{N\leq k\leq [nm]}|x_i-\hat{x}^{n+k}_{n+1}-d_{x,i}|=O_P(n^{-\delta_{2,x}}a_{x,N})$,*

as $n\to\infty$, followed by $N\to\infty$, where $a_{x,N}=o(1)$ as $N\to\infty$ and where the index set, which we maximize over in the last line, is replaced by $\{N,\dots\}$ in the open-end setting.

Additionally to the previous assumptions, we consider the cases where the parameters are allowed to have structural breaks. Again, we can separate them into two types of estimates, as already considered. However, in both cases we sequentially provide the observations. To understand what we have in mind we consider the following example:

Example 3.2.1. *Let $X_i=\epsilon_i+\mu_0+\sum_{j=1}^{N}\mathbb{1}_{i\leq k_j^*}\Delta_{\mu,j}$, $i=1,\dots,n(1+m)$, be a stochastic process with centered innovations ϵ_i, where $k_j^*\sim n$, $k_j^*\neq k_{j+1}^*$, $k_j^*\in\{1,\dots,[n(1+m)]\}$ and $|\Delta_{\mu,j}|>0$ for all $j=1,\dots,N$.*
Now, we observe one by one the samples $X_{n+1},X_{n+2}\cdots$. Then, there are at least two options to estimate the change-points in the mean of the process. On the one hand, we can split the data in the data group from 1 until n and from $n+1$ until $k=n+1,\dots$ to estimate the change-points in each group, respectively. It has the advantage that we estimate the change-points of the first group only one time. On the other hand, we do not split the data into two groups and estimate for each new observation the change-points of the whole data set once again. It has at least the advantage that a possible change-point near the nth observation could be estimated more precisely.

Assumption (PEE7). *Let the considered parameter x with sequence $\{x_i\}$, estimate*

$$\hat{x}^{(3)}_{i,k,n}=\hat{x}_{i,k,n}=\mathbb{1}_{i\leq n}\hat{x}_{i,n}+\mathbb{1}_{i>n}\hat{x}^{n+k}_{i,1+n} \quad (i=1,\dots,n+k;\ k=1,\dots)$$

and sequence $m_{2,x,n}$ fulfill the following conditions:

1. *the parameter x with estimates $\{\hat{x}_{i,n}\}_{i=1,\dots,n}$ fulfills Assumption (PEE3);*

2. *for each* $k = 1,\ldots,[mn]$ *there exists a decomposition* $\{\hat{I}_{x,j,n+k,n}\}_{j=1,\ldots,m_{x,2,n}}$ *of* $\{n+1,\ldots,n+k\}$ *and a random array* $e_{x,k,j}$ *so that*

$$e_{x,k,j} = \begin{cases} x_i - \hat{x}_{i,n+1}^{n+k} - d_{x,i}, & \text{for all } i \in \hat{I}_{x,j,n+k,n}, \\ 0, & \text{else;} \end{cases}$$

3. *there exists a sequence* $b_{x,2,j,n}$, $(j = 1,\ldots,m_{x,2,n};\ n = 1,\ldots)$ *so that*

$$\max_{1\leq j\leq m_{2,x,n}} \frac{\max_{1\leq k\leq [mn]} |e_{x,k,j}|}{b_{x,2,j,n}} = O_P(1), \tag{3.2.1}$$

4. *there exists a sequence of intervals* $\{I_{x,2,j}\}_{j=1,\ldots,m_{x,2,n}}$ *so that*

$$P\left(\bigcap_{j=1}^{m_{x,2,n}} \{\omega : \bigcup_{k=n+1}^{n+[mn]} \hat{I}_{x,j,k,n}(\omega) \subseteq I_{x,2,j}\}\right) \to 1. \tag{3.2.2}$$

The following assumption is similar to Assumption (PEE7). The only difference is that the parameter estimate always depends on the whole, already observed, data set.

Assumption (PEE8). *Let the considered parameter* x *with sequence* $\{x_i\}$ *and estimate* $\hat{x}_{i,k,n}^{(\psi)}$ $(i = 1,\ldots,n+k;\ k = 1,\ldots,[nm], j = 4)$ *fulfill the following conditions:*
There is a sequence $m_{x,n}$ *such that the following conditions hold:*

1. *for each* $k = 1,\ldots,[mn]$ *there is a decomposition* $\{\hat{I}_{x,j,k,n}\}_{j=1,\ldots,m_{x,n}}$ *of* $\{1,\ldots,n+k\}$ *so that* $e_{x,k,j} \equiv x_i - \hat{x}_{i,k,n}^{(4)} - d_{i,x}$ *for all* $i \in \hat{I}_{x,j,k,n}$;

2. *there is a sequence* $b_{x,j,n}$, $(j = 1,\ldots,m_{x,n};\ n = 1,\ldots)$ *so that*

$$\max_{1\leq j\leq m_{x,n}} \frac{\max_{k\in M_{n,m}} |e_{x,k,j}|}{b_{x,j,n}} = O_P(1); \tag{3.2.3}$$

3. *there is a sequence of intervals* $\{I_{x,j}\}_{j=1,\ldots,m_{x,n}}$ *so that*

$$P\left(\bigcap_{j=1}^{m_{x,n}} \{\omega : \bigcup_{k\in M_{n,m}} \hat{I}_{x,j,k,n}(\omega) \subseteq I_{x,j}\}\right) \to 1; \tag{3.2.4}$$

where $M_{n,m} = \{n,\ldots,[n(1+m)]\}$ *and* $M_{n,m} = \{n+1,\ldots\}$ *in the closed- and open-end setting, respectively.*

Remark 3.2.2. *Some of the sets* $I_{x,2,j}$, $j = 1,\ldots,m_{x,n}$, $n \in \mathbb{N}$ *and* $I_{x,j}$, $j = 1,\ldots,m_{x,n}$, $n \in \mathbb{N}$ *could be empty. In the open-end setting they could be even right-unbounded which later on produces some technical condition between the estimation rates* $b_{x,2,j,n}$, $b_{x,j,n}$ *and the cardinality of the trimmed sets* $I_{x,j}$.

4 Change-Point Analysis of the Correlation under Unknown Means and Known Variances

In addition to the main model of Section 1 we assume in this section that the parameters $\mu_{1,i}$ and $\mu_{2,i}$ are unknown so that we have to replace them by some estimates.

4.1 A Posteriori Analysis under a General Dependency Framework and General Mean Estimates

We define

$$\hat{\rho}_{\psi,k} = \frac{1}{k}\sum_{i=1}^{k}\frac{(X_i - \hat{\mu}^{(\psi)}_{1,i,k,n})(Y_i - \hat{\mu}^{(\psi)}_{2,i,k,n})}{\sigma_{1,i}\sigma_{2,i}} = \frac{1}{k}\sum_{i=1}^{k} Z^{(\psi)}_{i,k,n}, \tag{4.1.1}$$

where $\psi = 1,\dots,4$ is a design index to distinguish the four different mean estimates which fulfill (PEE1), (PEE2), (PEE3), and (PEE4), respectively.

Each of these parameter estimates results in an error and we are interested, as mentioned before, in the behavior of these errors so that the main results of Section 2 hold true. Hence, it is sufficient to consider the error terms $R_{linZ^{(\psi)}}$ for $i = 1,\dots,n$, $l = 1,2,3$, and $\psi = 1,\dots,4$ of the following decomposition

$$\begin{aligned} Z^{(\psi)}_{i,k,n} =& \frac{(X_i - \hat{\mu}^{(\psi)}_{1,i,k,n})(Y_i - \hat{\mu}^{(\psi)}_{2,i,k,n})}{\sigma_{1,i}\sigma_{2,i}} \\ =& Z_i^{(0)} + \frac{(X_i - \mu_{1,i})(\mu_{2,i} - \hat{\mu}^{(\psi)}_{2,i,k,n})}{\sigma_{1,i}\sigma_{2,i}} + \frac{(\mu_{1,i} - \hat{\mu}^{(\psi)}_{1,i,k,n})(Y_i - \mu_{2,i})}{\sigma_{1,i}\sigma_{2,i}} \\ & + \frac{(\mu_{1,i} - \hat{\mu}^{(\psi)}_{1,i,k,n})(\mu_{2,i} - \hat{\mu}^{(\psi)}_{2,i,k,n})}{\sigma_{1,i}\sigma_{2,i}} \\ =& Z_i^{(0)} + R_{1iknZ^{(\psi)}} + R_{2iknZ^{(\psi)}} + R_{3iknZ^{(\psi)}}, \end{aligned} \tag{4.1.2}$$

where we will suppress the unused indices of the parameter estimates, of the error terms, and of $Z^{(\psi)}_{i,k,n}$.

4.1.1 Testing under a Functional Central Limit Theorem and Unknown Means

In this sub-subsection, we consider sufficient properties of the mean estimation errors so that the main results of Subsection 2.1, Theorem 2.1.1, and Theorem 2.1.4, can be retained, replacing the expectations by their estimates. In doing so, we investigate the convergence behavior of the test statistics under the estimation assumptions (PEE1)–(PEE4).

Nearly Constant Means In this paragraph, we treat the situation of nearly constant unknown means, i.e., the expectations μ_1 and μ_2 fulfill either Assumption (PEE1) or Assumption (PEE2).

Lemma 4.1.1. *Let* $\{d_{in}\}_{i\in\mathbb{N}}$ *satisfy* $\sum_{i=1}^{n}|d_{in}| = o(\sqrt{n})$ *and* $\{Z_{in}\}$ *be a triangle array of random sequences with* $\max_{i,n}\mathbb{E}[|Z_{in}|] \le C < \infty$. *Then, it holds that*

$$\max_{1\le k\le n}\left|\sum_{i=1}^{k} Z_{in}d_{in}\right| = o_P(\sqrt{n}).$$

Proof. We obtain that $\max_{1\le k\le n}|\sum_{i=1}^{k} Z_{in}d_{in}| \le \sum_{i=1}^{n}|Z_{in}d_{in}|$ and by Markov's inequality

$$P\left(\max_{1\le k\le n}\left|\sum_{i=1}^{k} Z_{in}d_{in}\right| \ge \eta\right) \le \frac{C}{\eta}\sum_{i=1}^{n}|d_{in}| = o(\sqrt{n}).$$

□

Theorem 4.1.2. *Let the parameters* μ_1 *and* μ_2 *fulfill Assumption (PEE1) with* $\delta_{\mu,1}, \delta_{\mu,2} > 0$ *and* $\delta_{\mu,1} + \delta_{\mu,2} > \frac{1}{2}$. *Moreover, for* $l = 1, 2$, *let* $\{\epsilon_{l,n}/\sigma_{3-l,n}\}$ *satisfy* $(\mathcal{K}_{\mathbf{r}}^{(1)})$ *for* $r_l > (\frac{1}{2} + \delta_{\mu,3-l})^{-1}$. *Then, Theorem 2.1.1 holds true if we replace* $B_n^{0,0,0}$ *by* $B_n^{1,0,0}$.

Proof. By using the decomposition displayed in (4.1.2) we have

$$B_n^{1,0,0}(\cdot) = \hat{D}^{-1/2}\left[\hat{D}^{1/2}B_n^{0,0,0}(\cdot) + \sum_{l=1}^{3} R_{[n\cdot],l}\right], \tag{4.1.3}$$

where

$$R_{[n\cdot],l} = \frac{[n\cdot]}{\sqrt{n}}\left(\frac{1}{[n\cdot]}\sum_{i=1}^{[n\cdot]} R_{linZ^{(1)}} - \frac{1}{n}\sum_{i=1}^{n} R_{linZ^{(1)}}\right). \tag{4.1.4}$$

Hence, Theorem 2.1.1 holds true by Slutsky's Theorem if for each $l = 1,2,3$ the right–hand side equals $o_P(1)$. Furthermore, we obtain with the triangle inequality that

$$\|R_{[n\cdot],l}\|_{[0,1]} \leq \max_{1\leq k\leq n}\left|\frac{1}{\sqrt{n}}\sum_{i=1}^{k} R_{linZ^{(1)}}\right| + \max_{1\leq k\leq n}\left|\frac{k}{n}\frac{1}{\sqrt{n}}\sum_{i=1}^{n} R_{linZ^{(1)}}\right|. \tag{4.1.5}$$

We obtain that the first summand dominates the second. For $l = 1$, we have

$$\max_{1\leq k\leq n}\left|\frac{1}{\sqrt{n}}\sum_{i=1}^{k}\frac{(X_i-\mu_{1,i})(\mu_{2,i}-\hat{\mu}_{2,n})}{\sigma_{1,i}\sigma_{2,i}}\right| = O_P(n^{-\delta_{\mu,2}})\frac{1}{\sqrt{n}}\left\|\sum_{i=1}^{[n\cdot]}\frac{\epsilon_{1,i}}{\sigma_{2,i}}\right\| + o_P(1) = o_P(1),$$

where we use Lemma 4.1.1, the main model, the assumed rates of Assumption (PEE1), and $(\mathcal{K}_{\mathbf{r}}^{(1)})$. Hence, we get $\|R_{[n\cdot],1}\|_{[0,1]} = o_P(1)$ and similarly obtain $\|R_{[n\cdot],2}\|_{[0,1]} = o_P(1)$. It remains to consider $\|R_{[n\cdot],3}\|_{[0,1]}$:

$$\|R_{[n\cdot],3}\|_{[0,1]} \leq C\frac{n}{\sqrt{n}}|\mu_{1,i}-\hat{\mu}_{1,n}-d_{1,i}||\mu_{2,i}-\hat{\mu}_{2,n}-d_{2,i}| + o_P(1) + o(1) = o_P(1),$$

where we use the property of $\{d_{l,n}\}$, the uniform boundedness of the standard deviation, and the assumed rates of Assumption (PEE1) with $\delta_{\mu,1} + \delta_{\mu,2} > \frac{1}{2}$. Thus, Theorem 2.1.1 holds true if we replace $Z_i^{(0)}$ by $Z_{i,n}^{(1)}$. □

Remark 4.1.3. *Using the sample means as estimates under the Assumption (IID), (MIX), or (NED) yields a* $\delta_{\mu,1}, \delta_{\mu,2} = \frac{1}{2}$.

Under some additional technical assumptions we even get the weighted convergences of Theorem 2.1.4.

Corollary 4.1.4. *Under the assumptions of Theorem 4.1.2 let* $\{\epsilon_{l,n}/\sigma_{3-l,n}\}$ *satisfy* $(\mathcal{K}_{\mathbf{r}}^{(2)})$ *for* $r_l \geq (\frac{1}{2} + \delta_{\mu,3-l})^{-1}$ *and let* $\{\epsilon_{l,n}d_{\mu,3-l}/\sigma_{3-l,n}\}$ *satisfy* $(\mathcal{K}_{\mathbf{r}}^{(1)})$ *and* $(\mathcal{K}_{\mathbf{r}}^{(2)})$ *for* $r = 2$ *and* $l = 1,2$. *Then, Theorem 2.1.4 holds true if we replace* $B_n^{0,0,\gamma}$ *by* $B_n^{1,0,\gamma}$.

Proof. The proof follows the combination of the proofs' arguments of Theorem 2.1.4 and Theorem 4.1.2. Hence, it remains to prove

$$\begin{aligned}&\|R_{[n\cdot],l}\|_{[0,1]}\\ &= \max_{\substack{k\in[1,[\epsilon n]]\cup\\ [[(1-\epsilon)n],n]}}\left|\left(\frac{n^2}{k(n-k)}\right)^{\gamma}\frac{1}{\sqrt{n}}\left(\sum_{i=1}^{k} R_{linZ^{(1)}} - \frac{k}{n}\sum_{i=1}^{n} R_{linZ^{(1)}}\right)\right| = o_P(1)\end{aligned} \tag{4.1.6}$$

as $n\to\infty$, $\epsilon\to 0$, which is fulfilled if for $l = 1,2,3$

$$\max_{[(1-\epsilon)n]\leq k\leq n}\left|\left(\frac{n}{(n-k)}\right)^{\gamma}\frac{1}{\sqrt{n}}\sum_{i=k+1}^{n} R_{linZ^{(1)}}\right| = o_P(1) \tag{4.1.7}$$

and

$$\max_{1\leq k\leq[\epsilon n]}\left|\left(\frac{n}{k}\right)^{\gamma}\frac{1}{\sqrt{n}}\sum_{i=1}^{k}R_{linZ^{(1)}}\right|=o_P(1) \tag{4.1.8}$$

as $n\to\infty$, followed by $\epsilon\to 0$. We start with $l=1$ and the second term:

$$\max_{1\leq k\leq[\epsilon n]}\left|\frac{1}{\sqrt{n}}\left(\frac{n}{k}\right)^{\gamma}\sum_{i=1}^{k}\frac{(X_i-\mu_{1,i})(\mu_{2,i}-\hat{\mu}_{2,n})}{\sigma_{1,i}\sigma_{2,i}}\right|=o_P(1)+O_P(n^{-\delta_2-\frac{1}{2}})\left\|\left(\frac{n}{[n\cdot]}\right)^{\gamma}\sum_{i=1}^{[n\cdot]}\frac{\epsilon_{1,i}}{\sigma_{2,i}}\right\|_{[0,\epsilon]}$$

as $n\to\infty$, followed by $\epsilon\to 0$, where we add $\pm d_{\mu,2,i}$ and apply $(\mathcal{K}_{\mathrm{r}}^{(1)})$ to $\{\epsilon_{1,n}d_{\mu,2}/\sigma_{2,n}\}$. Hence, again by using $(\mathcal{K}_{\mathrm{r}}^{(1)})$ for $\{\epsilon_{1,n}/\sigma_{2,n}\}$, the right–hand side is equal to $o_P(1)$ as $n\to\infty$, followed by $\epsilon\to 0$. Analogously, we get the rates for $l=2,3$, where we additionally use

$$\left\|\left(\frac{n}{[n\cdot]}\right)^{\gamma}\sum_{i=1}^{[n\cdot]}|d_{\mu,l,i}|\right\|_{[0,\epsilon]}=o(\sqrt{n})\qquad\text{and}\qquad\left\|\left(\frac{n}{[n\cdot]}\right)^{\gamma}\sum_{i=1}^{[n\cdot]}|d_{\mu,l,i}d_{\mu,3-l,i}|\right\|_{[0,\epsilon]}=o(\sqrt{n})$$

as $n\to\infty$, which holds by the property of $\{d_{\mu,l,n}\}$. Analogously, the rate displayed in (4.1.7) can be proven such that (4.1.6) hold. □

Remark 4.1.5. *The preceding theorem and corollary present sufficient conditions on the mean estimates, which are constant for all $i=1,\ldots,n$, such that the convergences of Theorem 2.1.1 hold true. In particular, the estimates have to be consistent. However, if one expectation has a non-local structural break, these estimates are not consistent. For example let*

$$\sigma_{1,i}=\sigma_{2,i}\equiv 1,\quad \mu_{1,i}\equiv 0,\quad \textit{and}\quad \mu_{2,i}=\mu_{2,i,n}=g_\mu(i/n),\quad \hat{\mu}_{1,n}=\overline{X}_n,\quad \textit{and}\quad \hat{\mu}_{1,n}=\overline{Y}_n,$$

where $g_\mu\not\equiv 0$ is a suitable function. Under weak assumptions we obtain that

$$\left\|\left(\frac{n}{[n\cdot]}\right)^{\gamma}\sum_{i=1}^{[n\cdot]}\frac{\epsilon_{1,i}}{\sigma_{2,i}}(\mu_{2,i}-\hat{\mu}_{2,n})\right\|=O_P(\sqrt{n}),$$

but not $o_P(\sqrt{n})$. Notably, under certain assumptions, such as inter alia the asymptotic normality of the sample means, we obtain that

$$\frac{1}{\sqrt{n}}\left(\frac{n}{[n\cdot]}\right)^{\gamma}\sum_{i=1}^{[n\cdot]}\frac{\epsilon_{1,i}}{\sigma_{2,i}}(\mu_{2,i}-\hat{\mu}_{2,n})\overset{D[0,1]}{\longrightarrow}G(\cdot),$$

where G is a Gaussian process with a covariance structure depending on g_μ. Furthermore, under certain assumptions and under H_0 we could even obtain that $B_n^{1,0,\gamma}(\cdot)$ also converges towards a Gaussian process where its covariance structure depends on g_μ, too. Since g_μ is unknown, the test statistic would be unusable under these assumptions since the critical value would be unknown.
Suppose there are non-local structural breaks in the expectations of both time series. Then, in many cases $B_n^{1,0,0}(\cdot)$ will not be bounded by an order of $O_P(1)$.

Now, we consider the estimates which are only calculated by the first k observations so that they are still independent of the index i.

Theorem 4.1.6. *Let μ_1 and μ_2 fulfill Assumption (PEE2) and let for $l=1,2$*

$$\max_{1\leq k\leq n}\frac{k}{\sqrt{n}}|\mu_1-\hat{\mu}_{1,k}||\mu_2-\hat{\mu}_{2,k}|=o_P(1) \tag{4.1.9}$$

as $n\to\infty$. Moreover, for $l=1,2$, let $\{\epsilon_{l,n}/\sigma_{3-l,n}\}$ satisfy $(\mathcal{K}_{\mathrm{r}}^{(1)})$ for $r_l\geq(\frac{1}{2}+\delta_{\mu,3-l})^{-1}$. Then, Theorem 2.1.1 holds true if we replace $B_n^{0,0}$ by $B_n^{2,0,0}$.

Remark 4.1.7. *In (PEE2) we can even replace* N *by an increasing sequence* N_n *if* $\|\sum_{i=1}^{[\cdot N_n]}\epsilon_{l,i}\| = o_P(n^{1/2})$.

Proof of Theorem 4.1.6. The proof is essentially similar to the proof of Theorem 4.1.2 so that we do not reproduce every argument in detail. Due to (4.1.5), it is sufficient to show for $l = 1,2,3$ that

$$\max_{1\le k\le n}\left|\frac{1}{\sqrt{n}}\sum_{i=1}^{k} R_{likZ^{(2)}}\right| = o_P(1).$$

For $l = 1$ we obtain

$$\begin{aligned}\max_{1\le k\le n}\left|\frac{1}{\sqrt{n}}\sum_{i=1}^{k} R_{likZ^{(2)}}\right| \le &\max_{1\le k\le N}\left|\frac{1}{\sqrt{n}}(\mu_2-\hat\mu_{2,k})\sum_{i=1}^{k}\frac{\epsilon_{1,i}}{\sigma_{2,i}}\right| \\ &+\max_{N\le k\le n}\left|\frac{1}{\sqrt{n}}(\mu_2-\hat\mu_{2,k})\sum_{i=1}^{k}\frac{\epsilon_{1,i}}{\sigma_{2,i}}\right| + o_P(1)\end{aligned} \tag{4.1.10}$$

as $n\to\infty$, where we use the triangle inequality and Lemma 4.1.1. We obtain that the first summand is equal to $o_P(1)$ as $n\to\infty$ and that the second is $o_P(1)$ as $n\to\infty$, followed by $N\to\infty$ since $\max_{N\le k\le n}|\mu_2-\hat\mu_{2,k}| = O_P(n^{-\delta_{\mu,2}}a_{2,N})$ and $\max_{N\le k\le n}\left|\sum_{i=1}^{k}\frac{\epsilon_{1,i}}{\sigma_{2,i}}\right| = O_P(n^{1/r_1})$ with $1/r_1 \le 1/2+\delta_{\mu,2}$, where $a_{2,N} = o(1)$ as $N\to\infty$. Analogously, we get the desired rate $o_P(1)$ for $l = 2$. Hence, we consider the third error term and obtain

$$\|R_{[n\cdot],3}\| \le C\left\|\frac{[n\cdot]}{\sqrt{n}}|\mu_1-\hat\mu_{1,[n\cdot]}||\mu_2-\hat\mu_{2,[n\cdot]}|\right\| + o_P(1) = o_P(1),$$

where we use the triangle inequality, the property of the p.e.s., and the assumed rate displayed in (4.1.9). □

Corollary 4.1.8. *Under the assumptions of Corollary 4.1.4 we replace the assumptions of Theorem 4.1.2 by the ones of Theorem 4.1.6. Moreover, let*

$$\max_{1\le k\le[n\epsilon]}\left(\frac{n}{k}\right)^{\gamma}\frac{k}{\sqrt{n}}|\mu_1-\hat\mu_{1,k}||\mu_2-\hat\mu_{2,k}| = o_P(1),\quad \text{as } n\to\infty,\ \epsilon\to 0$$

and

$$\max_{n-[n\epsilon]\le k\le n-1}\left(\frac{n}{n-k}\right)^{\gamma}\frac{k}{\sqrt{n}}|\mu_1-\hat\mu_{1,k}||\mu_2-\hat\mu_{2,k}| = o_P(1),\quad \text{as } n\to\infty,\ \epsilon\to 0.$$

Then, Theorem 2.1.4 holds true if we replace $B_n^{0,0,\gamma}$ *by* $B_n^{2,0,\gamma}$.

Proof. The proof follows from the proofs' arguments of Theorem 2.1.4 and Theorem 4.1.6. □

Remark 4.1.9. *Similar to Remark 4.1.5 the process* $B_n^{2,0,\gamma}(\cdot)$ *can still converge in distribution towards a Gaussian process if there are non-local structural breaks in the expectations of one of the two time series* $X_1,\ldots,X_n$ *and* $Y_1,\ldots,Y_n$. *In this case, the covariance structure of this Gaussian process depends on the unknown change-function* g_μ, *meaning that the test is unusable.*

Non-constant Mean Estimates In this paragraph, we consider some general sufficient conditions on the mean estimation error where the mean estimates $\hat\mu_{l,i,\cdot}$ are non-constant.

Theorem 4.1.10. *Let the parameters* μ_1 *and* μ_1 *fulfill Assumption (PEE3). Additionally, let* $\{\epsilon_{l,n}/\sigma_{3-l,n}\}$ *fulfill* $(\mathcal{K}_{\mathbf{r}}^{(2)})$ *for* $r_l > 1$, $l = 1,2$, *and let the arrays* $b_{l,j,n} = b_{\mu_l,j,n}$ $(l = 1,2;\ j = 1,\ldots,m_{l,n};\ n = 1,\ldots)$ *satisfy*

$$\sum_{j=1}^{m_{l,n}} b_{l,j,n}^{r_{3-l}}\#I_{l,j} = o(n^{r_{3-l}/2}m_{l,n}^{-r_{3-l}}), \tag{4.1.11}$$

$$\sum_{v=1}^{m_{1,n}}\sum_{w=1}^{m_{2,n}} b_{1,v,n}b_{2,w,n}\#(I_{1,v}\cap I_{2,w}) = o(n^{1/2}),\quad \text{and}\quad \sum_{j=1}^{m_{l,n}} b_{l,j,n}\sum_{i\in I_{l,j}}|d_{3-l,i}| = o(\sqrt{n}). \tag{4.1.12}$$

Then, Theorem 2.1.1 holds true if we replace $B_n^{0,0}$ *by* $B_n^{3,0,0}$.

Proof. In analogy to the proofs of Theorem 4.1.2 and 4.1.6, it remains to show for $l = 1, 2, 3$ that

$$\max_{1 \le k \le n} \left| \frac{1}{\sqrt{n}} \sum_{i=1}^{k} R_{linZ^{(3)}} \right| = \|R_{[n\cdot],l}\| = o_P(1)$$

as $n \to \infty$. For $l = 1$ we obtain

$$\max_{1 \le k \le n} \left| \frac{1}{\sqrt{n}} \sum_{i=1}^{k} \frac{(X_i - \mu_{1,i})(\mu_{2,i} - \hat{\mu}_{2,i})}{\sigma_{1,i}\sigma_{2,i}} \right| = \max_{1 \le k \le n} \left| \frac{1}{\sqrt{n}} \sum_{i=1}^{k} \epsilon_{1,i} \frac{\mu_{2,i} - \hat{\mu}_{2,i} - d_{l,i}}{\sigma_{2,i}} \right| + o_P(1)$$
$$\le \sum_{j=1}^{m_{2,n}} \left| \frac{e_{2,j}}{\sqrt{n}} \right| \left\| \sum_{i \in \hat{I}_{2,j} \cap [0,[n\cdot]]} \frac{\epsilon_{1,i}}{\sigma_{2,i}} \right\| + o_P(1),$$

where we use Lemma 4.1.1, the main model, and the assumed property of the mean estimate.

Now, we use $\max_j |e_j|/b_{2,j,n} = O_P(1)$, i.e., equation (3.1.1), the assumed rate in (3.1.2), and the σ–additivity which implies

$$P\left(\sum_{j=1}^{m_{2,n}} \left| \frac{e_{2,j}}{\sqrt{n}} \right| \left\| \sum_{i \in \hat{I}_{2,j} \cap [0,[n\cdot]]} \frac{\epsilon_{1,i}}{\sigma_{2,i}} \right\| \ge \eta \right) \le P\left(\sum_{j=1}^{m_{2,n}} \frac{N b_{2,j,n}}{\sqrt{n}} \left\| \sum_{i \in \hat{I}_{2,j} \cap [0,[n\cdot]]} \frac{\epsilon_{1,i}}{\sigma_{2,i}} \right\| \ge \eta \right) + o(1)$$
$$\le P\left(\left\{ \sum_{j=1}^{m_{2,n}} \frac{N b_{2,j,n}}{\sqrt{n}} \left\| \sum_{i \in \hat{I}_{2,j} \cap [0,[n\cdot]]} \frac{\epsilon_{1,i}}{\sigma_{2,i}} \right\| \ge \eta \right\} \cap \bigcap_{j=1}^{m_{2,n}} \{\hat{I}_{2,j} \subseteq I_{2,j}\} \right) + o(1)$$
$$\le \sum_{j=1}^{m_{2,n}} P\left(\frac{N b_{2,j,n} m_{2,n}}{\sqrt{n}} \max_{k_1,k_2 \in I_{2,j}; k_1 \le k_2} \left| \sum_{i=k_1}^{k_2} \frac{\epsilon_{1,i}}{\sigma_{2,i}} \right| \ge \eta \right) + o(1)$$

as $n \to \infty$, followed by $N \to \infty$. The above maximum of the partial sums is of the type $\max_{0 \le k_1 \le k_2 \le M} |S_{N+k_2} - S_{N+k_1}|$ and can be estimated to $2 \max_{0 \le k \le M} |S_{N+M} - S_{N+M-k}|$ for $N, M \in \mathbb{N}$. Hence, by applying the second Kolmogorov-inequality it holds that

$$\sum_{j=1}^{m_{2,n}} P\left(\frac{N b_{2,j,n} m_{2,n}}{\sqrt{n}} \max_{k_1,k_2 \in I_{2,j}; k_1 \le k_2} \left| \sum_{i=k_1}^{k_2} \frac{\epsilon_{1,i}}{\sigma_{2,i}} \right| \ge \eta \right) \le C \frac{N^{r_1} m_{2,n}^{r_1}}{n^{r_1/2} \eta^{r_1}} \sum_{j=1}^{m_{2,n}} b_{2,j,n}^{r_1} \# I_{2,j} = o(1)$$

as $n \to \infty$, followed by $N \to \infty$, where we use the assumed equation (4.1.11). Combining the previous arguments provides $\|R_{[n\cdot],1}\| = o_P(1)$ and $\|R_{[n\cdot],2}\| = o_P(1)$ as $n \to \infty$. Hence, it remains to consider $\|R_{[n\cdot],3}\|$:

$$\|R_{[n\cdot],3}\|_{[0,1]} = \left\| \frac{1}{\sqrt{n}} \sum_{i=1}^{[n\cdot]} \frac{(\mu_{1,i} - \hat{\mu}_{1,i})(\mu_{2,i} - \hat{\mu}_{2,i})}{\sigma_{1,i}\sigma_{2,i}} \right\|$$
$$\le \frac{C}{\sqrt{n}} \sum_{v=1}^{m_{1,n}} \sum_{w=1}^{m_{2,n}} \#(\hat{I}_{1,v,n} \cap \hat{I}_{2,w,n}) |e_{1,v,n}||e_{2,w,n}| + \frac{C}{\sqrt{n}} \sum_{l=1}^{2} \sum_{v=1}^{m_{l,n}} \sum_{j \in \hat{I}_{l,v,n}} |e_{l,v,n}||d_{3-l,j}| + o(1)$$
$$\le \frac{CN^2}{\sqrt{n}} \sum_{v=1}^{m_{1,n}} \sum_{w=1}^{m_{2,n}} \#(I_{1,v,n} \cap I_{2,w,n}) b_{1,v,n} b_{2,w,n} + \frac{CN^2}{\sqrt{n}} \sum_{l=1}^{2} \sum_{v=1}^{m_{l,n}} b_{l,v,n} \sum_{j \in I_{l,v,n}} |d_{3-l,j}| + o_P(1)$$

as $n \to \infty$, followed by $N \to \infty$, where we use the upper and lower bounds of $\sigma_{l,i} \ge \epsilon > 0$, $l = 1, 2$. Hence, with the assumed equations (4.1.12) the proof's first display holds. □

Remark 4.1.11. *If we recall Example 3.1.3, see p. 48, then there is for each* $l = 1, 2$ *a sequence* $\{a_{l,n,j}\}$ *so that*

$$\max_{1 \le j \le m_{l,n}} \frac{\#(\tilde{J}_{l,j} \triangle J_{l,j})}{a_{l,j,n}} = o_P(1).$$

If we additionally assume that $\max_{1\le j\le m_{l,n}} \frac{a_{l,j,n}}{\#J_{l,j}} < 1$, *we set* $I_{l,j} = \{\min J_{l,j} - a_{l,j,n}, \ldots, \max J_{l,j} + a_{l,j,n}\}$, *which fulfills equation (3.1.2). Furthermore, we recognize that a good estimation of the mean is more important on big sets* $I_{l,i}$ *than on small ones.*

Corollary 4.1.12. *Under the assumptions of Theorem 4.1.10 let* $d_{\mu,l} \equiv 0$ *for* $l = 1,2$. *Moreover, as* $n \to \infty$, *followed by* $\epsilon \to 0$ *let*

$$\sum_{j=1}^{m_{l,n}} b_{l,j,n}^{r_{3-l}} \sum_{i\in I_{l,j}\cap((0,[n\epsilon]]\cup(n-[n\epsilon],n))} \frac{1}{(i\wedge(n-i))^{\gamma r_{3-l}}} = o(n^{r_{3-l}(1/2-\gamma)} m_{l,n}^{-r_{3-l}}), \tag{4.1.13}$$

$$\sup_{z\in(0,\epsilon]} \sum_{v=1}^{m_{1,n}} \sum_{w=1}^{m_{2,n}} b_{1,v,n} b_{2,w,n} \frac{\#(I_{1,v}\cap I_{2,w}\cap(0,[nz]])}{[nz]^\gamma} = o(n^{1/2-\gamma}), \tag{4.1.14}$$

and

$$\sup_{z\in[1-\epsilon,1)} \sum_{v=1}^{m_{1,n}} \sum_{w=1}^{m_{2,n}} b_{1,v,n} b_{2,w,n} \frac{\#(I_{1,v}\cap I_{2,w}\cap([nz],n])}{(n-[nz])^\gamma} = o(n^{1/2-\gamma}). \tag{4.1.15}$$

Then, Theorem 2.1.4 holds true if we replace $B_n^{0,0,\gamma}$ *by* $B_n^{3,0,\gamma}$.

Proof. The claim follows from the arguments used in the proofs of Theorem 2.1.4 and Theorem 4.1.10. □

Now, we consider the asymptotic behavior of the test statistics. The idea goes back to the maximum likelihood approach. Hence, we are interested in some parameter estimates which satisfy Assumption (PEE4).

Theorem 4.1.13. *Let the parameters* μ_1 *and* μ_2 *fulfill Assumption (PEE4), let* $\{\epsilon_{1,i}/\sigma_{2,i}\}$ *and* $\{\epsilon_{2,i}/\sigma_{1,i}\}$ *fulfill* $(\mathcal{K}_{\mathbf{r}}^{(2)})$ *for* $r_1,r_2 > 1$, *and let for* $l = 1,2$

$$\sum_{j_2=1}^{m_{3-l,2}} \left(m_{3-l,1}^{r_l} \sum_{j_1=1}^{m_{3-l,1}} b_{3-l,j_1,j_2}^{r_l} \#I_{3-l,1,j_1} \cap (0, \max I_{3-l,2,j_2}] \right)^{1/r_l} = o(n^{1/2}), \tag{4.1.16}$$

$$\max_{\substack{1\le j_1\le m_{2,1}\\ 1\le j_2\le m_{2,2}}} \sum_{i_1=1}^{m_{1,1}} \sum_{i_2=1}^{m_{1,2}} \#(I_{1,1,i_1}\cap I_{2,1,i_1}\cap(0,\max I_{1,2,j_1}\cap I_{2,2,j_2}]) b_{1,i_1,j_1} b_{2,i_2,j_2} = o(n^{1/2}), \tag{4.1.17}$$

and

$$\max_{1\le j_2\le m_{l,2}} \sum_{j_1=1}^{m_{l,1,n}} b_{l,j_1,j_2,n} \sum_{i\in I_{l,1,j_1}\cap(0,\max I_{l,2,j_2}]} |d_{3-l,i}| = o(\sqrt{n}). \tag{4.1.18}$$

as $n \to \infty$. *Then, Theorem 2.1.1 holds true if we replace* $B_n^{0,0}$ *by* $B_n^{4,0,0}$.

Proof. As in the proofs before it is sufficient that for $l = 1,2,3$

$$\max_{1\le k\le n} \left| \sum_{i=1}^{k} R_{likZ^{(4)}} \right| = o_P(n^{1/2}). \tag{4.1.19}$$

Firstly, we consider the error term with $l = 1$

$$\max_{1\le k\le n} \left| \sum_{i=1}^{k} R_{1ikZ^{(4)}} \right| = \max_{1\le k\le n} \left| \sum_{i=1}^{k} (\mu_{2,i} - \hat{\mu}_{2,i,k} - d_i)\epsilon_{1,i}/\sigma_{2,i} \right| + o_P(n^{1/2})$$

$$\le \max_{1\le j_2\le m_{2,2}} \sum_{j_1=1}^{m_{2,1}} \max_{k\in\hat{I}_{2,2,j_2}} |e_{2,j_1,j_2,k}| \max_{k\in\hat{I}_{2,2,j_2}} \left| \sum_{i\in\hat{I}_{2,1,j_1}\cap(0,k]} \frac{\epsilon_{1,i}}{\sigma_{2,i}} \right| + o_P(n^{1/2})$$

$$\leq O_P(1) \max_{1\leq j_2\leq m_{2,2}} \sum_{j_1=1}^{m_{2,1}} b_{2,j_1,j_2} \max_{k\in\hat{I}_{2,2,j_2}} \left| \sum_{i\in\hat{I}_{2,1,j_1}\cap(0,k]} \frac{\epsilon_{1,i}}{\sigma_{2,i}} \right| + o_P(n^{1/2})$$

$$= O_P\left(\sum_{j_2=1}^{m_{2,2}} \left(m_{2,1}^{r_1} \sum_{j_1=1}^{m_{2,1}} b_{2,j_1,j_2}^{r_1} \# I_{2,1,j_1} \cap (0, \max I_{2,2,j_2}] \right)^{1/r_1} \right) + o_P(n^{1/2})$$

as $n \to \infty$, where we apply Lemma 4.1.1, the triangle inequality, the additivity of a probability measure, and the Kolmogorov-type inequality. Hence, the equation displayed in (4.1.19) holds for $l = 1$ and is additionally fulfilled for $l = 2$ by using (4.1.16).

Now, we consider the case $l = 3$, use (4.1.18) and obtain that

$$\max_{1\leq k\leq n} \left| \sum_{i=1}^{k} R_{3ikZ^{(4)}} \right| \leq |R| + o_P(n^{1/2})$$

with

$$|R| = \max_{1\leq k\leq n} \left| \sum_{i=1}^{k} \frac{(\mu_{1,i} - \hat{\mu}_{1,i,k} - d_{1,i})(\mu_{2,i} - \hat{\mu}_{2,i,k} - d_{2,i})}{\sigma_{1,i}\sigma_{2,i}} \right|$$

$$= \max_{\substack{1\leq j_1\leq m_{1,2}\\ 1\leq j_2\leq m_{2,2}}} \max_{k\in\hat{I}_{1,2,j_1}\cap\hat{I}_{2,2,j_2}} \left| \sum_{i=1}^{k} \frac{(\mu_{1,i} - \hat{\mu}_{1,i,k} - d_{1,i})(\mu_{2,i} - \hat{\mu}_{2,i,k} - d_{2,i})}{\sigma_{1,i}\sigma_{2,i}} \right|$$

$$\leq c \max_{\substack{1\leq j_1\leq m_{2,1}\\ 1\leq j_2\leq m_{2,2}}} \max_{k\in\hat{I}_{1,2,j_1}\cap\hat{I}_{2,2,j_2}} \sum_{i_1=1}^{m_{1,1}} \sum_{i_2=1}^{m_{2,1}} \#(\hat{I}_{1,1,i_1} \cap \hat{I}_{2,1,i_1} \cap (0,k]) |e_{1,i_1,j_1,k}||e_{2,i_2,j_2,k}|$$

$$= O_P\left(\max_{\substack{1\leq j_1\leq m_{2,1}\\ 1\leq j_2\leq m_{2,2}}} \sum_{i_1=1}^{m_{1,1}} \sum_{i_2=1}^{m_{1,2}} \#(I_{1,1,i_1} \cap I_{2,1,i_1} \cap (0, \max I_{1,2,j_1} \cap I_{2,2,j_2}]) b_{1,i_1,j_1} b_{2,i_2,j_2} \right)$$

as $n \to \infty$. Hence, the claim follows by using (4.1.17). □

Corollary 4.1.14. *Under the assumptions of Theorem 4.1.13 let* $d_{l,i} \equiv 0$,

$$\sum_{j_2=1}^{m_{3-l,2}} \left(\sum_{j_1=1}^{m_{3-l,1}} b_{3-l,j_1,j_2}^{r_l} \sum_{i\in I_{3-l,1,j_1}\cap(0,\max I_{3-l,2,j_2}]\wedge[n\epsilon]]} \frac{1}{i^{\gamma r_{3-l}}} \right)^{1/r_l} = o\left(\frac{n^{1/2}}{m_{3-l,1}}\right), \tag{4.1.20}$$

$$\sum_{j_2=1}^{m_{3-l,2}} \left(\sum_{j_1=1}^{m_{3-l,1}} b_{3-l,j_1,j_2}^{r_l} \sum_{i\in I_{3-l,1,j_1}\cap([n-[n\epsilon],\max I_{3-l,2,j_2})} \frac{1}{(n-i)^{\gamma r_{3-l}}} \right)^{1/r_l} = o\left(\frac{n^{1/2}}{m_{3-l,1}}\right), \tag{4.1.21}$$

$$\sup_{z\in(0,\epsilon]} \max_{\substack{1\leq j_1\leq m_{2,1}\\ 1\leq j_2\leq m_{2,2}}} \sum_{i_1=1}^{m_{1,1}} \sum_{i_2=1}^{m_{1,2}} \frac{\#(I_{1,1,i_1} \cap I_{2,1,i_1} \cap (0,[nz] \vee \max I_{1,2,j_1} \cap I_{2,2,j_2}])}{[nz]^{\gamma}} \cdot b_{1,i_1,j_1} b_{2,i_2,j_2} = o(n^{1/2-\gamma}), \tag{4.1.22}$$

$$\sup_{z\in[1-\epsilon,1)} \max_{\substack{1\leq j_1\leq m_{2,1}\\ 1\leq j_2\leq m_{2,2}}} \sum_{i_1=1}^{m_{1,1}} \sum_{i_2=1}^{m_{1,2}} \frac{\#(I_{1,1,i_1} \cap I_{2,1,i_1} \cap ([nz], \max I_{1,2,j_1} \cap I_{2,2,j_2}])}{(n-[nz])^{\gamma}} \cdot b_{1,i_1,j_1} b_{2,i_2,j_2} = o(n^{1/2-\gamma}) \tag{4.1.23}$$

be fulfilled as $n \to \infty$, *followed by* $\epsilon \to 0$. *Then, Theorem 2.1.4 holds true if we replace* $B_n^{0,0,\gamma}$ *by* $B_n^{4,0,\gamma}$.

Proof. The claim follows from the arguments used in the proofs of Theorem 2.1.4 and Theorem 4.1.13. □

In this sub-subsection, we have seen that under suitable estimation rates the mean estimation has no influence on the limit of the test statistic. Note that, if these rate assumptions are not fulfilled, especially if the estimates are not consistent, the processes $B_n^{3,0,\gamma}(\cdot)$ and $B_n^{4,0,\gamma}(\cdot)$ can still converge towards a Gaussian process, cf. Remark 4.1.5.

4.1.2 Change-Point Estimation under Unknown Means

In this sub-subsection, we consider the setting of the multiple change-point problem for the correlation and under the conditions that the means are unknown. In the first paragraph, we consider the special case where we assume that the parameters $\mu_{1,i}$ and $\mu_{2,i}$ are constant. In the second paragraph, we allow for multiple change-points in these parameters.

Constant Means Firstly, we consider the special estimate for the change-points in an epidemic change-point setting.

Theorem 4.1.15. *Under the assumptions of Theorem 2.1.15 let*

1. $\{\epsilon_{1,n}/\sigma_{2n}\}$ *and* $\{\epsilon_{2,n}/\sigma_{1n}\}$ *fulfill* $(\mathcal{K}_{\mathbf{r}}^{(\mathbf{2})})$ *and* $(\mathcal{K}_{\mathbf{r}}^{(\mathbf{3})})$ *for* $r_1, r_2 > 1$;
2. *the parameters* μ_1 *and* μ_2 *satisfy Assumption (PEE1) with* $d_{\mu,l,i} \equiv 0$ *and* $\delta_1 \vee \delta_2 > 0$.

Then, it holds that

$$n\|\tilde{\theta}^{(1)} - \theta^*\| = O_P(1), \tag{4.1.24}$$

where $\tilde{\theta}^{(1)} = (\tilde{\theta}_1^{(1)},\tilde{\theta}_2^{(1)}) \in \arg\max\{\tilde{Q}_n^{(1)}(s,t) \,:\, 0 \leq s < t \leq 1\}$ *with*

$$\tilde{Q}_n^{(1)}(s,t) = \left(\sum_{i=1+[ns]}^{[nt]} \left(Z_i^{(1)} - \overline{Z^{(1)}}_n\right)\right)^T \left(\sum_{i=1+[ns]}^{[nt]} \left(Z_i^{(1)} - \overline{Z^{(1)}}_n\right)\right), \tag{4.1.25}$$

$\overline{Z^{(1)}}_n = n^{-1}\sum_{i=1}^n Z_i^{(1)}$, *and* $Z_i^{(1)} = (X_i - \hat{\mu}_{1,n})(Y_i - \hat{\mu}_{2,n})/(\sigma_{1,i}\sigma_{2,i})$.

Proof. Set $\tilde{Q}^{(1)}([ns],[nt]) = \tilde{Q}_n^{(1)}(s,t)$. Firstly, we prove that it is sufficient that the following rates hold true, where $Z_i^{(1)} = Z_i^{(0)} + R_i$ and $a_{2,N} = o(1)$ as $N \to \infty$,

$$\max_{\substack{k_1<k_2\\ \|k-k^*\|\geq N}} \frac{|\sum_{i=1+k_1}^{k_2} R_i - \sum_{i=1+k_1^*}^{k_2^*} R_i|}{\|k-k^*\|} = a_{N,2}O_P(1) \tag{4.1.26}$$

and

$$\max_{1\leq k\leq n}\left|\sum_{i=1}^k R_i\right| = o_P(n) \tag{4.1.27}$$

as $n \to \infty$, $N \to \infty$. It holds with $L_{n,k_1,k_2} = O(1)$ form the proof of Theorem 2.1.15 that

$$\mathbb{P}\left(n\|\hat{\boldsymbol{\theta}} - \boldsymbol{\theta}^0\| \geq N+1\right) \leq \mathbb{P}\left(\|\hat{k} - k^*\| \geq N\right) = \mathbb{P}\left(\max_{\substack{k_1<k_2\\ \|k-k^*\|\geq N}} (\tilde{Q}^{(1)}(k_1,k_2) - \tilde{Q}^{(1)}(k_1^*,k_2^*)) \geq 0\right)$$

$$= \mathbb{P}\Bigg(0 \leq \max_{\substack{k_1<k_2\\ \|k-k^*\|\geq N}} L_{n,k_1,k_2}\Bigg(\frac{\tilde{Q}(k_1,k_2) - \tilde{Q}(k_1^*,k_2^*)}{L_{n,k_1,k_2}}$$

$$+\frac{n\|k-k^*\|}{L_{n,k_1,k_2}}\cdot\Bigg[\frac{2\sum_{i=1+k_1}^{k_2}(Z_i^{(0)} - \overline{Z^{(0)}})\left[\sum_{i=1+k_1}^{k_2}(R_i - \overline{R}) - \sum_{i=1+k_1^*}^{k_2^*}(R_i - \overline{R})\right]}{\|k-k^*\|n}$$

$$+\frac{2\sum_{i=1+k_1^*}^{k_2^*}(R_i - \overline{R})\left[\sum_{i=1+k_1}^{k_2}(Z_i^{(0)} - \overline{Z^{(0)}}) - \sum_{i=1+k_1^*}^{k_2^*}(Z_i^{(0)} - \overline{Z^{(0)}})\right]}{\|k-k^*\|n}$$

$$+\frac{\left[\sum_{i=1+k_1}^{k_2}(R_i-\overline{R})+\sum_{i=1+k_1^*}^{k_2^*}(R_i-\overline{R})\right]^2}{\|k-k^*\|n}\Bigg]\Bigg)\Bigg)$$

$$\leq \mathbb{P}\left(O_P(1)a_N+\Delta_\rho^2+O_P(1)a_{N,2}\leq 0\right),$$

where we use the rates of the proof of Theorem 2.1.15 for the first summand and $n\|k-k^*\|/L_{n,k_1,k_2}=O(1)$. In the square brackets we use the Kolmogorov-type inequality and (4.1.26) for the first summand, (4.1.27) and Lemma B.0.2 for the second, and (4.1.26) and (4.1.27) for the third. Thus, it remains to prove (4.1.26) and (4.1.27):
With the structure of $R_i=\sum_{l=1}^{3}R_{i,l,Z^{(1)}}$ as displayed in (4.1.2) and by the triangle inequality, it is sufficient to prove the two necessary rates for each $R_{i,l,Z^{(1)}}$ by itself with $l=1,2,3$. We start with the one displayed in (4.1.27) :

Since $R_{i,1,Z^{(1)}}=\epsilon_{1,i}(\mu_2-\hat\mu_{2,n})/\sigma_{2,i}$, we apply the Kolmogorov-type inequality to obtain the rate displayed in (4.1.27) for $l=1$. Analogously, we get the rate for $l=2$. Now, we consider the case $l=3$ and recognize that $-\delta_1-\delta_2<0$ is sufficient.
Hence, it remains to prove (4.1.26). For $l=1$ we obtain that

$$\max_{\substack{k_1<k_2\\ \|k-k^*\|\geq N}}\frac{\left|\sum_{i=k_1+1}^{k_2}\epsilon_{1,i}(\mu_{2,i}-\hat\mu_{2,n})/\sigma_{2,i}-\sum_{i=k_1^*+1}^{k_2^*}\epsilon_{1,i}(\mu_{2,i}-\hat\mu_{2,n})/\sigma_{2,i}\right|}{\|k-k^*\|}$$

$$=o_P(1)\max_{\substack{k_1<k_2\\ \|k-k^*\|\geq N}}\frac{\left|\sum_{i=k_1+1}^{k_2}\epsilon_{1,i}/\sigma_{2,i}-\sum_{i=k_1^*+1}^{k_2^*}\epsilon_{1,i}/\sigma_{2,i}\right|}{\|k-k^*\|}=o_P(a_N)$$

by applying Lemma B.0.2, where $\{a_N\}$ is a sequence with $a_N\to 0$ as $N\to\infty$. Since for $l=2$ the rate similarly follows, we consider $l=3$

$$\max_{\substack{k_1<k_2\\ \|k-k^*\|\geq N}}\frac{\left|\sum_{i=k_1+1}^{k_2}R_{i,3,Z^{(1)}}-\sum_{i=k_1^*+1}^{k_2^*}R_{i,3,Z^{(1)}}\right|}{\|k-k^*\|}=o_P(1)\max_{\substack{k_1<k_2\\ \|k-k^*\|\geq N}}\frac{|k_2-k_2^*+k_1^*-k_1|}{\|k-k^*\|}=o_P(1),$$

where $-\delta_1-\delta_2<0$ is sufficient again. □

Remark 4.1.16. *Note that it is not necessary that both means have to be consistently estimated.*

In the last theorem we use a mean estimate depending on the whole sample. In the following theorem this aspect will be dropped.

Theorem 4.1.17. *Under the assumptions of Theorem 2.1.15 let for* $l=1,2$

1. $\{\epsilon_{l,n}/\sigma_{3-l,n}\}$ *fulfill* $(\mathcal{K}_{\mathbf{r}}^{(2)})$ *and* $(\mathcal{K}_{\mathbf{r}}^{(3)})$ *for* $r_l>1$;
2. $\mu_{l,i}-\hat\mu_{l,k_1,k_2,n}=e_{l,k_1,k_2}$ *be independent of* i *and there be an* $\epsilon\in(0,(\theta_2^*-\theta_1^*)/3]$ *and bounded sequences* $\{a_{l,n}\}$ *and* $\{b_{l,n,N}\}$ *so that*

$$|e_{l,k_1^*,k_2^*,n}|=O_P(a_{l,n}),\tag{4.1.28}$$

$$\max_{\substack{k_1<k_2\\ \lfloor\epsilon n\rfloor\geq\|k-k^*\|\geq N}}|e_{l,k_1,k_2}-e_{l,k_1,k_2}|=O_P(a_{l,n}),\tag{4.1.29}$$

$$\max_{\substack{k_1<k_2\\ \|k-k^*\|\geq N}}\frac{|e_{l,k_1^*,k_2^*,n}-e_{l,k_1,k_2,n}|}{\|k-k^*\|}=O_P(b_{l,n,N}),\tag{4.1.30}$$

$$\max_{\substack{k_1<k_2\\ \lfloor\epsilon n\rfloor\leq\|k-k^*\|}}(k_2-k_1)|e_{1,k_1,k_2,n}||e_{2,k_1,k_2,n}|=o_P(n),\tag{4.1.31}$$

$$a_{1,n}a_{2,n}=o(1),\qquad na_{l,n}b_{3-l,n,N}=o(1),\qquad \text{and}\qquad a_{3-l,n}=o(n^{1-1/r_l})\tag{4.1.32}$$

as $n\to\infty$, *followed by* $N\to\infty$.

Then, it holds that

$$n\|\tilde{\theta}^{(2)} - \theta^*\| = O_P(1), \tag{4.1.33}$$

where $\tilde{\theta}^{(2)} = (\tilde{\theta}_1^{(2)},\tilde{\theta}_2^{(2)}) \in \arg\max\{\tilde{Q}_n^{(2)}(s,t) \ : \ 0 \leq s < t \leq 1\}$ *with*

$$\tilde{Q}_n^{(2)}(s,t) = \left(\sum_{i=1+[ns]}^{[nt]} \left(Z^{(2)}_{i,[ns],[nt]} - \overline{Z^{(2)}}_n\right)\right)^T \left(\sum_{i=1+[ns]}^{[nt]} \left(Z^{(2)}_{i,[ns],[nt]} - \overline{Z^{(2)}}_n\right)\right), \tag{4.1.34}$$

where $\overline{Z^{(2)}}_n = n^{-1}\sum_{i=1}^n Z^{(2)}_{i,1,n}$ *and* $Z^{(2)}_{i,k_1,k_2} = (X_i - \hat{\mu}_{1,k_1,k_2})(Y_i - \hat{\mu}_{2,k_1,k_2})/(\sigma_{1,i}\sigma_{2,i})$.

Proof. The proof essentially follows the proof of Theorem 4.1.15. Hence, we have to prove the rate displayed in (4.1.26) with $R_{i,k_1,k_2} = \sum_{l=1}^3 R_{i,k_1,k_2,l,z^{(2)}}$ instead of R_i, where $R_{i,k_1,k_2} = Z^{(2)}_{i,k_1,k_2} - Z^{(0)}_i$. The second necessary rate displayed in (4.1.27) is replaced by the condition

$$\max_{1\leq k_1<k_2\leq n}\left|\sum_{i=k_1+1}^{k_2} R_{i,k_1,k_2}\right| = o_P(n).$$

The triangle inequality yields that it remains to prove for each $l = 1, 2, 3$

$$\max_{\substack{k_1<k_2\\ \|k-k^*\|\geq N}}\left|\sum_{i=k_1+1}^{k_2} R_{i,k_1,k_2,l,z^{(2)}}\right| = o_P(n). \tag{4.1.35}$$

We start with the previously described modification of (4.1.26). For $l = 1$ we obtain

$$\begin{aligned}
&\max_{\substack{k_1<k_2\\ \|k-k^*\|\geq N}} \frac{\left|e_{2,k_1,k_2}\sum_{i=k_1+1}^{k_2}\epsilon_{1,i}/\sigma_{2,i} - e_{2,k_1^*,k_2^*}\sum_{i=k_1^*+1}^{k_2^*}\epsilon_{1,i}/\sigma_{2,i}\right|}{\|k-k^*\|}\\
\leq &\max_{\substack{k_1<k_2\\ \|k-k^*\|\geq N}} \frac{\left|(e_{2,k_1,k_2} - e_{2,k_1^*,k_2^*})\sum_{i=k_1+1}^{k_2}\epsilon_{1,i}/\sigma_{2,i}\right|}{\|k-k^*\|}\\
+ &\max_{\substack{k_1<k_2\\ \|k-k^*\|\geq N}} \frac{\left|e_{2,k_1^*,k_2^*}\left(\sum_{i=k_1+1}^{k_2}\epsilon_{1,i}/\sigma_{2,i} - \sum_{i=k_1^*+1}^{k_2^*}\epsilon_{1,i}/\sigma_{2,i}\right)\right|}{\|k-k^*\|}\\
= &O_P(n^{1/r_1}b_{2,n,N}) + O_P(a_{2,n}a_N) = o_P(1)
\end{aligned}$$

as $n \to \infty$, followed by $N \to \infty$. Here, we use the rate assumption (4.1.30) combined with the Kolmogorov-type inequality for the first summand. For the second summand we use (4.1.28) and Lemma B.0.2, whereby we get a sequence $\{a_N\}$ with $a_N = o(1)$ as $N \to \infty$.

Analogously, we get the necessary rate for $l = 2$ so that we now consider the sequence for $l = 3$:

$$\begin{aligned}
&\max_{\substack{k_1<k_2\\ \|k-k^*\|\geq N}} \frac{\left|\sum_{i=k_1+1}^{k_2} R_{i,k_1,k_2,3,n} - \sum_{i=k_1^*+1}^{k_2^*} R_{i,k_1,k_2,3,n}\right|}{\|k-k^*\|}\\
= &\max_{\substack{k_1<k_2\\ \|k-k^*\|\geq N}} \frac{\left|e_{1,k_1,k_2}e_{2,k_1,k_2}\sum_{i=k_1+1}^{k_2}(\sigma_{1,i}\sigma_{2,i})^{-1} - e_{1,k_1^*,k_2^*}e_{2,k_1^*,k_2^*}\sum_{i=k_1^*+1}^{k_2^*}(\sigma_{1,i}\sigma_{2,i})^{-1}\right|}{\|k-k^*\|}\\
\leq c &\max_{\substack{k_1<k_2\\ [n\epsilon]\geq\|k-k^*\|\geq N}} \left[\prod_{l=1}^{2}|e_{l,k_1,k_2} - e_{l,k_1^*,k_2^*}| + |e_{1,k_1^*,k_2^*}||e_{2,k_1^*,k_2^*}|\right.\\
&\qquad + \sum_{l=1}^{2}|e_{l,k_1,k_2} - e_{l,k_1^*,k_2^*}||e_{3-l,k_1^*,k_2^*}|
\end{aligned}$$

$$+\frac{k_2^*-k_1^*}{\|k-k^*\|}\left(\prod_{l=1}^{2}|e_{l,k_1,k_2}-e_{l,k_1^*,k_2^*}|+\sum_{l=1}^{2}|e_{l,k_1,k_2}-e_{l,k_1^*,k_2^*}||e_{3-l,k_1^*,k_2^*}|\right)\Bigg]$$

$$+c\max_{\substack{k_1<k_2\\ \|k-k^*\|\geq[n\epsilon]}}\frac{\left|e_{1,k_1,k_2}e_{2,k_1,k_2}\sum_{i=k_1+1}^{k_2}(\sigma_{1,i}\sigma_{2,i})^{-1}\right|+\left|e_{1,k_1^*,k_2^*}e_{2,k_1^*,k_2^*}\sum_{i=k_1^*+1}^{k_2^*}(\sigma_{1,i}\sigma_{2,i})^{-1}\right|}{[n\epsilon]}$$

$$=O_P(a_{1,n}a_{2,n})+\sum_{l=1}^{2}O_P(nb_{l,n,N}a_{3-l,n})+o_P(1)=o_P(1),$$

where we use the assumed rates (4.1.28) and (4.1.29) for the summands in the first line, (4.1.28) and (4.1.30) in the second line, and (4.1.31) in the last line. Finally, we apply the assumed rates displayed in (4.1.32).

Now, we prove display (4.1.35) and obtain that

$$\max_{\substack{k_1<k_2\\ \|k-k^*\|\geq N}}\left|\sum_{i=k_1+1}^{k_2}R_{i,k_1,k_2,l,n}\right|\leq\max_{\substack{k_1<k_2\\ \|k-k^*\|\geq N}}n\frac{\left|\sum_{i=k_1+1}^{k_2}R_{i,k_1,k_2,l,n}-\sum_{i=k_1^*+1}^{k_2^*}R_{i,k_1^*,k_2^*,l,n}\right|}{\|k-k^*\|}$$
$$+\left|\sum_{i=k_1^*+1}^{k_2^*}R_{i,k_1^*,k_2^*,l,n}\right|$$
$$=o_P(n)+\mathbb{1}_{\{l=3\}}o_P(na_{1,n}a_{2,n})+\mathbb{1}_{\{l\neq3\}}O_P(a_{3-l,n}n^{1/r_l})=o_P(n)$$

as $n\to\infty$, followed by $N\to\infty$, where we use the previous arguments for the first summand and (4.1.32) for the two others. Hence, the two necessary rates are fulfilled and the claim finally follows. □

In the previous two theorems we have postulated an epidemic change-point setting. In the following part of this paragraph, we will focus on the general multiple change-point setting $\mathbf{H_A^{(M)}}$.

Theorem 4.1.18. *Define $Q_n^{(1)}$ as $Q_n^{(0)}$ with $Z_i^{(1)}$ instead of $Z_i^{(0)}$. Then, Theorem 2.1.22 holds true with $\hat\theta_n^{(1)}$ in place of $\hat\theta_n^{(0)}$ if the following conditions are additionally fulfilled:*

1. *the $\{\epsilon_{1,n}/\sigma_{2n}\}$ and $\{\epsilon_{2,n}/\sigma_{1n}\}$ satisfy $(\mathcal{K}_\mathbf{r}^{(\mathbf{2})})$ and $(\mathcal{K}_\mathbf{r}^{(\mathbf{3})})$ for $r_1>1$ and $r_2>1$, and have uniformly bounded r_1'th and r_2'th moments with $r_1',r_2'\geq2$, respectively;*
2. *the parameters μ_1 and μ_2 satisfy Assumption (PEE1) with $d_{\mu,l,i}\equiv0$;*
3. *the sequences $\underline{\Delta}_{k^*,n}$, $\Delta_{\rho,r,n}$, and a_n of Theorem 2.1.22 as well as the parameters δ_1 and δ_2 of (PEE1) satisfy as $n\to\infty$:*

$$\underline{\Delta}_{k^*,n}^{-1+(1/r_1'+1/r_1-2\delta_2)\vee(1/r_2'+1/r_2-2\delta_1)}=o(\min_{1\leq i\leq R}\Delta_{\rho,i,n}^2),\tag{4.1.36}$$

$$\left(\frac{a_n}{n}\right)^{((r_1-1)/r_1+\delta_2)\wedge((r_2-1)/r_2+\delta_1)}=O\left(\frac{\min_{1\leq i\leq R}\Delta_{\rho,i,n}^2}{\max_r|\Delta_{\rho,r,n}|}\right),\tag{4.1.37}$$

$$a_nn^{-1}\max_r|\Delta_{k^*,r,n}|^{(2/r_1-1-2\delta_2)\vee(2/r_2-1-2\delta_1)}=O(\min_{1\leq i\leq R}\Delta_{\rho,i,n}^2).\tag{4.1.38}$$

Proof. Firstly, with $k_0=0$, $k_{R+1}=n$, and $r_0\in\{1,\dots,R\}$ we obtain that

$$P(a_n|\hat\theta_{r_0}-\theta_{r_0}|\geq N+1)\leq P\left(\min_{1<k_1<\ldots<k_R<n;|k_{r_0}-k_{r_0}^*|\geq Nn/a_n}Q_n^{(1)}(k)\leq Q_n^{(1)}(k^*)\right)$$

and

$$\begin{aligned} &Q_n^{(1)}(k) - Q_n^{(1)}(k^*) \\ &= 2\sum_{r=1}^{R+1}\sum_{i=k_{r-1}+1}^{k_r} (\rho_i - \overline{\rho}(k_{r-1},k_r))\,\tilde{Z}_i^{(1)} + \sum_{r=1}^{R+1}\sum_{i=k_{r-1}+1}^{k_r} (\rho_i - \overline{\rho}(k_{r-1},k_r))^2 \\ &\quad + \sum_{r=1}^{R+1}\left[(k_r^* - k_{r-1}^*)\left(\overline{\tilde{Z}^{(1)}}(k_{r-1}^*,k_r^*)\right)^2 - (k_r - k_{r-1})\left(\overline{\tilde{Z}^{(1)}}(k_{r-1},k_r)\right)^2\right] \end{aligned}$$

with

$$\tilde{Z}_i^{(1)} = Z_i^{(0)} - \rho_i + \frac{\epsilon_{1,i}}{\sigma_{2,i}} O_P(n^{-\delta_2}) + \frac{\epsilon_{2,i}}{\sigma_{1,i}} O_P(n^{-\delta_1}) + O_P(n^{-\delta_2-\delta_1})$$

and where the $O_P(\cdot)$-terms are independent of the index i.

Now, we follow the idea of the proof of Theorem 2.1.22. In the first case, if there exists a $r^* \in \{1,\ldots,R\}$ and an arbitrarily small $\epsilon > 0$ so that $|k_{r^*} - k_{r^*}^*| \geq \epsilon \min_{1\leq i\leq R+1} \Delta_{k^*,i,n} = \epsilon \underline{\Delta}_{k^*,n}$ for all sufficient large n, we obtain that

$$\max_{\|k-k^*\|\geq\epsilon\underline{\Delta}_{k^*,n}} \left|\sum_{r=1}^{R+1}\sum_{i=k_{r-1}+1}^{k_r} (\rho_i - \overline{\rho}(k_{r-1},k_r))\,\tilde{Z}_i^{(1)}\right| = O_P((n^{1/r_z\vee(1/r_1-\delta_2)\vee(1/r_2-\delta_1)}) \max_{1\leq i\leq R} |\Delta_{\rho,i,n}|),$$

where we use the arguments of the proof of Theorem 2.1.22. Using the same arguments of the proof of Theorem 2.1.22 for the last row yields that there is a lower bound which is equal to

$$O_P\left(n^{1/r+1/r_z} + n^{1/r_1'+1/r_1-2\delta_2} + n^{1/r_2'+1/r_2-2\delta_1}\right).$$

Hence, by applying (4.1.36) and the rates of Theorem 2.1.22 we observe that

$$P\left(\min_{1<k_1<\ldots<k_R<n;\|k-k^*\|\geq\epsilon\underline{\Delta}_{k^*,n}} Q_n^{(1)}(k) - Q_n^{(1)}(k^*) \leq 0\right) \to 0.$$

In the second case, we minimize over each k_r which is inside an ϵ-neighborhood of k_r^* with a radius of $\epsilon\underline{\Delta}_{k^*,n}$. Then, we get

$$\begin{aligned} &P(n|\hat{\theta}_{r_0} - \theta_{r_0}| \geq N+1) \leq o(1) \\ &+ P\Bigg(c \min_{1\leq i\leq R} \Delta_{\rho,i}^2 - O_P\left(\max_r |\Delta_{\rho,r,n}| \underline{\Delta}_{k^*,n}^{-1+1/r_z\vee(1/r_1-\delta_2)\vee(1/r_2-\delta_1)}\right) \\ &\quad - O_P\left(\frac{\max_r |\Delta_{k^*,r,n}|^{(2/r_z-1)\vee(2/r_1-1-2\delta_2)\vee(2/r_2-1-2\delta_1)}}{Nn/a_n}\right) \\ &\quad - C(R+1) \max_{1\leq i\leq R} \Delta_{\rho,i,n} \max_{1\leq r\leq R} \max_{1\leq|k_r-k_r^*|\leq\epsilon n} \frac{|\sum_{i=k_r\wedge k_r^*+1}^{k_r\vee k_r^*} \tilde{Z}_i^{(1)}|}{|k_r-k_r^*| \vee (Nn/a_n)} \geq 0\Bigg) \end{aligned}$$

as $n \to \infty$, followed by $N \to \infty$. Here, we can now use the Hájek-Rényi-type inequalities to obtain

$$\max_{1\leq|k_r-k_r^*|\leq\epsilon n} \frac{|\sum_{i=k_r\wedge k_r^*+1}^{k_r\vee k_r^*} \tilde{Z}_i^{(1)}|}{|k_r-k_r^*| \vee (Nn/a_n)} = O_P\left((Nn/a_n)^{-[(r_z-1)/r_z\wedge((r_1-1)/r_1+\delta_2)\wedge((r_2-1)/r_2+\delta_1)]}\right).$$

Due to the assumed (2.1.28) of Theorem 2.1.22, the rate assumption (2.1.29) of Theorem 2.1.22, (4.1.36), (4.1.37), and (4.1.38) it holds that the previous probability converges towards zero as $n \to \infty$, followed $N \to \infty$. This finally implies the claim. □

Remark 4.1.19. *If* $1/r_z + 1/r' \geq (1/r_1 + 1/r_1' - 2\delta_2) \vee (1/r_2 + 1/r_2' - 2\delta_1)$, *the rate assumption in* (4.1.36) *is implicitly fulfilled by the second condition of Theorem 2.1.22. If even*

$$1/r_z \geq (1/r_1 - \delta_2) \vee (1/r_2 - \delta_1) \quad \text{and} \quad 1/r_z \geq (1/r_2 - \delta_1) \vee (1/r_1 - \delta_2),$$

the rate assumptions in (4.1.37) *and* (4.1.38) *are implicitly fulfilled by the assumption of Theorem 2.1.22, too. Thus, if* $r_z \leq r_1 \wedge r_2$, $r'_z \leq r'_1 \wedge r'_2$, *and* $\delta_l > 0$, *the estimation rate* a_n *of the change-points in the correlations is independent of the mean estimation.*

Theorem 4.1.20. *Define* $\hat{R}_n^{(1)}$ *as* $\hat{R}$ *with* $Z_i^{(1)}$ *instead of* $Z_i^{(0)}$. *Under the assumptions of Theorem 4.1.18 let*

$$d_n^{(1)} \ll \beta_n^{(1)} \leq \frac{1}{4C^*} \min_{1\leq i\leq m} \Delta_{\rho,i,n}^2 \underline{\Delta}_{k^*,n} \tag{4.1.39}$$

with

$$d_n^{(1)} = d_n^{(0)} \vee n^{(1/r'_1+1/r_1-2\delta_2)\vee(1/r'_2+1/r_2-2\delta_1)}$$

and with $d_n^{(0)}$ *from Theorem 2.1.25. Then, the estimate* $\hat{R}^{(1)}$ *is a consistent estimate for the number of change-points* R^*.

Proof. Set $\beta_n = \beta_n^{(1)}$. This proof follows the arguments of the proof of Theorem 2.1.25. Hence, it is sufficient to show that the sets $\{\hat{R}^{(1)} < R^*\}$ and $\{\hat{R}^{(1)} > R^*\}$ are asymptotically empty. The asymptotic behavior of $\{\hat{R}^{(1)} < R^*\}$ follows in the same way as $\{\hat{R} < R^*\}$ in the proof of Theorem 2.1.25 with the arguments of Theorem 4.1.18 instead of those of Theorem 2.1.22.

Now, we consider $\{\hat{R}^{(1)} > R^*\}$ and obtain by using the same arguments as in the proof of Theorem 2.1.25 the lower bounds

$$O_P\left(d_n^{(0)} \vee n^{(1/r'_1+1/r_1-2\delta_2)\vee(1/r'_2+1/r_2-2\delta_1)} \vee \max_{1\leq i\leq R^*} |\Delta_{k^*,r,n}|^{(2/r_1-1-2\delta_2)\vee(2/r_2-1-2\delta_1)}\right)$$

and

$$|\Delta_{k^*,r,n}|^{(2-r_z)/r_z\vee[(2-r_1)/r_1-2\delta_2]\vee[(2-r_2)/r_2-2\delta_1]}$$

instead of

$$O_P\left(n^{1/r'_z+1/r}\right) \quad \text{and} \quad |\Delta_{k^*,r,n}|^{(2-r_z)/r_z},$$

respectively. Using $\max_{1\leq i\leq R^*} |\Delta_{k^*,r,n}| \sim n$ and $1/r_1 - 1 \leq 1/r'_1$ yields the claim. □

Non-constant Means In this paragraph, we present change-point estimates for the change-points in the correlation under the condition that the mean estimates are non-constant. Again, we consider first the special case of epidemic changes before we focus on the general multiple change-point setting.

The first change-point estimator postulates an epidemic change in the correlation and uses for the estimation of the unknown means the whole sample. Here, we allow structural breaks in the mean.

Theorem 4.1.21. *Under the assumptions of Theorem 2.1.15 let*

1. $\{\epsilon_{1,n}/\sigma_{2n}\}$ *and* $\{\epsilon_{2,n}/\sigma_{1n}\}$ *fulfill* $(\mathcal{K}_{\mathbf{r}}^{(\mathbf{2})})$ *and* $(\mathcal{K}_{\mathbf{r}}^{(\mathbf{3})})$ *for* $r_1, r_2 = 2$;

2. μ_1 *and* μ_2 *fulfill Assumption (PEE3) with* $d_{l,i} \equiv 0$ *such that it holds that*

$$\sum_{j=1}^{m_{l,n}} b_{l,j}^{r_{3-l}} \# I_{l,j} = o((n/m_{l,n})^{r_l}) \tag{4.1.40}$$

and

$$\sum_{j_1=1}^{m_{1,n}} \sum_{j_2=1}^{m_{2,n}} b_{1,j_1} b_{2,j_2} \# I_{1,j_1} \cap I_{2,j_2} = o(n); \tag{4.1.41}$$

3. *it holds for*

$$K_{1,k}^* = (k_1^* - k,k_1^*],\ K_{2,k}^* = (k_1^*,k_1^* + k],\ K_{3,k}^* = (k_2^* - k,k_2^*],\ K_{4,k}^* = (k_2^*,k_2^* + k]$$

that

$$\max_{v\in\{1,\dots,4\}} \sum_{j_1=1}^{m_{1,n}} \sum_{j_2=1}^{m_{2,n}} |b_{1,j_1}||b_{2,j_2}| \max_{1\le k\le n} \frac{1}{k \vee N} \# \left(K_{v,k}^* \cap I_{1,j_1} \cap I_{2,j_2}\right) = o(1); \tag{4.1.42}$$

4. *for each* $f = 1,2$ *exists* $K_f < \infty$ *sequences of natural numbers* $N = a_{l,f,0,N} < a_{l,f,1,n} < \dots < a_{l,f,K,n} = k_0^f - 1$ *with* $a_{l,f,1,n} \to \infty$ *so that for each* $v = 1,\dots,K_f$, $f = 1,2$, *and* $l = 1,2$ *it holds that*

$$(\#M_{l,f,0,N})^2 \sum_{j\in M_{l,f,0,N}} b_{l,j}^2 \frac{(\max I_{l,j} \wedge k_f^* - \min I_{l,j}) \wedge N}{N^2} = o(1), \tag{4.1.43}$$

$$(\#M_{l,f,0,N})^2 \sum_{j\in M_{l,f,0,N}} b_{l,j}^2 \frac{(k_f^* - \min I_{l,j} + 1 - N) \vee 0}{N((k_f^* - \min I_{l,j} + 1) \vee N)} = o(1), \tag{4.1.44}$$

$$(\#M_{l,f,v})^2 \sum_{j\in M_{l,f,v}} b_{l,j}^2 \frac{\#I_{l,j} + 1}{(k_f^* - \min I_{l,j} + 1)[(k_f^* - \max I_{l,j}) \vee N]} = o(1), \tag{4.1.45}$$

as $n \to \infty$ *followed by* $N \to \infty$, *where*

$$M_{l,f,0,N} = \left\{1 \le j \le m_{l,n} \ :\ \min I_{l,j} \le k_f^*,\ k_f^* - N < \max I_{l,j}\right\},$$
$$M_{l,f,v} = \left\{1 \le j \le m_{l,n} \ :\ \max I_{l,j} \in (k_f^* - a_{v,l,f,n},k_f^* - a_{v-1,l,f,n}]\right\};$$

5. *for each* $f = 1,2$ *exist* $K_f < \infty$ *sequences of natural numbers* $N = b_{l,f,0,N} < b_{l,f,1,n} < \dots < b_{l,f,K,n} = k_0^f - 1$ *with* $b_{l,f,1,n} \to \infty$ *so that for each* $v = 1,\dots,K$, $f = 1,2$, *and* $l = 1,2$ *it holds that*

$$(\#M_{l,f,0,N})^2 \sum_{j\in M_{l,f,0,N}} b_{l,j}^2 \frac{(\max I_{l,j} - \min I_{l,j} \vee k_f^*) \wedge N}{N^2} = o(1), \tag{4.1.46}$$

$$(\#M_{l,f,0,N})^2 \sum_{j\in M_{l,f,0,N}} b_{l,j}^2 \frac{(\max I_{l,j} - k_f^* + 1 - N) \vee 0}{N((\max I_{l,j} - k_f^* + 1) \vee N)} = o(1), \tag{4.1.47}$$

$$(\#M_{l,f,v})^2 \sum_{j\in M_{l,f,v}} b_{l,j}^2 \frac{\#I_{l,j} + 1}{(\max I_{l,j} - k_f^* + 1)[(\min I_{l,j} - k_f^*) \vee N]} = o(1), \tag{4.1.48}$$

as $n \to \infty$ *followed by* $N \to \infty$, *where*

$$M_{l,f,0,N} = \left\{1 \le j \le m_{l,n} \ :\ \min I_{l,j} \le k_f^* + N,\ k_f^* \le \max I_{l,j}\right\},$$
$$M_{l,f,v} = \left\{1 \le j \le m_{l,n} \ :\ \min I_{l,j} \in (k_f^* + b_{v-1,l,f,n},k_f^* + b_{v,l,f,n}]\right\}.$$

Then, it holds that

$$n\|\tilde{\theta}^{(3)} - \theta^*\| = O_P(1), \tag{4.1.49}$$

where $\tilde{\theta}^{(3)} = (\tilde{\theta}_1^{(3)},\tilde{\theta}_2^{(3)}) \in \arg\max\{\tilde{Q}_n^{(3)}(s,t) \ :\ 0 \le s < t \le 1\}$ *with*

$$\tilde{Q}_n^{(3)}(s,t) = \left(\sum_{i=1+[ns]}^{[nt]} \left(Z_i^{(3)} - \overline{Z^{(3)}}_n\right)\right)^T \left(\sum_{i=1+[ns]}^{[nt]} \left(Z_i^{(3)} - \overline{Z^{(3)}}_n\right)\right) \tag{4.1.50}$$

and where $\overline{Z^{(3)}}_n = n^{-1}\sum_{i=1}^n Z_{i,1,n}^{(2)}$ *and* $Z_{i,k_1,k_2}^{(3)} = (X_i - \hat{\mu}_{1,i,n})(Y_i - \hat{\mu}_{2,i,n})/(\sigma_{1,i}\sigma_{2,i})$.

Proof. The proof essentially follows the proof of Theorem 4.1.15. Hence, we prove that the equations displayed in (4.1.26) and (4.1.27) hold for $Z_i^{(3)} = Z_i^{(0)} + R_i$ and $R_i = \sum_{l=1}^3 R_{i,l,Z^{(3)}}$. Again, it is sufficient to show the rates for each l separately. We start with the proof of (4.1.27) and obtain that

$$\max_{1\le k_1<k_2\le n}\left|\sum_{i=k_1+1}^{k_2} R_{i,l,Z^{(3)}}\right| \le 2\max_{1\le k\le n}\left|\sum_{i=1}^{k} R_{i,l,Z^{(3)}}\right|,$$

which has already been treated in the proof of Theorem 4.1.10. Using (4.1.40) and (4.1.41) yields that (4.1.27) is fulfilled. Now, we consider (4.1.26) and obtain that it is sufficient to prove the rate for each $l = 1,2,3$. Moreover, we obtain that we can fragment the set $\{1 \le k_1 < k_2 \le n\}$ cut with $\|k-k^*\| \ge N$, over which the maximum is taken, in the sets

$$\{k_1 < k_2 \le k_1^* < k_2^*\},\ \{k_1 \le k_1^* \le k_2 < k_2^*\},\ \{k_1^* < k_1 < k_2 < k_2^*\},$$
$$\{k_1^* < k_1 \le k_2^* < k_2\},\ \{k_1 \le k_1^* < k_2^* \le k_2\},\ \{k_1^* < k_2^* \le k_1 < k_2\},$$

where each set is cut with $\{1,\ldots,n\}$ and $\|k-k^*\| \ge N$. Hence, the primary maximum can be estimated by the sum of the maxima on each of the above sets. Since $k_2^* - k_1^* \sim n$, it holds on the first and last set that $\|k-k^*\| \sim n$. Furthermore, this implies that

$$\max_{1\le k_1<k_2\le k_1^*<k_2^*,\|k-k^*\|\ge N}\frac{\left|\sum_{i=k_1+1}^{k_2} R_{i,l,Z^{(3)}} - \sum_{i=k_1^*+1}^{k_2^*} R_{i,l,Z^{(3)}}\right|}{\|k-k^*\|}$$
$$\le \frac{C}{n}\max_{1\le k_1<k_2\le n}\left|\sum_{i=k_1+1}^{k_2} R_{i,l,Z^{(3)}}\right| \le \frac{2C}{n}\max_{1\le k\le n}\left|\sum_{i=1}^{k} R_{i,l,Z^{(3)}}\right|,$$

where we have already treated these maximums in the proof of Theorem 4.1.10. The maximum on the three other sets are estimated to a linear combination of the following types

$$\begin{aligned}&\max_{1\le k\le k_1^*}\frac{c}{k\vee N}\left|\sum_{i=k_1^*+1-k}^{k_1^*} R_{i,l,Z^{(3)}}\right|,\ \max_{1\le k\le k_2^*-k_1^*}\frac{c}{k\vee N}\left|\sum_{i=k_1^*+1}^{k_1^*+k} R_{i,l,Z^{(3)}}\right|,\\ &\max_{1\le k\le k_2^*-k_1^*}\frac{c}{k\vee N}\left|\sum_{i=k_2^*+1-k}^{k_2^*} R_{i,l,Z^{(3)}}\right|,\ \max_{1\le k\le n-k_2^*}\frac{c}{k\vee N}\left|\sum_{i=k_2^*+1}^{k_2^*+k} R_{i,l,Z^{(3)}}\right|,\end{aligned} \tag{4.1.51}$$

where c is a suitable constant. Since each of this maximum can be treated in the same way, we just consider the first one. We start with $l = 1$:

$$\max_{1\le k\le k_1^*}\frac{c}{k\vee N}\left|\sum_{i=k_1^*+1-k}^{k_1^*} R_{i,l,Z^{(3)}}\right| \le \sum_{j=1}^{m_{2,n}}\max_{1\le k\le k_1^*}\frac{c|e_{2,j}|}{(k_1^*-k+1)\vee N}\left|\sum_{i\in[k,k_1^*]\cap\hat{I}_{l,j}}\epsilon_{1,i}/\sigma_{2,i}\right|$$
$$\le \sum_{j=1}^{m_{2,n}}\max_{\min\hat{I}_{l,j}\le k\le(k_1^*\wedge\max\hat{I}_{l,j})}\frac{c|e_{2,j}|}{(k_1^*-k+1)\vee N}\left|\sum_{i\in[k,k_1^*]\cap\hat{I}_{l,j}}\epsilon_{1,i}/\sigma_{2,i}\right|$$
$$\le \sum_{j=1}^{m_{2,n}}\max_{\min I_j\le k\le(k_1^*\wedge\max I_j),s_1,s_2\in I_j}\frac{c|e_{2,j}|}{(k_1^*-k+1)\vee N}\left|\sum_{i\in[k,k_1^*]\cap I_j\cap[s_1,s_2]}\epsilon_{1,i}/\sigma_{2,i}\right| + o_P(1)$$
$$= \sum_{j=1}^{m_{2,n}}\max_{\min I_j\le k\le s\le(k_1^*\wedge\max I_j)}\frac{c|e_{2,j}|}{(k_1^*-k+1)\vee N}\left|\sum_{i\in[k,s]}\epsilon_{1,i}/\sigma_{2,i}\right| + o_P(1)$$
$$\le 2\sum_{j=1}^{m_{2,n}}\max_{\min I_j\le k\le(k_1^*\wedge\max I_j)}\frac{c|e_{2,j}|}{(k_1^*-k+1)\vee N}\left|\sum_{i=k}^{(k_1^*\wedge\max I_j)}\epsilon_{1,i}/\sigma_{2,i}\right| + o_P(1)$$

$$= O_P\left(\sum_{j=1}^{m_{2,n}} \max_{\min I_j \le k \le (k_1^* \wedge \max I_j)} \frac{c|b_{2,j}|}{(k_1^* - k + 1) \vee N} \left| \sum_{i=k}^{(k_1^* \wedge \max I_j)} \epsilon_{1,i}/\sigma_{2,i} \right| \right) + o_P(1)$$
$$= O_P\left(\sum_{j=1}^{m_{2,n}} \max_{1 \le k \le k_1^*} \frac{c|b_{2,j}|}{(k_1^* - k + 1) \vee N} \left| \sum_{i \in [k,k_1^*] \cap I_j} \epsilon_{1,i}/\sigma_{2,i} \right| \right) + o_P(1),$$

where we use the estimation rates of Assumption (PEE3). Now, we decompose $\{1,\ldots,m_{2,n}\}$ in

$$\begin{aligned}
M_1 &= \{1 \le j \le m_{2,n} \ : \ \max I_j \in (0,k_1^* - a_{K-1,n}]\}, \\
M_2 &= \{1 \le j \le m_{2,n} \ : \ \max I_j \in (k_1^* - a_{K-1,n},k_1^* - a_{K-2,n}]\}, \\
&\vdots \\
M_{K,N} &= \{1 \le j \le m_{2,n} \ : \ \max I_j \in (k_1^* - a_{1,n},k_1^* - N]\}, \\
M_{K+1,N} &= \{1 \le j \le m_{2,n} \ : \ \min I_j \le k_1^*,\ k_1^* - N < \max I_j\}, \\
M_{K+2,N} &= \{1,\ldots,m_{2,n}\} \setminus \bigcup_{i=1}^{K+1} M_i
\end{aligned}$$

and obtain that the above estimated term is zero for each $j \in M_{K+2}$. Furthermore, we define $a_{K+1,n} = 1$ to obtain that

$$\begin{aligned}
&P\left(2\sum_{j=1}^{m_{2,n}} \max_{\min I_j \le k \le (k_1^* \wedge \max I_j)} \frac{c|b_{2,j}|}{(k_1^* - k + 1) \vee N} \left| \sum_{i=k}^{(k_1^* \wedge \max I_j)} \epsilon_{1,i}/\sigma_{2,i} \right| \ge \eta \right) \\
&\le C \sum_{v=1}^{K-1} \frac{(\#M_v)^2}{\eta^2} \sum_{j \in M_v} b_{2,j}^2 \sum_{i=\min I_j}^{\max I_j} \frac{1}{(k_1^* - k + 1)^2} \\
&\quad + C\frac{(\#M_{K,N})^2}{\eta^2} \sum_{j \in M_{K,N}} b_{2,j}^2 \sum_{i=\min I_j}^{\max I_j} \frac{1}{(k_1^* - k + 1)^2 \vee N^2} \\
&\quad + C\frac{(\#M_{K+1,N})^2}{\eta^2} \sum_{j \in M_{K+1,N}} b_{2,j}^2 \sum_{i=\min I_j}^{k_1^* \wedge \max I_j} \frac{1}{(k_1^* - k + 1)^2 \vee N^2} \\
&\le C \sum_{v=1}^{K-1} \frac{(\#M_v)^2}{\eta^2} \sum_{j \in M_v} b_{2,j}^2 \frac{\#I_j + 1}{(k_1^* - \min I_j + 1)(k_1^* - \max I_j)} \\
&\quad + C\frac{(\#M_{K,N})^2}{\eta^2} \sum_{j \in M_{K,N}} b_{2,j}^2 \frac{\#I_j + 1}{(k_1^* - \min I_j + 1)[(k_1^* - \max I_j) \vee N]} \\
&\quad + C\frac{(\#M_{K,N})^2}{\eta^2} \sum_{j \in M_{K,N}} b_{2,j}^2 \Bigg[\frac{(\max I_j \wedge k_1^* - \min I_j) \wedge N}{N^2} \\
&\qquad\qquad + \frac{(k_1^* - \min I_j + 1 - N) \vee 0}{N((k_1^* - \min I_j + 1) \vee N)}\Bigg],
\end{aligned}$$

where we use the above decomposition, the σ-additivity, and the second Kolmogorov-type inequality for the first inequality and the estimation by the integral method for the second. For $l = 2$ we get a similar rate so that we consider the case $l = 3$ now:

$$\begin{aligned}
&\max_{1 \le k \le k_1^*} \frac{c}{k \vee N} \left| \sum_{i=k_1^*+1-k}^{k_1^*} R_{i,l,Z^{(3)}} \right| \\
&\quad = O_P\left(\sum_{j_1=1}^{m_{1,n}} \sum_{j_2=1}^{m_{2,n}} |b_{1,j_1}||b_{2,j_2}| \max_{1 \le k \le k_1^*} \frac{1}{k \vee N} \# \left([k_1^* - k + 1,k_1^*] \cap I_{1,j_1} \cap I_{2,j_2}\right)\right),
\end{aligned}$$

where we use Assumption (PEE3). Hence, the first maximum of (4.1.51) fulfills the sufficient rate. The third maximum essentially follows in the same way. For the second and the last one we use the same arguments, where we apply the shifted Kolmogorov inequality. Hence, the claim is proven. □

Theorem 4.1.22. *Define* $Q_n^{(3)}$ *as* $Q_n^{(0)}$ *with* $Z_i^{(3)}$ *instead of* $Z_i^{(0)}$. *Then, Theorem 2.1.22 holds true with* $\hat{\theta}_n^{(3)}$ *in place of* $\hat{\theta}_n^{(0)}$ *if the following conditions are additionally fulfilled:*

1. *the sequences* $\{\epsilon_{1,i}/\sigma_{2i}\}$ *and* $\{\epsilon_{1,i}/\sigma_{2i}\}$ *satisfy* $(\mathcal{K}_{\mathrm{r}}^{(2)})$ *and* $(\mathcal{K}_{\mathrm{r}}^{(3)})$ *for* $r_1, r_2 > 1$, *respectively;*
2. *the parameters* μ_1 *and* μ_2 *satisfy Assumption (PEE3) with* $d_{\mu,l,i} \equiv 0$;
3. *the sequences* $\underline{\Delta}_{k^*,n}$, $\Delta_{\rho,r,n}$, *and* a_n *of Theorem 2.1.22, as well as the sequences* $b_{l,j}$ *and* $I_{l,j}$ *of (PEE3) satisfy for* $l = 1,2$ *and as* $n \to \infty$:

$$m_{l,n}^{r_{3-l}} \sum_{j=1}^{m_{l,n}} b_{l,j}^{r_{3-l}} \# I_{l,j} = o\left(\left(\frac{\underline{\Delta}_{k^*,n} \underline{\Delta}_{\rho,n}^2}{\max_{1\le r\le R} |\Delta_{\rho,r}|} \right)^{r_{3-l}} \wedge (\underline{\Delta}_{k^*,n} \underline{\Delta}_{\rho,n}^2)^{r_{3-l}/2} \right), \tag{4.1.52}$$

$$\sum_{j=1}^{m_{2,n}} \sum_{j=1}^{m_{1,n}} b_{1,j} b_{2,j} \# I_{1,j} \cap I_{2,j} = o\left(\frac{\underline{\Delta}_{k^*,n} \underline{\Delta}_{\rho,n}^2}{\max_{1\le r\le R} |\Delta_{\rho,r}|} \wedge \sqrt{\underline{\Delta}_{k^*,n} \underline{\Delta}_{\rho,n}^2} \right), \tag{4.1.53}$$

$$\max_{1\le r\le R} (\# A_{l,r,n})^{r_{3-l}} \sum_{j\in A_{l,r,n}} b_{l,j}^{r_{3-l}} (Nn/a_n)^{-(r_{3-l}-1)} = o\left(\frac{\underline{\Delta}_{\rho,n}^2}{\max_{1\le r\le R} |\Delta_{\rho,r}|} \right), \tag{4.1.54}$$

$$\begin{aligned} &\max_{1\le r\le R} \sum_{j_1\in A_{1,r,n}, j_2 \in A_{2,r,n}} \frac{b_{1,j} b_{2,j} \# ((k_r^* - \epsilon \underline{\Delta}_{k^*,n}, k_r^* + \epsilon \underline{\Delta}_{k^*,n}] \cap I_{1,j_1} \cap I_{2,j_2})}{\# ((k_r^* - \epsilon \underline{\Delta}_{k^*,n}, k_r^* + \epsilon \underline{\Delta}_{k^*,n}] \cap I_{1,j_1} \cap I_{2,j_2}) \vee (Nn/a_n)} \\ &= o\left(\frac{\underline{\Delta}_{\rho,n}^2}{\|\Delta_{\rho,r}\|} \right), \end{aligned} \tag{4.1.55}$$

for an arbitrarily small $\epsilon > 0$, *where*

$$A_{l,r,n} = \left\{ 1 \le j \le m_{l,n} \ : \ I_{l,j} \cap (k_r^* - \epsilon \underline{\Delta}_{k^*,n}, k_r^* + \epsilon \underline{\Delta}_{k^*,n}] \neq \emptyset \right\}.$$

Remark 4.1.23. *The assumed rates in* (4.1.52) *and* (4.1.53) *are sufficient for the change-point estimation of the correlation to guarantee that there are no big mean estimation errors over the whole time. Additionally, in the direct neighborhood of the change-points of the correlations the mean estimation should in particular be not too unfavorable in the sense of* (4.1.54) *and* (4.1.55).

Proof of Theorem 4.1.22. Set $a_{1,n,N} = Nn/a_n$. Firstly, we obtain with $k_0 = 0$, $k_{R+1} = n$, and $r_0 \in \{1,\ldots,R\}$ that

$$P(a_n |\hat{\theta}_{r_0} - \theta_{r_0}| \ge N+1) \le P\left(\min_{1<k_1<\ldots<k_R<n; |k_{r_0} - k_{r_0}^*| \ge a_{1,n,N}} Q_n^{(3)}(k) \le Q_n^{(3)}(k^*) \right)$$

and that

$$\begin{aligned} Q_n^{(3)}(k) - Q_n^{(3)}(k^*) &= 2 \sum_{r=1}^{R+1} \sum_{i=k_{r-1}+1}^{k_r} (\rho_i - \overline{\rho}(k_{r-1},k_r)) \tilde{Z}_i^{(3)} + \sum_{r=1}^{R+1} \sum_{i=k_{r-1}+1}^{k_r} (\rho_i - \overline{\rho}(k_{r-1},k_r))^2 \\ &+ \sum_{r=1}^{R+1} \left[(k_r^* - k_{r-1}^*) \left(\overline{\tilde{Z}^{(3)}}(k_{r-1}^*,k_r^*) \right)^2 - (k_r - k_{r-1}) \left(\overline{\tilde{Z}^{(3)}}(k_{r-1},k_r) \right)^2 \right] \end{aligned}$$

with

$$\tilde{Z}_i^{(3)} = Z_i^{(0)} - \rho_i + \frac{\epsilon_{1,i}}{\sigma_{2,i}} (\mu_{2,i} - \hat{\mu}_{2,i}) + \frac{\epsilon_{2,i}}{\sigma_{1,i}} (\mu_{1,i} - \hat{\mu}_{1,i}) + (\mu_{1,i} - \hat{\mu}_{1,i})(\mu_{2,i} - \hat{\mu}_{2,i}).$$

Now, we follow the idea of the proof of Theorem 2.1.22. In the first case, if there is a $r^* \in \{1, \ldots, R\}$ and an arbitrarily small $\epsilon > 0$ so that $|k_{r^*} - k^*_{r^*}| \geq \epsilon \underline{\Delta}_{k^*,n} \vee a_{1,n,N}$ for all sufficient large n, we obtain that

$$\max_{\|k-k^*\| \geq \epsilon \underline{\Delta}_{k^*,n} \vee a_{1,n,N}} \left| \sum_{r=1}^{R+1} \sum_{i=k_{r-1}+1}^{k_r} (\rho_i - \overline{\rho}(k_{r-1},k_r)) \tilde{Z}_i^{(3)} \right| = O_P(n^{1/r_z} \max_{1 \leq i \leq R} |\Delta_{\rho,i,n}|)$$

$$+ \max_{\|k-k^*\| \geq \epsilon \underline{\Delta}_{k^*,n} \vee a_{1,n,N}} \left| \sum_{j=1}^{m_{2,n}} e_{2,j} \sum_{r=1}^{R+1} \sum_{i \in (k_{r-1},k_r] \cap \hat{I}_{2,j}} (\rho_i - \overline{\rho}(k_{r-1},k_r)) \frac{\epsilon_{1,i}}{\sigma_{2,i}} \right|$$

$$+ \max_{\|k-k^*\| \geq \epsilon \underline{\Delta}_{k^*,n} \vee a_{1,n,N}} \left| \sum_{j=1}^{m_{1,n}} e_{1,j} \sum_{r=1}^{R+1} \sum_{i \in (k_{r-1},k_r] \cap \hat{I}_{1,j}} (\rho_i - \overline{\rho}(k_{r-1},k_r)) \frac{\epsilon_{2,i}}{\sigma_{1,i}} \right|$$

$$+ \max_{\|k-k^*\| \geq \epsilon \underline{\Delta}_{k^*,n} \vee a_{1,n,N}} \left| \sum_{j=1}^{m_{2,n}} \sum_{j=1}^{m_{1,n}} e_{1,j} e_{2,j} \sum_{r=1}^{R+1} \sum_{i \in (k_{r-1},k_r] \cap \hat{I}_{1,j} \cap \hat{I}_{2,j}} (\rho_i - \overline{\rho}(k_{r-1},k_r)) \right|.$$

Here, we use in the first row the arguments of the proof of Theorem 2.1.22. Furthermore, with the triangle inequality, σ-additivity, and the Kolmogorov-type inequality (and $|\Delta_{\rho,r}| \leq C$) we get that

$$P\left(\max_{\|k-k^*\| \geq \epsilon \underline{\Delta}_{k^*,n} \vee a_{1,n,N}} \left| \sum_{j=1}^{m_{2,n}} e_{2,j} \sum_{r=1}^{R+1} \sum_{i \in (k_{r-1},k_r] \cap \hat{I}_{2,j}} (\rho_i - \overline{\rho}(k_{r-1},k_r)) \frac{\epsilon_{1,i}}{\sigma_{2,i}} \right| \geq \eta \underline{\Delta}_{k^*,n} \underline{\Delta}^2_{\rho,n} \right)$$
$$\leq C \left(\frac{R m_{1,n} \max_{1 \leq r \leq R} |\Delta_{\rho,r}|}{\eta \underline{\Delta}_{k^*,n} \underline{\Delta}^2_{\rho,n}} \right)^{r_1} \sum_{j=1}^{m_{2,n}} b_{2,j}^{r_1} \# I_{2,j} + o(1)$$

as $n \to \infty$, followed by $N \to \infty$. Due to (4.1.52), this tends to zero. Similarly, we can treat the above third summand. For the fourth we obtain

$$\max_{\|k-k^*\| \geq \epsilon \underline{\Delta}_{k^*,n}} \left| \sum_{j_1=1}^{m_{2,n}} \sum_{j_1=1}^{m_{1,n}} e_{1,j_1} e_{2,j_2} \sum_{r=1}^{R+1} \sum_{i \in (k_{r-1},k_r] \cap \hat{I}_{1,j_1} \cap \hat{I}_{2,j_2}} (\rho_i - \overline{\rho}(k_{r-1},k_r)) \right|$$
$$= O_P \left(R \sum_{j_2=1}^{m_{2,n}} \sum_{j_1=1}^{m_{1,n}} b_{1,j_1} b_{2,j_2} \# I_{1,j_1} \cap I_{2,j_2} \max_{1 \leq r \leq R} |\Delta_{\rho,r}| \right),$$

which tends to zero, if it is divided by $\underline{\Delta}_{k^*,n} \underline{\Delta}^2_{\rho,n}$, see (4.1.53). For the last summand of the second display in this proof we use the same arguments as in the proof of Theorem 2.1.22 which yield a lower bound of an order of $O_P(b_n^{(3)})$ with

$$b_n^{(3)} = n^{1/r + 1/r_z} + \sum_{l=1}^{2} m_{l,n}^2 \left(\sum_{j=1}^{m_{l,n}} b_{l,j}^{r_{3-l}} \# I_{3-l,j} \right)^{2/r_{3-l}} + \left(\sum_{j_1=1}^{m_{1,n}} \sum_{j_2=1}^{m_{2,n}} b_{1,j_1} b_{2,j_2} \# I_{1,j_1} \cap I_{2,j_2} \right)^2. \tag{4.1.56}$$

This tends to zero if it is divided by $\underline{\Delta}_{k^*,n} \underline{\Delta}^2_{\rho,n}$, see (4.1.52) and (4.1.53). Analogously to the proof of Theorem 2.1.22, it follows that

$$P\left(\min_{1 < k_1 < \ldots < k_R < n; \|k-k^*\| \geq \epsilon \underline{\Delta}_{k^*,n} \vee a_{1,n,N}} Q_n^{(3)}(k) - Q_n^{(3)}(k^*) \leq 0 \right) \to 0.$$

In the second case, we minimize over each k_r which is inside an ϵ-neighborhood of k^*_r with a radius of $\epsilon \underline{\Delta}_{k^*,n}$. Then, we get

$$P(n|\hat{\theta}_{r_0} - \theta_{r_0}| \geq N + 1) \leq o(1)$$

$$+P\Bigg(c\min_{1\le i\le R}\Delta_{\rho,i}^2-O_P\left(\frac{\max_r|\Delta_{k^*,r,n}|^{2/r_z-1}}{Nn/a_n}\right)-O_P\left(\frac{\sum_{l=1}^2 m_{l,n}^2\left(\sum_{j=1}^{m_{l,n}}b_{l,j}^{r_{3-l}}\#I_{3-l,j}\right)^{2/r_{3-l}}}{\underline{\Delta}_{k^*,n}Nn/a_n}\right.$$

$$\left.+\frac{\left(\sum_{j_1=1}^{m_{1,n}}\sum_{j_1=1}^{m_{1,n}}b_{1,j_1}b_{2,j_2}\#I_{1,j_1}\cap I_{2,j_2}\right)^2}{\underline{\Delta}_{k^*,n}Nn/a_n}\right)$$

$$-C(R+1)\max_{1\le i\le R}\Delta_{\rho,i,n}\max_{1\le r\le R}\max_{1\le|k_r-k_r^*|\le\epsilon\underline{\Delta}_{k^*,n}}\frac{|\sum_{i=k_r\wedge k_r^*+1}^{k_r\vee k_r^*}\tilde{Z}_i^{(3)}|}{|k_r-k_r^*|\vee(Nn/a_n)}\ge 0\Bigg)$$

as $n\to\infty$ followed by $N\to\infty$, where we obtain that the first term dominates the second and third, by using the fourth condition of Theorem 2.1.22, (4.1.52), and (4.1.53). For the last term we use the Hájek-Rényi-type inequalities to obtain

$$\max_{1\le|k_r-k_r^*|\le\epsilon\underline{\Delta}_{k^*,n}}\frac{|\sum_{i=k_r\wedge k_r^*+1}^{k_r\vee k_r^*}\tilde{Z}_i^{(3)}|}{|k_r-k_r^*|\vee(Nn/a_n)}=O_P((Nn/a_n)^{-(r_z-1)/r_z})$$

$$+\sum_{l=1}^{2}\max_{1\le|k_r-k_r^*|\le\epsilon\underline{\Delta}_{k^*,n}}\left|\sum_{j=1}^{m_{l,n}}e_{l,j}\frac{\sum_{i\in(k_r\wedge k_r^*,k_r\vee k_r^*]\cap\hat{I}_{l,j}}\frac{\epsilon_{3-l,i}}{\sigma_{l,i}}}{|k_r-k_r^*|\vee(Nn/a_n)}\right|$$

$$+\max_{1\le|k_r-k_r^*|\le\epsilon\underline{\Delta}_{k^*,n}}\left|\sum_{j=1}^{m_{2,n}}\sum_{j=1}^{m_{1,n}}e_{1,j}e_{2,j}\frac{\#(k_r\wedge k_r^*,k_r\vee k_r^*]\cap\hat{I}_{1,j}\cap\hat{I}_{2,j}}{|k_r-k_r^*|\vee(Nn/a_n)}\right|$$

$$=O_P((Nn/a_n)^{-(r_z-1)/r_z})+\sum_{l=1}^{2}O_P\left(\#A_{l,r,n}\left(\sum_{j\in A_{l,r,n}}b_{l,j}^{r_{3-l}}(Nn/a_n)^{-(r_{3-l}-1)}\right)^{1/r_{3-l}}\right)$$

$$+O_P\left(\sum_{j_1\in A_{1,r,n},j_2\in A_{2,r,n}}b_{1,j}b_{2,j}\frac{\#((k_r^*-\epsilon\underline{\Delta}_{k^*,n},k_r^*+\epsilon\underline{\Delta}_{k^*,n}]\cap I_{1,j_1}\cap I_{2,j_2})}{\#((k_r^*-\epsilon\underline{\Delta}_{k^*,n},k_r^*+\epsilon\underline{\Delta}_{k^*,n}]\cap I_{1,j_1}\cap I_{2,j_2})\vee(Nn/a_n)}\right),$$

where

$$A_{l,r,n}=\left\{1\le j\le m_{l,n}\ :\ I_{l,j}\cap(k_r^*-\epsilon\underline{\Delta}_{k^*,n},k_r^*+\epsilon\underline{\Delta}_{k^*,n}]\neq\emptyset\right\}.$$

Due to the assumed rates in (4.1.54) and (4.1.55), the claim finally follows. □

Theorem 4.1.24. *Define* $\hat{R}_n^{(3)}$ *as* $\hat{R}$ *with* $Z_i^{(3)}$ *instead of* $Z_i^{(0)}$. *Let the assumptions of Theorem 4.1.22 be fulfilled and suppose that*

$$d_n^{(3)}\ll\beta_n^{(3)}\le\frac{1}{4C^*}\min_{1\le i\le m}\Delta_{\rho,i,n}^2\underline{\Delta}_{k^*,n}\quad\textit{with}\quad d_n^{(3)}=b_n^{(3)}\vee d_n^{(0)},\tag{4.1.57}$$

where $b_n^{(3)}$ *is from* (4.1.56), $d_n^{(0)}$ *and* C^* *are defined as in Theorem 2.1.25. Then,* $\hat{R}^{(3)}$ *consistently estimates the number of change-points* R^*.

Proof. Set $\beta_n=\beta_n^{(3)}$. This proof follows the arguments of the one of Theorem 2.1.25. Hence, it is sufficient to show that the sets $\{\hat{R}^{(3)}<R^*\}$ and $\{\hat{R}^{(3)}>R^*\}$ are asymptotically empty. The asymptotic behavior of $\{\hat{R}^{(3)}<R^*\}$ follows in the same way as $\{\hat{R}<R^*\}$ in the proof of Theorem 2.1.25 with the arguments of Theorem 4.1.22 instead of Theorem 2.1.22.

Now, we consider $\{\hat{R}^{(3)}>R^*\}$ by using the same arguments as in the proof of Theorem 2.1.25

and obtain the lower bounds:

$$\min_k Q_n^{(3)}(k) - Q_n^{(3)}(k^*) + \beta_n$$
$$= \min Q_n(k) - Q_n(k^*) + \beta_n - O_P\left(a_n \vee \left(\sum_{l=1}^{2} m_{l,n}\left(\sum_{j=1}^{m_{l,n}} b_{l,j}^{r_{3-l}} \# I_{3-l,j}\right)\right)^{1/r_{3-l}}\right.$$
$$\left.\vee \left(\sum_{j_1=1}^{m_{1,n}}\sum_{j_1=1}^{m_{1,n}} b_{1,j_1} b_{2,j_2} \# I_{1,j_1} \cap I_{2,j_2}\right)\right),$$

where $a_n = n^{1/r_z + 1/r} \vee b_n^{(3)}$ with $b_n^{(3)}$ defined in (4.1.56). Since a_n is the dominating sequence of the last two lines, the claim is proven. □

4.1.3 Long-run Variance Estimation under Unknown Means

In this sub-subsection, we present some LRV estimates which can be used for the Theorems 4.1.2, 4.1.6, 4.1.10, and 4.1.13. Since both LRVs are the same in the constant mean setting, i.e. under Theorems 4.1.2 and 4.1.6, and since both LRVs are the same in the non-constant mean setting, i.e. under Theorems 4.1.10 and 4.1.13, we will present the two LRV estimates $\hat{D}_{1,n}$ and $\hat{D}_{3,n}$ defined by (2.1.34). One LRV estimate uses the mean estimate-type which is presented in Assumption (PEE1) and the other one uses a mean estimate-type as presented in Assumption (PEE3).

Nearly constant means In this paragraph, we consider the LRV estimate type using a mean estimate. Since we are interested in the consequences of nonconsistent estimates, we display the potential error in each case (A)–(H); cf. Subsection 2.1.3.

Theorem 4.1.25. *Let the assumptions of Theorem 2.1.33 and the following conditions hold true:*

1. *the sequences* $\{\epsilon_{1,n}/\sigma_{2l,n}\}$ *and* $\{\epsilon_{2,n}/\sigma_{1,n}\}$ *fulfill* $(\mathcal{K}_{\mathbf{r}}^{(1)})$ *for* $r_1,r_2 > 1$;

2. *the parameters* μ_1 *and* μ_2 *satisfy Assumption (PEE1);*

3. *for each* $l,l_1,l_2 = 1,2,$ $k = 1,2,3,$ *and with* $d_{\mu,l}$ *from Assumption (PEE1) and* $d_{jn}^{(k)}$ *from Theorem 2.1.33 let*

$$\frac{1}{n}\left|\sum_{i,j=1}^{n} \mathfrak{f}\left(\frac{i-j}{q_n}\right) d_{jn}^{(k)} \epsilon_{l,i}\right| = O_P(b_{l,n}^{(k)}), \tag{4.1.58}$$

$$\sum_{i=1}^{n}\sum_{j=1}^{n} \mathfrak{f}\left(\frac{i-j}{q_n}\right) \epsilon_{l,i}(Z_j^{(0)} - \rho_j) = o_P(n^{1+\delta_{\mu,3-l}}), \tag{4.1.59}$$

$$\sum_{i=1}^{n}\sum_{j=1}^{n} \mathfrak{f}\left(\frac{i-j}{q_n}\right) \epsilon_{l_1,i}\epsilon_{l_2,j} = o_P(n^{1+\delta_{\mu,3-l_1}+\delta_{\mu,3-l_2}}). \tag{4.1.60}$$

Then, it holds that $\hat{D}_{1,n} = D + \hat{R}_n^{(0)} + \hat{R}_n^{(1)}$, *where* $\hat{R}_n^{(0)}$ *is defined as in Theorem 2.1.33 and*

$$\begin{aligned}
\hat{R}_n^{(1)} =& o_P(1) + O_P\left(q_n n^{-\delta_{\mu,1}-\delta_{\mu,2}}(n^{-\delta_{\mu,1}-\delta_{\mu,2}} \vee n^{-1+1/r} \vee b_{1,n} n^{-1+1/r_2} \vee b_{2,n} n^{-1+1/r_1})\right) \\
&+ o_P(q_n n^{-\delta_1}) + o_P(q_n n^{-1/2-(\delta_{\mu,1}\wedge\delta_{\mu,2})}) + o_P(1) \\
&+ \begin{cases}
O_P(q_n n^{-\delta_1}(n^{-\delta_{\mu,1}-\delta_{\mu,2}} + n^{1/r_{3-l}-1-(\delta_{\mu,1}\wedge\delta_{\mu,2})})), & \text{under (A)}, \\
T_n^{(A)} + O_P(\max_{l=1,2} b_{3-l,n}^{(1)} n^{-\delta_{\mu,l}}), & \text{under (B)}, \\
T_n^{(A)} + O_P(\max_{l=1,2} b_{3-l,n}^{(2)} n^{-\delta_{\mu,l}}), & \text{under (C)}, \\
T_n^{(A)} + O(q_n n^{-\delta_{\mu,1}-\delta_{\mu,2}}) + O_P \max_{l=1,2} b_{3-l,n}^{(3)} n^{-\delta_{\mu,l}}), & \text{under (D)}, \\
\begin{matrix} O_P(q_n \max_k n^{-1-\delta_{\mu,1}-\delta_{\mu,2}-\delta_k} \# C_k) \\ +O_P(\sum_{l=1}^2 q_n n^{-1+1/r_{3-l}-\delta_{\mu,l}-\min \delta_k}) \end{matrix}, & \text{under (E)}, \\
T_n^{(E)} + O_P(\max_{l=1,2} b_{3-l,n}^{(1)} n^{-\delta_{\mu,l}}), & \text{under (F)}, \\
T_n^{(E)} + O_P(\max_{l=1,2} b_{3-l,n}^{(2)} n^{-\delta_{\mu,l}}), & \text{under (G)}, \\
T_n^{(E)} + O(q_n n^{-\delta_{\mu,1}-\delta_{\mu,2}}) + O_P(\max_{l=1,2} b_{3-l,n}^{(3)} n^{-\delta_{\mu,l}}), & \text{under (H)},
\end{cases}
\end{aligned}$$

as $n \to \infty$, *where* $T_n^{(A)}$ *and* $T_n^{(E)}$ *denote the rate terms in the cases (A) and (E), respectively.*

Proof. Define $b_{l,n} = n^{-\delta_{\mu,l}}$. Firstly, we obtain that

$$\begin{aligned}
&\frac{1}{n}\sum_{i=1}^n \sum_{j=1}^n \mathfrak{f}\left(\frac{i-j}{q_n}\right)(Z_i^{(1)} - \tilde{\rho}_i)(Z_j^{(1)} - \tilde{\rho}_j) \\
&= \frac{1}{n}\sum_{i=1}^n \sum_{j=1}^n \mathfrak{f}\left(\frac{i-j}{q_n}\right)(Z_i^{(0)} - \tilde{\rho}_i)(Z_j^{(0)} - \tilde{\rho}_j) \\
&+ \frac{1}{n}\sum_{i=1}^n \sum_{j=1}^n \mathfrak{f}\left(\frac{i-j}{q_n}\right)\left[R_i(Z_j^{(0)} - \tilde{\rho}_j) + R_i R_j + (Z_i^{(0)} - \tilde{\rho}_i) R_j\right],
\end{aligned}$$

where $R_i = \sum_{l=1}^3 R_{l,i,Z^{(1)}}$ and the first summand is equal to $D + \tilde{R}_n$ and $\tilde{R}_n$ is the estimation error of the LRV defined in Theorem 2.1.33. Hence, it is necessary and sufficient to consider that the second summand is equal to $o_P(1)$:
Firstly, we obtain that

$$\begin{aligned}
\frac{1}{n}\sum_{i=1}^n \sum_{j=1}^n \mathfrak{f}\left(\frac{i-j}{q_n}\right) R_i R_j = \frac{1}{n}\sum_{i=1}^n \sum_{j=1}^n \mathfrak{f}\left(\frac{i-j}{q_n}\right) \Bigg[& R_{3,i,Z^{(1)}} R_{3,j,Z^{(1)}} + R_{3,i,Z^{(1)}} \sum_{l_2=1}^2 R_{l_2,j,Z^{(1)}} \\
& + R_{3,j,Z^{(1)}} \sum_{l_1=1}^2 R_{l_1,i,Z^{(1)}} + \sum_{l_1=1}^2 \sum_{l_2=1}^2 R_{l_1,i,Z^{(1)}} R_{l_2,j,Z^{(1)}} \Bigg].
\end{aligned}$$

Using Assumption (PEE1), we obtain terms of the following form

$$\begin{aligned}
&\frac{1}{n}\sum_{i=1}^n \sum_{j=1}^n \mathfrak{f}\left(\frac{i-j}{q_n}\right)[(b_{1,n}b_{2,n} + b_{1,n}d_{2,i} + b_{2,n}d_{1,i})(b_{1,n}b_{2,n} + b_{1,n}d_{2,j} + b_{2,n}d_{1,j}) \\
&+ (b_{1,n}b_{2,n} + b_{1,n}d_{2,i} + b_{2,n}d_{1,i})(b_{1,n}\epsilon_{2,j} + b_{2,n}\epsilon_{1,j} + d_{1,j}\epsilon_{2,j} + d_{2,j}\epsilon_{1,j}) \\
&+ (b_{1,n}\epsilon_{2,i} + b_{2,n}\epsilon_{1,i} + d_{1,i}\epsilon_{2,i} + d_{2,i}\epsilon_{1,i})(b_{1,n}b_{2,n} + b_{1,n}d_{2,j} + b_{2,n}d_{1,j}) \\
&+ (b_{1,n}\epsilon_{2,i} + b_{2,n}\epsilon_{1,i} + d_{1,i}\epsilon_{2,i} + d_{2,i}\epsilon_{1,i})(b_{1,n}\epsilon_{2,j} + b_{2,n}\epsilon_{1,j} + d_{1,j}\epsilon_{2,j} + d_{2,j}\epsilon_{1,j})] \\
&= O(q_n b_{1,n}^2 b_{2,n}^2) + o\left((q_n b_{1,n}^2 b_{2,n} n^{-1/2} + q_n b_{2,n}^2 b_{1,n} n^{-1/2} + b_{1,n}^2 + b_{1,n}b_{2,n} + b_{2,n}^2)\right) \\
&+ O_P(q_n b_{1,n} b_{2,n} n^{-1}(b_{1,n} n^{1/r_2} + b_{2,n} n^{1/r_1})) \\
&+ o_P(q_n n^{-1/2}(1 + b_{1,n} + b_{2,n})(b_{1,n}b_{2,n} + b_{1,n} + b_{2,n})) \\
&+ o_P(b_{1,n} + b_{2,n}) + o_P(1)
\end{aligned}$$

$$+\frac{1}{n}\left|\sum_{i=1}^{n}\sum_{j=1}^{n}\mathfrak{f}\left(\frac{i-j}{q_n}\right)\left[b_{2,n}^2\epsilon_{1,i}\epsilon_{1,j}+b_{1,n}b_{2,n}\epsilon_{1,i}\epsilon_{2,j}+b_{1,n}^2\epsilon_{2,i}\epsilon_{2,j}\right]\right|$$
$$=O(q_n b_{1,n}^2 b_{2,n}^2)+O_P(q_n b_{1,n}b_{2,n}(b_{1,n}n^{-1+1/r_2}+b_{2,n}n^{-1+1/r_1}))$$
$$+o_P(q_n n^{-1/2}(b_{1,n}+b_{2,n}))$$
$$+\frac{1}{n}\left|\sum_{i=1}^{n}\sum_{j=1}^{n}\mathfrak{f}\left(\frac{i-j}{q_n}\right)\left[b_{2,n}^2\epsilon_{1,i}\epsilon_{1,j}+b_{1,n}b_{2,n}\epsilon_{1,i}\epsilon_{2,j}+b_{1,n}^2\epsilon_{2,i}\epsilon_{2,j}\right]\right|+o_P(1)$$

as $n\to\infty$, followed by $N\to\infty$, where we use $b_{l,n}=O(1)$ and

$$\sum_{i=1}^{n}\sum_{j=1}^{n}\mathfrak{f}\left(\frac{i-j}{q_n}\right)\epsilon_{l,i}=O_P(q_n n^{1/r_l})\qquad\text{and}\qquad\sum_{i=1}^{n}\sum_{j=1}^{n}\mathfrak{f}\left(\frac{i-j}{q_n}\right)d_{3-l,i}\epsilon_{l,i}=o_P(q_n n^{1/2}),$$
$$\sum_{i=1}^{n}\sum_{j=1}^{n}\mathfrak{f}\left(\frac{i-j}{q_n}\right)d_{l_1,j}\epsilon_{l_2,i}=o_P(q_n n^{1/2})\qquad\text{and}\qquad\sum_{i=1}^{n}\sum_{j=1}^{n}\mathfrak{f}\left(\frac{i-j}{q_n}\right)d_{l_1,j}d_{3-l_2,i}\epsilon_{l_2,i}=o_P(n)$$

for $l,l_1,l_2=1,2$. Here, we use Markov's and Kolmogorov's inequality, as well as the kernel and p.e.s. properties. Hence, the double sum with R_iR_j is of order $o_P(1)$.

Furthermore, we obtain

$$\frac{1}{n}\sum_{i=1}^{n}\sum_{j=1}^{n}\mathfrak{f}\left(\frac{i-j}{q_n}\right)R_i(Z_j^{(0)}-\tilde\rho_j)=\frac{1}{n}\sum_{i=1}^{n}\sum_{j=1}^{n}\mathfrak{f}\left(\frac{i-j}{q_n}\right)R_i(\rho_j-\tilde\rho_j)$$
$$+\frac{1}{n}\sum_{i=1}^{n}\sum_{j=1}^{n}\mathfrak{f}\left(\frac{i-j}{q_n}\right)R_i(Z_j^{(0)}-\rho_j).$$

Then, in the case of (A) to (H) as defined in Theorem 2.1.33 we get that

$$O_P\Bigg(\frac{1}{n}\sum_{i=1}^{n}\sum_{j=1}^{n}\mathfrak{f}\left(\frac{i-j}{q_n}\right)(b_{1,n}b_{2,n}+b_{1,n}d_{2,i}+b_{2,n}d_{1,i}+b_{1,n}\epsilon_{2,i}$$
$$+b_{2,n}\epsilon_{1,i}+d_{1,i}\epsilon_{2,i}+d_{2,i}\epsilon_{1,i})(\rho_j-\tilde\rho_j)\Bigg)$$
$$=\begin{cases}O_P(q_n n^{-\delta_1}(b_{1,n}b_{2,n}+\sum_{l=1}^2 b_{l,n}n^{1/r_{3-l}-1}))+o_P(q_n n^{-\delta_1-1/2}(b_{1,n}+b_{2,n}+1)), & \text{case }(A),\\ T_n^{(A)}+o(q_n n^{-1/2}(b_{1,n}b_{2,n}+b_{1,n}+b_{2.n}))+O_P(b_{1,n}b_{2,n}^{(1)}+b_{2,n}b_{1,n}^{(1)}), & \text{case }(B),\\ T_n^{(A)}+o(q_n n^{-1/2}(b_{1,n}+b_{2.n}))+O(q_n n^{-1/2}b_{1,n}b_{2,n})+O_P(b_{1,n}b_{2,n}^{(2)}+b_{2,n}b_{1,n}^{(2)}), & \text{case }(C),\\ T_n^{(A)}+O(q_n b_{1,n}b_{2,n})+o(q_n(b_{1,n}+b_{2,n})n^{-1/2})+O_P(b_{1,n}b_{2,n}^{(3)}+b_{2,n}b_{1,n}^{(3)}), & \text{case }(D),\\ O_P(q_n b_{1,n}b_{2,n}\max_k n^{-1-\delta_k}\#C_k)+O_P(\sum_{l=1}^2 b_{l,n}q_n n^{-1+1/r_{3-l}-\min\delta_k}), & \text{case }(E),\\ T_n^{(E)}+o(q_n n^{-1/2}(b_{1,n}b_{2,n}+b_{1,n}+b_{2.n}))+O_P(b_{1,n}b_{2,n}^{(1)}+b_{2,n}b_{1,n}^{(1)}), & \text{case }(F),\\ T_n^{(E)}+o(q_n n^{-1/2}(b_{1,n}+b_{2.n}))+O(q_n n^{-1/2}b_{1,n}b_{2,n})+O_P(b_{1,n}b_{2,n}^{(2)}+b_{2,n}b_{1,n}^{(2)}), & \text{case }(G),\\ T_n^{(E)}+O(q_n b_{1,n}b_{2,n})+o(q_n(b_{1,n}+b_{2,n})n^{-1/2})+O_P(b_{1,n}b_{2,n}^{(3)}+b_{2,n}b_{1,n}^{(3)}), & \text{case }(H)\end{cases}$$

as $n\to\infty$ by using Assumption (PEE1). Here, we can reduce the second rate in the case of (A) to $o_P(q_n n^{-\delta_1-1/2})$ and the second rate in the case of (B) to $o(q_n n^{-1/2}(b_{1,n}\vee b_{2,n}))$ due to $b_{l,n}=O(1)$. Here, we denote by $T_n^{(A)}$ and $T_n^{(E)}$ the rates in the case of (A) and of (E), respectively.

Finally, we obtain by using Assumption (PEE1) that

$$\frac{1}{n}\sum_{i=1}^{n}\sum_{j=1}^{n}\mathfrak{f}\left(\frac{i-j}{q_n}\right)R_i(Z_j^{(0)}-\rho_j)$$
$$=O_P\Bigg(\frac{1}{n}\sum_{i,j=1}^{n}\mathfrak{f}\left(\frac{i-j}{q_n}\right)(b_{1,n}b_{2,n}+b_{1,n}d_{2,i}+b_{2,n}d_{1,i}+b_{1,n}\epsilon_{2,i}+b_{2,n}\epsilon_{1,i}+d_{1,i}\epsilon_{2,i}+d_{2,i}\epsilon_{1,i})$$

$$
\begin{aligned}
&\qquad \cdot (Z_j^{(0)} - \rho_j)\Big) \\
&= O_P(q_n b_{1,n} b_{2,n} n^{-1+1/r}) + o_P(q_n(b_{1,n}+b_{2,n})n^{-1/2}) + o_P(q_n n^{-1/2}) \\
&\quad + \frac{1}{n}\sum_{i=1}^n\sum_{j=1}^n \mathfrak{f}\left(\frac{i-j}{q_n}\right)(b_{1,n}\epsilon_{2,i} + b_{2,n}\epsilon_{1,i})\,(Z_j^{(0)} - \rho_j).
\end{aligned}
$$

Now, we use the 3rd assumption and combine the previous rates. This yields the claimed rates. □

Non-constant Means In contrast to the last theorem we assume that the p.e.s. is zero, i.e., $d_{\mu,l,i} \equiv 0$. Furthermore, we present sufficient conditions such that the estimation of the mean asymptotically has no influence on the LRV estimate. Therefore, we have to add some assumption on the moments. It would also be possible to display the additional rates $\hat{R}_n^{(3)}$, but they are more technical than $\hat{R}_n^{(1)}$ of Theorem 4.1.25.

Theorem 4.1.26. *Let the assumptions of Theorem 2.1.33 and the following conditions hold true:*

1. $\max_{i,j} \mathbb{E}[|V_i W_j|] \le C \in \mathbb{R}$ *for all* $\{V_n\},\{W_n\} \in \{\{\epsilon_{1,n}\},\{\epsilon_{2,n}\},\{Z_n^{(0)} - \rho_n\}\}$;
2. *the sequences* $\{\epsilon_{1,n}/\sigma_{2,n}\}$ *and* $\{\epsilon_{2,n}/\sigma_{1,n}\}$ *fulfill* $(\mathcal{K}_{\mathrm{r}}^{(2)})$ *for* $r_1,r_2 > 1$;
3. *the parameters* μ_1 *and* μ_2 *fulfill Assumption (PEE3) with* $d_{\mu,l,i} \equiv 0$;
4. *let the bandwidth* q_n *and the sequences of Assumption (PEE3) fulfill*

$$
\begin{aligned}
&\Bigg[\sum_{l_1,v_1=1}^{m_{1,n}}\sum_{l_2,v_2=1}^{m_{2,n}} (b_{1,l_1}b_{2,l_2}b_{1,v_1}b_{2,v_2})[\max_{w\in\{l,v\}} \#I_{1,w_1}\cap I_{2,w_2} \wedge q_n] \min_{w\in\{l,v\}} \#I_{1,w_1}\cap I_{2,w_2} \\
&\quad + \sum_{l=1}^{2}\sum_{v_{3-l}=1}^{m_{3-l,n}}\sum_{i_l,v_l=1}^{m_{l,n}} (b_{l,i_l}b_{l,v_l}b_{3-l,v_{3-l}})[(\#I_{l,i_l} \vee \#I_{l,v_l}\cap I_{3-l,v_{3l}}) \wedge q_n] \\
&\quad \cdot (\#I_{l,i_l} \wedge \#I_{l,v_l}\cap I_{3-l,v_{3-l}}) \\
&\quad + \sum_{l=1}^{2}\sum_{v=1}^{2}\sum_{i=1}^{m_{l,n}}\sum_{j=1}^{m_{v,n}} b_{l,i_l}b_{v,j}[(\#I_{l,i} \vee \#I_{v,j}) \wedge q_n](\#I_{l,i} \wedge \#I_{v,j})\Bigg] = o(n),
\end{aligned}
\tag{4.1.61}
$$

$$
\begin{aligned}
&\Bigg[\sum_{l_1=1}^{m_{1,n}}\sum_{l_2=1}^{m_{2,n}}\sum_{j=1}^{m}(b_{1,l_1}b_{2,l_2}n^{-\delta_j})[(\#C_j \vee \#I_{1,l_1}\cap I_{2,l_2}) \wedge q_n](\#C_j \wedge \#I_{1,l_1}\cap I_{2,l_2}) \\
&\quad + \sum_{l=1}^{2}\sum_{i=1}^{m_{l,n}}\sum_{j=1}^{m} b_{l,i}n^{-\delta_j}[(\#I_{l,i} \vee \#C_j) \wedge q_n](\#I_{l,i} \wedge \#C_j)\Bigg] = o(n),
\end{aligned}
\tag{4.1.62}
$$

$$
\sum_{l_1=1}^{m_{1,n}}\sum_{l_2=1}^{m_{2,n}} b_{1,l_1,n}b_{2,l_2,n}\#(I_{1,l_1}\cap I_{2,l_2}) + \sum_{l=1}^{2}\sum_{i=1}^{m_{l,n}} b_{l,i,n}\#I_{l,i} = o\left(\frac{n}{q_n}\right),
\tag{4.1.63}
$$

$$
\sum_{l=1}^{2} m_{l,n}\left(\sum_{i=1}^{m_{l,n}} b_{l,i,n}^{r_{3-l}}\#I_{l,i}\right)^{1/r_{3-l}} = o\left(\frac{n^{1+\delta_1}}{q_n}\right).
\tag{4.1.64}
$$

Then, it holds as $n \to \infty$ *that*

$$
\hat{D}_{3,n} = \hat{D}_n + \hat{R}_n^{(0)} + o_P(1).
$$

Proof. Firstly, we obtain that

$$
\frac{1}{n}\sum_{i=1}^n\sum_{j=1}^n \mathfrak{f}\left(\frac{i-j}{q_n}\right)(Z_i^{(3)} - \tilde{\rho}_i)(Z_j^{(3)} - \tilde{\rho}_j) = \frac{1}{n}\sum_{i=1}^n\sum_{j=1}^n \mathfrak{f}\left(\frac{i-j}{q_n}\right)(Z_i^{(0)} - \tilde{\rho}_i)(Z_j^{(0)} - \tilde{\rho}_j)
$$

$$+\frac{1}{n}\sum_{i=1}^{n}\sum_{j=1}^{n}\mathfrak{f}\left(\frac{i-j}{q_n}\right)\left[R_i(Z_j^{(0)}-\tilde{\rho}_j)+R_iR_j+R_j(Z_i^{(0)}-\tilde{\rho}_i)\right],$$

where $R_i=\sum_{l=1}^{3}R_{l,i,Z^{(3)}}$ and the first summand is equal to $D+\hat{R}_n^{(0)}$, where $\hat{R}_n^{(0)}$ is the estimation error of the LRV defined in Theorem 2.1.33. Hence, it is necessary and sufficient to consider the second summand, which can be split in three double sums:

Firstly, we obtain as $n\to\infty$, followed by $N\to\infty$,

$$\begin{aligned}
&\left|\frac{1}{n}\sum_{i=1}^{n}\sum_{j=1}^{n}\mathfrak{f}\left(\frac{i-j}{q_n}\right)R_iR_j\right|=\left|\frac{1}{n}\sum_{k=-q_n}^{q_n}\mathfrak{f}\left(\frac{k}{q_n}\right)\sum_{i=1}^{n}\mathbb{1}_{\{i+k\in(0,n]\}}R_iR_{i+k}\right|\\
&\leq\frac{1}{n}\sum_{k=-q_n}^{q_n}\left|\mathfrak{f}\left(\frac{k}{q_n}\right)\right|\sum_{i=1}^{n}\mathbb{1}_{\{i+k\in(0,n]\}}\left|R_iR_{i+k}\right|\\
&=O_P\Bigg(n^{-1}\Bigg[\sum_{l_1,v_1=1}^{m_{1,n}}\sum_{l_2,v_2=1}^{m_{2,n}}(b_{1,l_1}b_{2,l_2}b_{1,v_1}b_{2,v_2})[\max_{w\in\{l,v\}}\#I_{1,w_1}\cap I_{2,w_2}\wedge q_n]\min_{w\in\{l,v\}}\#I_{1,w_1}\cap I_{2,w_2}\\
&\quad+\sum_{l_1,v_1=1}^{m_{1,n}}\sum_{v_2=1}^{m_{2,n}}(b_{1,l_1}b_{1,v_1}b_{2,v_2})[(\#I_{1,l_1}\vee\#I_{1,v_1}\cap I_{2,v_2})\wedge q_n](\#I_{1,l_1}\wedge\#I_{1,v_1}\cap I_{2,v_2})\\
&\quad+\sum_{l=1}^{2}\sum_{v_1=1}^{m_{1,n}}\sum_{i_l,v_2=1}^{m_{2,n}}(b_{1,i_l}b_{1,v_1}b_{2,v_2})[(\#I_{l,i_2}\vee\#I_{1,v_1}\cap I_{2,v_2})\wedge q_n](\#I_{l,i_2}\wedge\#I_{1,v_1}\cap I_{2,v_2})\\
&\quad+\sum_{l=1}^{2}\sum_{v=1}^{2}\sum_{i=1}^{m_{l,n}}\sum_{j=1}^{m_{v,n}}b_{l,i_l}b_{v,j}[(\#I_{l,i}\vee\#I_{v,j})\wedge q_n](\#I_{l,i}\wedge\#I_{v,j})\Bigg]\Bigg),
\end{aligned}$$

where we use the triangle inequality, Assumption (PEE3), the upper bound of the kernel, Markov's inequality, and the uniform boundedness of the joint moments.

Now, we obtain that

$$\frac{1}{n}\sum_{i=1}^{n}\sum_{j=1}^{n}\mathfrak{f}\left(\frac{i-j}{q_n}\right)R_i(Z_j^{(0)}-\tilde{\rho}_j)=\frac{1}{n}\sum_{i=1}^{n}\sum_{j=1}^{n}\mathfrak{f}\left(\frac{i-j}{q_n}\right)R_i\left[(Z_j^{(0)}-\rho_j)+(\rho_j-\tilde{\rho}_j)\right],$$

where it holds that

$$\begin{aligned}
&\frac{1}{n}\sum_{i=1}^{n}\sum_{j=1}^{n}\mathfrak{f}\left(\frac{i-j}{q_n}\right)R_i(Z_j^{(0)}-\rho_j)\\
&=O_P\Bigg(\frac{1}{n}\sum_{k=-q_n}^{q_n}\sum_{l_1=1}^{m_{1,n}}\sum_{l_2=1}^{m_{2,n}}\mathfrak{f}\left(\frac{k}{q_n}\right)\sum_{i\in\hat{I}_{1,l_1}\cap\hat{I}_{2,l_2};i+k\in(0,n]}\Big[b_{1,l_1,n}b_{2,l_2,n}(Z_{i+k}^{(0)}-\rho_{i+k})\\
&\qquad+(b_{1,l_1,n}\epsilon_{2,i}+b_{2,l_2,n}\epsilon_{1,i})(Z_{i+k}^{(0)}-\rho_{i+k})\Big]\Bigg)\\
&=O_P\left(\frac{q_n}{n}\left[\sum_{l_1=1}^{m_{1,n}}\sum_{l_2=1}^{m_{2,n}}b_{1,l_1,n}b_{2,l_2,n}\#(I_{1,l_1}\cap I_{2,l_2})+\sum_{l=1}^{2}\sum_{i=1}^{m_{l,n}}b_{l,i,n}\#I_{l,i}\right]\right)=o_P(1),
\end{aligned}$$

by using similar arguments as above as well as in the last step (4.1.61) and (4.1.63). Similarly, for case (A) we obtain that

$$\frac{1}{n}\sum_{i=1}^{n}\sum_{j=1}^{n}\mathfrak{f}\left(\frac{i-j}{q_n}\right)R_i(\rho_j-\tilde{\rho}_j)=O_P\Bigg(\frac{q_n}{n}\sum_{l_1=1}^{m_{1,n}}\sum_{l_2=1}^{m_{2,n}}b_{1,l_1,n}b_{2,l_2,n}(\#I_{1,l_1}\cap I_{2,l_2})n^{-\delta_1}$$

$$+ \frac{q_n}{n} \sum_{l=1}^{2} m_{l,n} \left(\sum_{i=1}^{m_{l,n}} b_{l,i,n}^{r_{3-l}} \# I_{l,i} \right)^{1/r_{3-l}} n^{-\delta_1} \Bigg)$$

which is equal to $o_P(1)$ due to (4.1.61) and (4.1.63). In the cases (B), (C), and (D) we replace δ_1 by zero and have to add

$$O_P\left(\frac{q_n}{n} \sum_{l=1}^{2} \sum_{i=1}^{m_{l,n}} b_{l,i,n} \# I_{l,i} \right).$$

In case of (E) we obtain that

$$\frac{1}{n} \sum_{i=1}^{n} \sum_{j=1}^{n} \mathfrak{f}\left(\frac{i-j}{q_n} \right) R_i(\rho_j - \tilde{\rho}_j)$$
$$= O_P\Bigg(n^{-1} \Bigg[\sum_{l_1=1}^{m_{1,n}} \sum_{l_2=1}^{m_{2,n}} \sum_{j=1}^{m} (b_{1,l_1} b_{2,l_2} n^{-\delta_j}) [(\#C_j \vee \# I_{1,l_1} \cap I_{2,l_2}) \wedge q_n](\#C_j \wedge \# I_{1,l_1} \cap I_{2,l_2})$$
$$+ \sum_{l=1}^{2} \sum_{i=1}^{m_{l,n}} \sum_{j=1}^{m} b_{l,i} n^{-\delta_j} [(\# I_{l,i} \vee \# C_j) \wedge q_n](\# I_{l,i} \wedge \# C_j) \Bigg] \Bigg),$$

where the last step follows from (4.1.61) and (4.1.63). In the cases (F) to (H) we have to add the same rates as in the cases (B) to (D). Analogously, we estimate the double sum with $R_j(Z_i^{(0)} - \tilde{\rho}_i)$. □

4.2 Sequential Analysis under a General Dependency Framework and General Mean Estimates

In this subsection, we consider the asymptotic behavior of the stopping time when the means are unknown. Hence, we have to estimate them sequentially. We define for $k = 1, 2, \dots$

$$\hat{\rho}^n_{j,k,1} = \frac{1}{n}\sum_{i=1}^{n} Z^{(\psi)}_{i,k,n} \quad \text{and} \quad \hat{\rho}^{n+k}_{j,k,n+1} = \frac{1}{k}\sum_{i=n+1}^{n+k} Z^{(\psi)}_{i,k,n} \tag{4.2.1}$$

with

$$Z^{(\psi)}_{i,k,n} = \frac{(X_i - \hat{\mu}^{(\psi)}_{1,i,n+k,n})(Y_i - \hat{\mu}^{(\psi)}_{2,i,n+k,n})}{\sigma_{1,i}\sigma_{2,i}} = Z^{(0)}_i + \sum_{l=1}^{3} R_{li(n+k)nZ^{(\psi)}}, \tag{4.2.2}$$

where $\psi = 1,\dots,4$ is a design index for different mean estimate types fulfilling (PEE5), (PEE6), (PEE7), and (PEE8), respectively. Moreover, we use the decomposition of (4.1.2).

4.2.1 Closed-end Procedure under Unknown Means

Nearly Constant Means

Theorem 4.2.1. *Let the parameters μ_1 and μ_2 fulfill Assumption (PEE5) with $\delta_{\mu,1}, \delta_{\mu,2} > 0$, $\delta_{\mu,1} + \delta_{\mu,2} > \frac{1}{2}$. Moreover, for $l = 1, 2$, let $\{\epsilon_{l,n}/\sigma_{3-l,n}\}$ satisfy $(\mathcal{K}^{(1)}_{\mathbf{r}})$ for $r_l > (\frac{1}{2} + \delta_{\mu,3-l})^{-1}$. Then, Theorem 2.2.1 holds true if we replace $\tau^{(c)}_{n,\iota,0,0}$ by $\tau^{(c)}_{n,\iota,1,0}$.*

Proof. Firstly, we obtain that

$$\tilde{B}^{1,0,0}_n(k) = \hat{D}_0^{-1/2}\left[\hat{D}_0^{1/2}\tilde{B}^{0,0,0}_n(k) + \frac{\sqrt{n}k}{n+k}\sum_{l=1}^{3}\left(\frac{1}{k}\sum_{i=1+n}^{n+k} R_{li(n+k)nZ^{(1)}} - \frac{1}{n}\sum_{i=1}^{n} R_{li(n+k)nZ^{(1)}}\right)\right]. \tag{4.2.3}$$

From Theorem 2.2.1 we already know the asymptotic behavior of the first summand inside the brackets. Hence, by Slutsky's Theorem it is sufficient to prove that the second summand vanishes in probability since $\hat{D}^{-1/2} = O_P(1)$ under $H_0^{(2)}$ and $\mathbf{H}^{(\mathbf{c})}_{\mathbf{LA}}$. Under Assumption $\mathbf{H}^{(\mathbf{c})}_{\mathbf{A}}$ this is sufficient since then, the first summand is the dominating term. Hence, it remains to show that

$$\sum_{l=1}^{3} \max_{1\le k\le nm}\left|\frac{1}{\sqrt{n}}\sum_{i=1+n}^{n+k} R_{li(n+k)nZ^{(1)}}\right| = o_P(1) \tag{4.2.4}$$

and

$$\sum_{l=1}^{3} \max_{1\le k\le nm}\frac{k}{n}\left|\frac{1}{\sqrt{n}}\sum_{i=1}^{n} R_{li(n+k)nZ^{(1)}}\right| = o_P(1). \tag{4.2.5}$$

Inserting the definition of R_{lin} for $l = 1$ in (4.2.4) yields that

$$\max_{1\le k\le nm}\left|\frac{1}{\sqrt{n}}\sum_{i=1+n}^{n+k} R_{1i(n+k)nZ^{(1)}}\right| = \max_{1\le k\le nm}|e_{2,k,n}|\left|\frac{1}{\sqrt{n}}\sum_{i=1+n}^{n+k}\frac{\epsilon_{1,i}}{\sigma_{2,i}}\right| + o_P(1) = o_P(1),$$

where we use Lemma 4.1.1, Assumption (PEE5), and $(\mathcal{K}^{(1)}_{\mathbf{r}})$ for

$$\max_{1\le k\le nm}\left|\frac{1}{\sqrt{n}}\sum_{i=1+n}^{n+k}\frac{\epsilon_{1,i}}{\sigma_{2,i}}\right| = \frac{\sqrt{n(1+m)}}{\sqrt{n}}\max_{1\le k\le n(m+1)}\left|\frac{1}{\sqrt{n(1+m)}}\sum_{i=1}^{k}\frac{\epsilon_{1,i}}{\sigma_{2,i}}\right| + \left|\frac{1}{\sqrt{n}}\sum_{i=1}^{n}\frac{\epsilon_{1,i}}{\sigma_{2,i}}\right|,$$

which is equal to $o_P(n^{\delta_{\mu,2}})$ as $n \to \infty$. Analogously, we get the rate for $l = 2$. For $l = 3$ and with the same arguments as before, we obtain that

$$\max_{1\le k\le nm}\left|\frac{1}{\sqrt{n}}\sum_{i=1+n}^{n+k} R_{3i(n+k)nZ^{(1)}}\right| \le \frac{n+[nm]}{\sqrt{n}}\max_{1\le k\le nm}|e_{2,k,n}e_{1,k,n}| + o_P(1) = o_P(1).$$

Since rate as displayed in (4.2.5) follows in the same way, the three convergence results of Theorem 2.2.1 hold true. □

Corollary 4.2.2. *Under the assumptions of Theorem 4.2.1 let* $d_{\mu,l,n} \equiv 0$ *for* $l = 1, 2$. *Moreover, for* $l = 1, 2$, *let* $\{\epsilon_{l,n}/\sigma_{3-l,n}\}$ *satisfy* $(\mathcal{K}_{\mathbf{r}}^{(\mathbf{3})})$ *for* $r_l \geq (\frac{1}{2} + \delta_{\mu,3-l})^{-1}$. *Then, Theorem 2.2.3 holds true if we replace* $\tau^{(c)}_{n,\iota,0,0,\gamma}$ *by* $\tau^{(c)}_{n,\iota,1,0,\gamma}$.

Proof. Firstly, we obtain that we have to include the weighting function $w_\gamma(k/n)$ in (4.2.3). Due to the continuity of $w_\gamma(\cdot)$ on (ϵ,∞) for each arbitrarily small $\epsilon > 0$, it remains to consider the error term on $(0,\epsilon]$ or for all $k \in [1,[\epsilon nm]]$. Since $(kn^{-1})^\gamma w_\gamma(k/n)$ is uniformly bounded on $[1,[\epsilon nm]]$, it remains to prove

$$\sum_{l=1}^{3} \max_{1\leq k\leq[\epsilon nm]} \left(\frac{n}{k}\right)^\gamma \left|\frac{1}{\sqrt{n}} \sum_{i=1+n}^{n+k} R_{li(n+k)Z^{(1)}}\right| = o_P(1)$$

as $n \to \infty$, followed by $\epsilon \to 0$. If we insert the definition of $R_{linZ^{(1)}}$, we can apply $(\mathcal{K}_{\mathbf{r}}^{(\mathbf{3})})$ in case of $l = 1,2$ and use $\delta_{\mu,1} + \delta_{\mu,2} > \frac{1}{2}$ in case of $l = 3$. Thereby, the claim directly follows. □

Theorem 4.2.3. *Let the parameters* μ_1 *and* μ_2 *fulfill Assumption (PEE6) and let*

$$\max_{1\leq k\leq mn} \frac{k}{\sqrt{n}} |\mu_1 - \hat{\mu}^{n+k}_{1,n+1}||\mu_2 - \hat{\mu}^{n+k}_{2,n+1}| = o_P(1) \tag{4.2.6}$$

as $n \to \infty$. *Additionally, let* δ_{l_1,μ,l_2}, $l_1, l_2 \in \{1, 2\}$, *of Assumption (PEE6) fulfill* $\delta_{1,\mu,1} + \delta_{1,\mu,2} > \frac{1}{2}$. *Moreover, for* $l_2 = 1, 2$, *let* $\{\epsilon_{l_2,n}/\sigma_{3-l_2,n}\}$ *satisfy* $(\mathcal{K}_{\mathbf{r}}^{(\mathbf{3})})$ *for* $r_{l_2} > (\frac{1}{2} + \min_{l_1} \delta_{l_1,\mu,3-l_2})^{-1}$. *Then, Theorem 2.2.1 holds true if we replace* $\tau^{(c)}_{n,\iota,0,0}$ *by* $\tau^{(c)}_{n,\iota,2,0}$.

Proof. Similarly to the proof of Theorem 4.2.1 it is sufficient to show

$$\sum_{l=1}^{3} \max_{1\leq k\leq nm} \left|\frac{1}{\sqrt{n}} \sum_{i=1+n}^{n+k} R_{li(n+k)Z^{(2)}}\right| = o_P(1) \quad \text{and} \quad \sum_{l=1}^{3} \max_{1\leq k\leq nm} \frac{k}{n} \left|\frac{1}{\sqrt{n}} \sum_{i=1}^{n} R_{linZ^{(2)}}\right| = o_P(1), \tag{4.2.7}$$

where the latter rate has already been proven in the proof of Theorem 4.1.2. Inserting the definition of $R_{1i(n+k)Z^{(1)}}$ in the first term yields that

$$\max_{1\leq k\leq nm} \left|\frac{1}{\sqrt{n}} \sum_{i=1+n}^{n+k} R_{1i(n+k)Z^{(2)}}\right| = \max_{1\leq k\leq nm} |e_{2,k,n}| \left|\frac{1}{\sqrt{n}} \sum_{i=1+n}^{n+k} \frac{\epsilon_{1,i}}{\sigma_{2,i}}\right| + o_P(1),$$

where we use the property of the p.e.s. To obtain that the first summand is equal to $o_P(1)$, we just have to split the index set over which the maximum is taken, into $\{1, \ldots ,N\}$ and $\{N+1, \ldots ,[nm]\}$. For the second index set we apply Assumption (PEE6) and $(\mathcal{K}_{\mathbf{r}}^{(\mathbf{3})})$. For the first one we use

$$\max_{k\in\{1,\ldots,N\}} \left|\frac{1}{\sqrt{n}} \sum_{i=1+n}^{n+k} \frac{\epsilon_{1,i}}{\sigma_{2,i}}\right| = o_P(1)$$

as $n \to \infty$, which clearly follows by Markov's inequality. Analogously, we get the rate for $l = 2$. For $l = 3$ we obtain that

$$\max_{1\leq k\leq nm} \left|\frac{1}{\sqrt{n}} \sum_{i=1+n}^{n+k} R_{3i(n+k)Z^{(2)}}\right| = \max_{1\leq k\leq nm} \frac{k}{\sqrt{n}} |e_{2,k,n} e_{1,k,n}| + o_P(1) = o_P(1)$$

with the the property of the p.e.s. and the assumed rate displayed in (4.2.6). Hence, the three convergence results of Theorem 2.2.1 hold. □

Corollary 4.2.4. *Corollary 4.2.2 holds true if we replace the assumption of Theorem 4.2.1 by Theorem 4.2.3 and* $\tau^{(c)}_{n,\iota,1,0,\gamma}$ *by* $\tau^{(c)}_{n,\iota,2,0,\gamma}$.

Non-constant Means

Theorem 4.2.5. *Let the parameters* μ_1 *and* μ_2 *fulfill Assumption (PEE7). Additionally, for* $l = 1, 2$ *let* (4.1.11) *and* (4.1.12) *hold true, let* $\{\epsilon_{l,n}/\sigma_{3-l,n}\}$ *fulfill* $(\mathcal{K}_{\mathbf{r}}^{(2)})$ *for* $r_l > 1$*, and let the arrays* $b_{l,2,j,n} = b_{\mu_l,2,j,n}$ $(j = 1,\dots,m_{l,n};\ n = 1,\dots)$ *hold that*

$$\sum_{j=1}^{m_{l,n}} b_{l,2,j,n}^{r_{3-l}} \# I_{l,2,j} = o(n^{r_{3-l}/2} m_{l,2,n}^{-r_{3-l}}), \tag{4.2.8}$$

$$\sum_{v=1}^{m_{1,n}} \sum_{w=1}^{m_{2,n}} b_{1,2,v,n} b_{2,2,w,n} \#(I_{1,2,v} \cap I_{2,2,w}) = o(n^{1/2}), \tag{4.2.9}$$

and

$$\sum_{j=1}^{m_{l,2,n}} b_{l,2,j,n} \sum_{i \in I_{l,2,j}} |d_{3-l,i}| = o(\sqrt{n}) \tag{4.2.10}$$

as $n \to \infty$*. Then, Theorem 2.2.1 holds true under the assumption of Theorem 4.1.10 if we replace* $\tau_{n,\iota,0,0}^{(c)}$ *by* $\tau_{n,\iota,3,0}^{(c)}$*.*

Proof. As in the proof of Theorem 4.2.1 it is sufficient to prove the rates as displayed in (4.2.4) and (4.2.5) with $Z^{(3)}$ instead of $Z^{(1)}$. Since the modified (4.2.5) is satisfied by (4.1.11) and (4.1.12), see proof of Theorem 4.1.21, it remains to prove the modified (4.2.4).

Firstly, we consider the summand with $l = 1$ and obtain that

$$\begin{aligned}
&\frac{1}{\sqrt{n}} \max_{1\le k\le nm} \left| \sum_{i=1+n}^{n+k} R_{1i(n+k)nZ^{(3)}} \right| = \frac{1}{\sqrt{n}} \max_{1\le k\le nm} \left| \sum_{j=1}^{m_{2,2,n}} e_{2,j,k,n} \sum_{i\in \hat{I}_{2,2,j,n+k,n}} \epsilon_{1,i}/\sigma_{2,i} \right| + o_P(1) \\
&\le \frac{1}{\sqrt{n}} m_{2,2,n} \max_{1\le j\le m_{2,n}} \max_{1\le k\le nm} |e_{2,j,k,n}| \left| \sum_{i\in \hat{I}_{2,2,j,n+k,n}} \epsilon_{1,i}/\sigma_{2,i} \right| + o_P(1) \\
&= O_P(1) \frac{m_{2,2,n}}{\sqrt{n}} \max_{1\le j\le m_{2,2,n}} b_{2,2,j} \max_{1\le k\le nm} \left| \sum_{i\in \hat{I}_{2,j,n+k,n}} \epsilon_{1,i}/\sigma_{2,i} \right| + o_P(1) \\
&= O_P\left(\frac{m_{2,2,n}}{\sqrt{n}} \left(\sum_{1\le j\le m_{2,2,n}} b_{2,2,j}^{r_1} \# I_{2,2,j,n} \right)^{1/r_1} \right) + o_P(1) = o_P(1)
\end{aligned}$$

as $n \to \infty$, where we use Lemma 4.1.1, the triangle inequality, the 3rd and 4th assumptions on (PEE7), as well as the Kolmogorov-type inequalities. Analogously, the summand with $l = 2$ vanishes in probability. Hence, it remains to consider $l = 3$:

$$\begin{aligned}
&\frac{1}{\sqrt{n}} \max_{1\le k\le nm} \left| \sum_{i=1+n}^{n+k} R_{3i(n+k)nZ^{(3)}} \right| \\
&= \frac{1}{\sqrt{n}} \max_{1\le k\le nm} \left| \sum_{j_1=1}^{m_{1,n}} \sum_{j_2=1}^{m_{2,n}} e_{1,j_1,k,n} e_{2,j_2,k,n} \# \hat{I}_{1,j_1,n+k,n} \cap \hat{I}_{2,j_2,n+k,n} \right| + o_P(1) \\
&= O_P\left(\frac{1}{\sqrt{n}} \sum_{j_1=1}^{m_{1,n}} \sum_{j_2=1}^{m_{2,n}} b_{1,2,j_1} b_{2,2,j_2} \# I_{1,2,j_1,n} \cap I_{2,2,j_2,n} \right) + o_P(1)
\end{aligned}$$

as $n \to \infty$, where we use the property of the p.e.s. and (4.2.10) in the second row. Thus, by (4.2.9) the summand with $l = 3$ vanishes in probability as well. Then, the claim finally follows. □

Corollary 4.2.6. *Corollary 4.2.2 holds true if we replace the assumptions of Theorem 4.2.1 by the ones of Theorem 4.2.5,* $\tau^{(c)}_{n,\iota,1,0,\gamma}$ *by* $\tau^{(c)}_{n,\iota,3,0,\gamma}$*, and if we additionally assume that* $\max_{j\in M_{l,\epsilon,n}} b_{l,2,j} = O(1)$ *and* $\#M_{l,\epsilon,n} = O(1)$ *as* $n \to \infty$*, followed by* $\epsilon \to 0$*, where*

$$M_{l,\epsilon,n} = \{1 \le j \le m_{l,n} \ : \ I_{l,2,j,n} \cap [n+1,n+[\epsilon mn]] \neq \emptyset\}.$$

Proof. As in the proof of Corollary 4.2.2 it is sufficient to prove

$$\sum_{l=1}^{3} \max_{1\le k\le[\epsilon nm]} w_\gamma(k/n) \left| \frac{1}{\sqrt{n}} \sum_{i=1+n}^{n+k} R_{li(n+k)nZ^{(3)}} \right| = o_P(1).$$

If we use the assumed rates of Corollary 4.2.2 and the estimation as used in the proof of Theorem 4.2.5 for $l=1$, it is easy to see that

$$\max_{1\le k\le[\epsilon nm]} w_\gamma(k/n) \left| \frac{1}{\sqrt{n}} \sum_{i=1+n}^{n+k} R_{1i(n+k)nZ^{(3)}} \right|$$
$$= O_P(1) \max_{j\in M_{l,\epsilon,n}} \max_{k_1,k_2\in[n+1,n+[\epsilon nm]]\cap I_{2,j,n}} w_\gamma(k_2/n) \frac{1}{\sqrt{n}} \left| \sum_{i=k_1}^{k_2} \epsilon_{1,i}/\sigma_{2,i} \right| + o_P(1)$$

as $n\to\infty$, $\epsilon \to 0$. This tends towards zero in probability by using $(\mathcal{K}_{\mathbf{r}}^{(\mathbf{3})})$. □

Theorem 4.2.7. *Let the parameters* μ_1 *and* μ_2 *fulfill Assumption (PEE8). Additionally, for* $l=1,2$ *let* $\{\epsilon_{l,n}/\sigma_{3-l,n}\}$ *fulfill* $(\mathcal{K}_{\mathbf{r}}^{(\mathbf{2})})$ *for* $r_l > 1$ *and let* (4.2.8), (4.2.9), *and* (4.2.10) *be fulfilled with* $\{b_{l,j,n}\}$ *and* $\{I_{l,j,n}\}$ *instead of* $\{b_{l,2,j,n}\}$ *and* $\{I_{l,2,j,n}\}$*, respectively. Then, Theorem 2.2.1 holds true if we replace* $\tau^{(c)}_{n,\iota,0,0,0}$ *by* $\tau^{(c)}_{n,\iota,4,0,0}$.

Proof. It is sufficient to prove

$$\sum_{l=1}^{3} \max_{1\le k\le nm} \left| \frac{1}{\sqrt{n}} \sum_{i=1+n}^{n+k} R_{li(n+k)Z^{(4)}} \right| = o_P(1) \quad \text{and} \quad \sum_{l=1}^{3} \max_{1\le k\le nm} \left| \frac{1}{\sqrt{n}} \sum_{i=1}^{n} R_{li(k+n)Z^{(4)}} \right| = o_P(1).$$

Using Lemma 4.1.1, $(\mathcal{K}_{\mathbf{r}}^{(\mathbf{2})})$, and the rate assumptions of Assumption (PEE8), we observe that

$$\max_{1\le k\le nm} \left| \sum_{i=1+n}^{n+k} R_{li(n+k)Z^{(4)}} \right| = O_P\left(m_{l,n} \left(\sum_{j=1}^{m_{l,n}} b_{l,j,n}^{r_{3-l}} \# I_{l,j} \cap (n,n+nm] \right)^{1/r_{3-l}} \right) + o_P(1),$$

as $n\to\infty$ and for $l=1,2$. For $l=3$ we get

$$\frac{1}{\sqrt{n}} \max_{1\le k\le nm} \left| \sum_{i=n+1}^{n+k} R_{3i(n+k)Z^{(4)}} \right| = O_P\left(\frac{1}{\sqrt{n}} \sum_{j_1=1}^{m_{1,n}} \sum_{j_2=1}^{m_{2,n}} b_{1,j_1,k} b_{2,j_2,k} I_{1,j_1,n} \cap I_{2,j_2,n} \cap (n,n+nm] \right) + o_P(1),$$

where we use the rate assumptions of Assumption (PEE8) and the modified (4.2.10). Since both rates are $o_P(1)$ by applying the modified (4.2.8) and (4.2.9), the first rate of the first displayed holds true. Analogously, the second rate of the first display can be proven. This completes the proof. □

Now, we use the assumptions of Corollary 4.2.6 and Theorem 4.2.5 to get the results for the weighted stopping times.

Corollary 4.2.8. *Corollary 4.2.6 holds true if we replace the assumptions of Theorem 4.2.5 by the ones of Theorem 4.2.7,* $b_{l,2,j}$ *by* $b_{l,j}$*,* $I_{l,2,j,n}$ *by* $I_{l,j,n}$*, and* $\tau^{(c)}_{n,\iota,3,0,\gamma}$ *by* $\tau^{(c)}_{n,\iota,4,0,\gamma}$.

Proof. The claim directly follows from the combination of the arguments used in the proofs of Corollary 4.2.2 and Corollary 4.2.6. □

4.2.2 Open-end under Unknown Means

Nearly Constant Means

Theorem 4.2.9. *Under the assumptions of Corollary 4.2.2 let* $r_l > 1 \vee (\frac{1}{2}+\delta_{\mu,3-l})^{-1}$. *Then, we can replace* $\tau^{(o)}_{n,\iota,0,0,\gamma}$ *by* $\tau^{(o)}_{n,\iota,1,0,\gamma}$ *and Theorem 2.2.5 holds true.*

Proof. This proof follows the proof of Theorem 2.2.5. We already know that

$$\tilde{B}_n^{1,0,\gamma}(k) = \tilde{B}_n^{0,0,\gamma}(k) + \hat{D}_0^{-1/2} w_\gamma(k/n)\frac{n}{n+k}\frac{k}{\sqrt{n}}\sum_{l=1}^{3}\left(\frac{1}{k}\sum_{i=1+n}^{n+k} R_{li(n+k)Z^{(1)}} - \frac{1}{n}\sum_{i=1}^{n} R_{li(n+k)Z^{(1)}}\right).$$

On the one hand, we know from the proof of Corollary 4.2.2 that the maximum over $k \le [mn]$ of the second summand converges in probability towards zero. Hence, if, on the other hand, we show that the maximum over $k > [mn]$ of the second summand convergences in probability towards zero as $n \to \infty$, followed by $m \to \infty$, we get the claim by the same arguments as in the proof of Theorem 2.2.5.

Since $\max_{k>nm} |w_\gamma(k/n)k/(n+k) - 1|$ converges towards zero as $n \to \infty$, followed by $m \to \infty$, and $\hat{D}_0^{-1/2} = O_P(1)$ it remains to consider

$$\max_{k>nm}\sum_{l=1}^{3}\left|\frac{\sqrt{n}}{k}\sum_{i=1+n}^{n+k} R_{li(n+k)Z^{(1)}}\right| \quad \text{and} \quad \max_{k>nm}\sum_{l=1}^{3}\left|\frac{1}{\sqrt{n}}\sum_{i=1}^{n} R_{li(n+k)Z^{(1)}}\right|, \tag{4.2.11}$$

where for the second term we directly obtain that k has just influence on the estimation of the parameter. So due Assumption (PEE5) and $(\mathcal{K}_{\mathrm{r}}^{(\mathbf{3})})$ it is equal to an order of $O_P(n^{1/r_l-1/2-\delta_{\mu,3-l}}) = o_P(1)$. Hence, it remains to estimate the first term. Due to Assumption (PEE5), we get for the summand with $l = 1$

$$\max_{k>nm}\frac{\sqrt{n}}{k}\left|\sum_{i=1+n}^{n+k} R_{1i(n+k)Z^{(1)}}\right| = O_P\left(n^{-\delta_{\mu,2}}\max_{k>nm}\frac{\sqrt{n}}{k}\left|\sum_{i=1+n}^{n+k}\epsilon_{2i}/\sigma_{1,i}\right|\right)$$

as $n \to \infty$ and

$$P\left(Nn^{-\delta_{\mu,2}}\max_{k>nm}\frac{\sqrt{n}}{k}\left|\sum_{i=1+n}^{n+k}\epsilon_{2i}/\sigma_{1,i}\right| \ge \eta\right) \le \frac{CN^{r_2}}{\eta^{r_2}}\sum_{j=0}^{\infty}\frac{nm2^{j+1}}{(2^j mn^{\delta_{\mu,2}+1/2})^{r_2}} = o(1)$$

as $n \to \infty$ by the use of the Kolmogorov-type inequality. Here, we use $r_l > 1 \vee (\frac{1}{2}+\delta_{\mu,3-l})^{-1}$. Analogously, we get an upper bound for $l = 2$. For $l = 3$ the convergence rate directly follows from the estimation rates with $\delta_{\mu,1} + \delta_{\mu,2} > \frac{1}{2}$. Hence, the second summand of the right–hand side in the first display vanishes in probability as $n \to \infty$, followed by $m \to \infty$, so that we can apply the same arguments as used in the proof of Theorem 4.2.1 to obtain the claim. □

Theorem 4.2.10. *Under the assumptions of Corollary 4.2.4 let* $r_l > 1 \vee (\frac{1}{2}+\delta_{\mu,3-l})^{-1}$. *Then, we can replace* $\tau^{(o)}_{n,\iota,0,0,\gamma}$ *by* $\tau^{(o)}_{n,\iota,2,0,\gamma}$ *and Theorem 2.2.5 holds true.*

Proof. The proof essentially follows the way of the one of Theorem 4.2.9. We just have to replace the arguments of Corollary 4.2.2 by the ones of Corollary 4.2.4 and the used Assumption (PEE5) by (PEE6). □

Non-constant Means

Theorem 4.2.11. *Under the slightly modified assumptions of Corollary 4.2.6, i.e., replacing* $I_{l,2,j}$ *by* $I_{l,2,j} \cap (0, n+nm]$ *in (4.2.8) and in (4.2.9), we can replace* $\tau^{(o)}_{n,\iota,0,0,\gamma}$ *by* $\tau^{(o)}_{n,\iota,3,0,\gamma}$ *and Theorem 2.2.5 holds true if additionally*

$$\sum_{j=1}^{m_{l,n}}\sum_{i=0}^{\infty} b_{l,2,j}^{r_{3-l}}\frac{\#(I_{l,2,j,n}\cap[1,n+nm2^{i+1}])}{(2^{r_{3-l}})^i} = o(n^{r_{3-l}/2} m_{l,2,n}^{-r_{3-l}}) \tag{4.2.12}$$

and

$$\sum_{j_1=1}^{m_{1,2,n}} \sum_{j_2=1}^{m_{1,2,n}} b_{1,2,j_1} b_{2,2,j_2} \max_{k \geq mn} \frac{\# I_{1,2,j_1} \cap I_{2,2,j_2} \cap [n,n+k]}{k} = o(n^{-1/2}) \tag{4.2.13}$$

as $n \to \infty$*, followed by* $m \to \infty$.

Proof. As in the proof of Theorem 4.2.9 it is easy to see that the second term is of order $o_P(1)$ such that it remains to show that the first term displayed in (4.2.11) vanishes in probability as $n \to \infty$, followed by $m \to \infty$. Under the slightly modified assumptions of Corollary 4.2.6 we obtain for the summand with $l = 1$:

$$\max_{k > nm} \frac{\sqrt{n}}{k} \left| \sum_{i=1+n}^{n+k} R_{1i(n+k)Z^{(1)}} \right| = \max_{k > nm} \frac{\sqrt{n}}{k} \left| \sum_{j=1}^{m_{2,n}} e_{2,j,k,n} \sum_{i \in \hat{I}_{2,j,n+k,n}} \epsilon_{1,i}/\sigma_{2,i} \right| + o_P(1)$$

$$\leq m_{2,n} \max_{1 \leq j \leq m_{2,n}} \max_{k > nm} \frac{\sqrt{n}}{k} |e_{2,j,k,n}| \left| \sum_{i \in \hat{I}_{2,j,n+k,n}} \epsilon_{1,i}/\sigma_{2,i} \right| + o_P(1)$$

$$\leq N m_{2,n} \max_{1 \leq j \leq m_{2,n}} b_{2,j} \max_{k > nm} \frac{\sqrt{n}}{k} \left| \sum_{i \in \hat{I}_{2,j,n+k,n}} \epsilon_{1,i}/\sigma_{2,i} \right| + o_P(1) + O_P(a_N).$$

For the first summand we obtain

$$P\left(N m_{2,n} \max_{1 \leq j \leq m_{2,n}} b_{2,j} \max_{k > nm} \frac{\sqrt{n}}{k} \left| \sum_{i \in \hat{I}_{2,j,n+k,n}} \epsilon_{1,i}/\sigma_{2,i} \right| \geq \eta \right)$$

$$\leq P\left(N m_{2,n} \max_{1 \leq j \leq m_{2,n}} b_{2,j} \max_{k > nm} \frac{\sqrt{n}}{k} \max_{k_1,k_2 \in \hat{I}_{2,j,n+k,n}} \left| \sum_{i=k_1}^{k_2} \epsilon_{1,i}/\sigma_{2,i} \right| \geq \eta \right)$$

$$\leq P\left(N m_{2,n} \max_{1 \leq j \leq m_{2,n}} b_{2,j} \max_{k > nm} \frac{\sqrt{n}}{k} \max_{k_1,k_2 \in I_{2,j,n} \cap [1,n+k]} \left| \sum_{i=k_1}^{k_2} \epsilon_{1,i}/\sigma_{2,i} \right| \geq \eta \right) + o(1)$$

$$\leq \sum_{j=1}^{m_{2,n}} \sum_{i=0}^{\infty} P\left(N m_{2,n} b_{2,j} \frac{1}{\sqrt{n} m 2^i} \max_{nm2^i \leq k \leq nm2^{i+1}} \max_{k_1,k_2 \in I_{2,j,n} \cap [1,n+k]} \left| \sum_{i=k_1}^{k_2} \epsilon_{1,i}/\sigma_{2,i} \right| \geq \eta \right) + o(1)$$

$$\leq C \frac{N^{r_1} m_{2,n}^{r_1}}{\eta^{r_1} m^{r_1} n^{r_1/2}} \sum_{j=1}^{m_{2,n}} \sum_{i=0}^{\infty} b_{2,j}^{r_1} \frac{\#(I_{2,j,n} \cap [1,n+nm2^{i+1}])}{(2^{r_1})^i} + o(1)$$

as $n \to \infty$, $m \to \infty$, and $N \to \infty$. Analogously, we get a similar estimation for the summand with $l = 2$. Hence, it remains to consider the one with $l = 3$:

$$\max_{k > nm} \frac{\sqrt{n}}{k} \left| \sum_{i=1+n}^{n+k} R_{3i(n+k)Z^{(1)}} \right| = O_P\left(\sqrt{n} \sum_{j_1=1}^{m_{1,n}} \sum_{j_2=1}^{m_{1,n}} b_{1,j_1} b_{2,j_2} \max_{k \geq mn} \frac{\# I_{1,j_1} \cap I_{2,j_2} \cap [n,n+k]}{k} \right)$$

as $n \to \infty$, followed by $m \to \infty$. □

Theorem 4.2.12. *Under the assumptions of Corollary 4.2.8 let* (4.2.8) *and* (4.2.9) *be fulfilled with* $I_{l,j} \cap (0,n+nm]$ *instead of* $I_{l,2,j}$. *If* (4.2.12) *and* (4.2.13) *are satisfied with* $I_{l,j}$ *and* $m_{l,n}$ *instead of* $I_{l,2,j}$ *and* $m_{l,2,n}$, *respectively, we can replace* $\tau_{n,\iota,0,0}^{(o)}$ *by* $\tau_{n,\iota,4,0}^{(o)}$ *and Theorem 2.2.5 holds true.*

Proof. As in the proof of Theorem 4.2.11 we have to prove that both rates displayed in (4.2.11) are fulfilled, where the second follows by the same arguments used in the proof of Theorem 4.2.11. Hence, it remains to prove

$$\max_{k > nm} \frac{1}{\sqrt{n}} \left| \sum_{l=1}^{3} \sum_{i=1}^{n} R_{li(n+k)Z^{(1)}} \right| = o_P(1).$$

For the summand with $l = 1$ we obtain that

$$\max_{k>nm} \frac{1}{\sqrt{n}} \left| \sum_{i=1}^{n} R_{1i(n+k)Z^{(1)}} \right| = \max_{k>nm} \frac{1}{\sqrt{n}} \left| \sum_{j=1}^{m_{2,n}} e_{2,j,k,n} \sum_{i \in \hat{I}_{2,j,n+k,n} \cap (0,n]} \epsilon_{1,i}/\sigma_{2,i} \right| + o_P(1)$$

$$\leq N m_{2,n} \max_{1 \leq j \leq m_{2,n}} b_{2,j} \max_{k>nm} \frac{1}{\sqrt{n}} \left| \sum_{i \in \hat{I}_{2,j,n+k,n} \cap (0,n]} \epsilon_{1,i}/\sigma_{2,i} \right| + o_P(1) + O_P(a_N)$$

and therefore by the Kolmogorov-type inequalities that

$$P\left(\max_{k>nm} \frac{1}{\sqrt{n}} \left| \sum_{i=1}^{n} R_{1i(n+k)Z^{(1)}} \right| \geq \eta \right) \leq \frac{(N m_{2,n})^{r_1}}{\eta^{r_1} n^{r_1/2}} \sum_{j=1}^{m_{2,n}} b_{2,j}^{r_1} \# \left(I_{2,j,n} \cap (0,n] \right) + o(1)$$

as $n \to \infty$, $m \to \infty$ and followed by $N \to \infty$ which is equal to $o(1)$ by the modified (4.2.8). The summand with $l = 2$ can be estimated in the same way. The one with $l = 3$ can be estimated as in the proof of Theorem 4.2.5, where we intersect the sets $I_{l,\cdot}$ with $(0,n]$ and use the assumed modified rate assumption (4.2.9). □

4.3 Examples

In this subsection, we continue the three example assumptions (IID), (MIX), and (NED) and present scenarios resulting in the availability of the different main results of Section 4. Therefore, we will not demonstrate in detail that every assumption of the considered results is fulfilled. Furthermore, in this subsection we will consider the following specification of Assumption (MIX) and Assumption (NED):

Assumption (MIX1). *Under Assumption (MIX) set* $r_l' = \inf_n r_{l,n}'$ *and let* $r_1', r_2' > 2$.

Assumption (NED1). *Under Assumption (NED) let* $r_1', r_2' > 2$ *and* $p_1, p_2 = 2$.

Firstly, we obtain that under each of the Assumption (IID), (MIX1), or (NED1) the sequences $\{Z_n^{(0)} - \rho_n\}$, $\{\epsilon_{1,n}\}$, and $\{\epsilon_{2,n}\}$ (see Subsection 2.3) fulfill the Kolmogorov-type inequalities for $r = 2$, respectively. Thus, we get $r_z = r_1 = r_2 = 2$.
Since the dependency structure still holds after multiplication with any uniformly bounded, deterministic sequence $\{c_n\}$, $\{\epsilon_{l,n}\sigma_{3-l,n}^{-1}\}$ satisfies the Kolmogorov-type inequalities for $r_1 = r_2 = 2$, too.
Hence, it remains to prove the assumed properties of the estimates for the following parameter model:

$$\mu_{l,i} = \mu_{l0} + \sum_{j=1}^{R_{l,n}} \Delta_{\mu,l,j} \mathbb{1}_{\{i \le k^*_{\mu,l,j}\}} \tag{4.3.1}$$

for $i = 1,\dots,n$ and $r = 1,\dots,R_{l,n}+1$, where $R_{l,n} \ge 0$, $\Delta_{\mu,l,j} \ne 0$ and $0 = k^*_{\mu,l,0} < k^*_{\mu,l,1} < \dots < k^*_{\mu,l,R_{l,n}} < k^*_{\mu,l,R_{l,n}+1} = N_n$. Here $N_n \in \{n, n(1+m), \infty\}$ depends on the considered procedure: a posteriori, closed-end or open-end. In addition, we assume that $k^*_{\mu,l,i+1} - k^*_{\mu,l,i} \to \infty$ as $n \to \infty$ for all $i = 1,\dots,R_{l,n}-1$.

Remark 4.3.1. *In fact, under (NED) it is not necessary to set* $p_l = 2$. *In general, (NED) implies that* $\{Z_{l,n}\}$ *is an* L_{p_l}*-mixingale of size* $-v_l = -\min\{a_l, a_V(1p_l - 1/r_l)\}$ *which implies*

$$\left\| \sum_{i=1}^{[n\cdot]} \epsilon_{l,i} \right\| = \begin{cases} O_P(n^{1/2}), & \text{if } p_l = 2 \text{ and } -v_l \le -\frac{1}{2}, \\ O_P(n^{1/p_l}), & \text{if } p_l \in (1,2) \text{ and } -v_l \le -1. \end{cases}$$

It would also be possible to choose $p_1, p_2 \in (1,2)$ *such that some of the following results hold true. However, we will drop this additional case since the arguments are similar but more technical.*

4.3.1 Constant Means

In this sub-subsection, we consider the special case of $R_{l,n} \equiv 0$.

A Posteriori Analysis We define $\hat\mu_{l,n}^{(1)}$ and $\{\hat\mu_{l,k}^{(2)}\}_{k=1,\dots,n}$ as the general weighted sample mean

$$\hat\mu_{l,n}^{(1)} = \overline{Z}_{lw,1}^{n} = n^{-1} \sum_{i=1}^{n} w_{i,n} Z_{l,i} \tag{4.3.2}$$

and general cumulative average

$$\hat\mu_{l,k}^{(2)} = \hat\mu_{l,k}^{(1)} \quad \text{for } k = 1,\dots,n, \tag{4.3.3}$$

where the deterministic, positive weights $\{w_{i,n}\}$ fulfill $\sum_{i=1}^n w_{i,n} = n$ for all $n \in \mathbb{N}$ and $\sum_{i=1}^n w_{i,n}^2 \sim n$. Then, from the Kolmogorov-type inequality and the Hájek-Rényi-type inequalities we observe that

$$|\hat\mu_{l,n}^{(1)} - \mu_{l,0}| = O_P(n^{-1/2}) \quad \text{and} \quad \max_{N \le k \le n} |\hat\mu_{l,k}^{(2)} - \mu_{l,0}| = O_P(N^{-1/2}).$$

Hence, we can apply Theorem 4.1.2, Theorem 4.1.6, and the weighted results, Corollary 4.1.4 and Corollary 4.1.8, to obtain the asymptotic limits of the tests $\phi_{\iota,1}^{\gamma}$ and $\phi_{\iota,2}^{\gamma}$. The corresponding LRV estimate from Theorem 4.1.25, case (A), is a consistent estimate under H_0 and $H_0^{(2)}$ for each $q_n = o(n^{1/2})$. Here, we use Markov's inequality to obtain the rate conditions of this theorem.

Remark 4.3.2. *For example, if under Assumption (NED)* $p_l \in (1,2)$, $-\min\{a_l, a_V(1p_l - 1/r_l)\} \leq -1$, *and* $2 - 1/p_1 - 1/p_2 \geq 1/2$, *the assumptions of Theorem 4.1.2 are still fulfilled.*

Furthermore, we can estimate the change-point $k^*_{\rho,1}$ ($k^*_2 = n$) with Theorem 4.1.18 under Assumption $\mathbf{H_A}$ and obtain the same estimation rate as in Section 2 if Assumption (IID), (MIX1), or (NED1) and the rate conditions

$$(k^*_1 \wedge (n - k^*_1))^{-1} = o\left(|\Delta_{\rho,n}|^2\right) \quad \text{and} \quad |\Delta_{\rho,n}|^{-1} = O(n^{1(2})$$

hold true. Then we get $n|\Delta_{\rho,n}||\hat{\theta}^*_{\rho,1} - \theta^*_{\rho,1}| = O_P(1)$. With the same theorem we estimate multiple change-points with identical rates if the distance between the pair of change-points increases linear with n and if each change size is of the same order. More generally, if the Kolmogorov-type inequality is fulfilled for $r_1, r_2 = 2$, we estimate the change-points in the correlation with the same rates as in the situation of known means. In particular, we use this estimate for the LRV estimation by applying Theorem 4.1.25, case (E), with some $q_n \to \infty$ and $q_n = o(n^{-1/2})$ as $n \to \infty$. This might provide a higher power under the alternative, cf. Bucchia and Heuser (2015).

Sequential Analysis Applying $\{\hat{\mu}^{(1)}_{l,n+k}\}_{k=1,\ldots}$ as estimates we could use the closed-end and open-end procedures, Corollary 4.2.2 and Theorem 4.2.9, to test sequentially with the stopping times $\tau^{(c)}_{\iota,1,1,\gamma}$ and $\tau^{(o)}_{\iota,1,1,\gamma}$ whether there appears a change in the correlation. Otherwise, we could use $\hat{\mu}^{(1)}_{l,n}$ for the training phase and $\{\hat{\mu}^{(1)}_{l,k}\}_{k=1,\ldots}$ depending on the data set $Z_{l,n+1}, \ldots, Z_{l,n+k}$ for the new observed data. Then, we would apply Corollary 4.2.4 and Theorem 4.2.10 to obtain the limit distribution of the stopping times $\tau^{(c)}_{\iota,2,1,\gamma}$ and $\tau^{(o)}_{\iota,2,1,\gamma}$.

In addition, it is possible to use a piecewise weighted sample mean

$$\hat{\mu}^{(3)}_{l,i,n} = \overline{Z}_{l_{w,\hat{I}_{l,j}}} = (\#\hat{I}_{l,j})^{-1} \sum_{k \in \hat{I}_{l,j}} w_{i,\#\hat{I}_{l,j}} Z_{l,k}, \qquad \forall i \in \hat{I}_{l,j},\, j = 1, \ldots, m$$

with disjoint random sets $\hat{I}_{l,j} \subset (0,n]$ and $\min_j \#\hat{I}_{l,j} \geq \epsilon n$ almost surely, for a fixed $\epsilon > 0$. On the one hand, we can test a posteriori and sequentially. On the other hand, we can estimate the change-points in the correlation. For this reason, one can now apply the results for the non-constant mean.

4.3.2 Non-constant Means

In this sub-subsection, we consider the case of $R_{l,n} \in \mathbb{N}_{>0}$. Here, we first focus on the case of known locations of the change-points.

A Posteriori Testing under Known Change-Points in the Mean If the change-points in the mean, $k_{\mu,l,1}, \ldots, k_{\mu,l,R}$ are known, we can use the piecewise weighted sample mean, i.e.,

$$\hat{\mu}^{(3')}_{l,i,n} = \overline{Z}^{k^*_{\mu,l,r}}_{l_{w,k^*_{\mu,l,r-1}}} = (k^*_{\mu,l,r} - k^*_{\mu,l,r-1})^{-1} \sum_{k=k^*_{\mu,l,r-1}+1}^{k^*_{\mu,l,r}} w_{k,k_{\mu,l,r}-k_{\mu,l,r-1}} Z_{l,k},$$

$\forall i \in (k^*_{\mu,l,r-1}, k^*_{\mu,l,r}]$, $r = 1, \ldots, R_{l,n} + 1$, where the weights fulfill the same properties as in the previous sub-subsection. Then, under Assumption (IID), (MIX1), or (NED1) we observe that

$$|\hat{\mu}^{(3')}_{l,i,n} - \mu_{l,i}| = O_P((k^*_{\mu,l,r} - k^*_{\mu,l,r-1})^{-1/2}) \qquad \forall i \in (k^*_{\mu,l,r-1}, k^*_{\mu,l,r}],\, r = 1, \ldots, R_{l,n} + 1.$$

Denote $\mu^*_{l,r}$ as the exact mean on $(k^*_{\mu,l,r-1}, k^*_{\mu,l,r}]$. Then, we obtain

$$\max_{r \in \{1,\ldots,R_{l,n}+1\}} \frac{|\overline{Z}^{k^*_{\mu,l,r}}_{l_{w,k^*_{\mu,l,r-1}}} - \mu^*_{l,r}|}{(k^*_{\mu,l,r} - k^*_{\mu,l,r-1})^{1/2}} = O_P(R^{1/2}_{l,n}).$$

Moreover, we define

$$\begin{aligned} I_{\mu,l,i} &= \{(k^*_{l,i-1,n}, k^*_{l,i,n}] \qquad \text{for } i = 1,\ldots,R_{l,n}+1, \\ b_{\mu,l,i} &= \{R_{l,n}^{1/2}(k^*_{\mu,l,i} - k^*_{\mu,l,i-1})^{-1/2} \qquad \text{for } i = 1,\ldots,R_{l,n}+1. \end{aligned} \tag{4.3.4}$$

Then, we notice that the rate assumptions of Theorem 4.1.10 are fulfilled if

$$\sum_{j=1}^{R_{l,n}} b^2_{\mu,l,j} \# I_{l,j} = R^2_{l,n} \stackrel{!}{=} o(nR^{-2}_{l,n}) \text{ and } \sum_{j=1}^{R_{1,n}} \sum_{i=1}^{R_{2,n}} b_{\mu,1,j} b_{\mu,2,i} \# I_{1,j} \cap I_{2,i} \leq cR^2_{l,n} \stackrel{!}{=} o(n^{1/2}).$$

Thus, these two rates are fulfilled if $R_{l,n} = o(n^{1/4})$.

Then, if $2r_1/(r_1-1) \leq r_2$, we can test whether or not there is a change in the correlation. Here, we use the LRV estimate from Theorem 4.1.26 with well chosen q_n. The condition $2r_1/(r_1-1) \leq r_2$ implies that the moment condition of Theorem 4.1.26 is fulfilled. Furthermore, we can apply the closed-end and open-end procedures under the same settings as used in Subsection 2.3, because the change-points of the mean are known and thereby the mean estimation is sufficient.

Now, we consider the second testing procedure based on the second type of estimates where we assume that the change-points in the means are known once again and $R_{l,n} \equiv R_l$. We define

$$\hat{\mu}^{(4')}_{l,i,k,n} = (k \wedge k^*_{\mu,l,r} - k^*_{\mu,l,r-1})^{-1} \sum_{k=k^*_{\mu,l,r-1}+1}^{k^*_{\mu,l,r}\wedge k} w_{i,k\wedge k^*_{\mu,l,r} - k^*_{\mu,l,r-1}} Z_{l,k}$$

$$\forall i \in (k^*_{\mu,l,r-1}, k \wedge k^*_{\mu,l,r}],\ k = 2,\ldots,n,\ r = 1,\ldots,R_l+1$$

and note that Assumption (PEE4) is fulfilled for $d_{\mu,l,n} \equiv 0$,

$$\begin{aligned} \hat{I}_{\mu,l,1,r} &= I_{\mu,l,1,r} = (k^*_{\mu,l,r-1}, k^*_{\mu,l,r}], \\ \hat{I}_{\mu,l,2,2r+1} &= I_{\mu,l,2,2r+1} = (k^*_{\mu,l,r}, k^*_{\mu,l,r+1} \wedge (k^*_{\mu,l,r} + c_{r,n})], \\ \hat{I}_{\mu,l,2,2r+2} &= I_{\mu,l,2,2r+2} = (k^*_{\mu,l,r+1} \wedge (k^*_{\mu,l,r} + c_{r,n}), k^*_{\mu,l,r+1}], \end{aligned}$$

for $r = 0,\ldots,R$, and

$$(b_{\mu,l,j_1,j_2})_{\substack{j_1=1,\ldots,R+1 \\ j_2=1,\ldots,2(R+1)}} = \begin{pmatrix} 1 & c_{0,n}^{-1/2} & (k^*_{\mu,l,1})^{-1/2} & \cdots & & & \\ 0 & 0 & 1 & c_{1,n}^{-1/2} & (k^*_{\mu,l,2} - k^*_{\mu,l,1})^{-1/2} & \cdots & \\ 0 & 0 & 0 & 0 & \ddots & \ddots & \\ 0 & 0 & 0 & \cdots & 0 & 1 & c_{R,n}^{-1/2} \end{pmatrix}^T,$$

where $c_{1,n},\ldots,c_{R+1,n}$ are suitable chosen sequences which tend towards infinity, e.g., $c_{r,n} = (k^*_{\mu,l,r+1} - k^*_{\mu,l,r})^\delta$ with $\delta \in (\frac{1}{4},\frac{1}{2})$. Then, the rate assumptions (4.1.16) and (4.1.17) are fulfilled so that we can apply Theorem 4.1.13 to test whether or not there is a change in the correlation. Therefore, we use the consistent LRV estimate of Theorem 4.1.26.

Change-Point Estimation under Known Change-Points in the Mean Now, let $R_{l,n} \equiv R \in \mathbb{N}_{>0}$. The estimation of the changes in a multiple change-point setting of the correlations is influenced by the mean estimation if at least two change-points in the mean lie in every $n\epsilon$-ball around a change-point in the correlation. Let, for example, $k^*_{\rho,1} = [0.5n]$, $k^*_{\mu,1} = [0.5n - \log(n)]$, and $k^*_{\mu,2} = [0.5n + \log(n)]$ which implies that the mean estimation error between the two change-points is equal to $O_P(\log(n)^{-1})$. Hence, the rate assumption (4.1.54) is only fulfilled if $a_n = O(n \log(n)^{1/2} \min_i \Delta_{\rho,i,n} / \max_i \Delta_{\rho,i,n})$, which could provide the change-point estimation rates of the correlation if the change sizes of the correlation vanish with different rates.

A Posteriori Testing under Unknown Change-Points in the Mean Now, we consider the case where the change-points in the mean, $k^*_{\mu,l,i}$, are unknown. Thus, we estimate them and apply Theorem 2.1.22 to get the estimation rate of these change-points. Assuming the change size does not vanish and the distance between the change-points increases linearly with rate n, we obtain $\max_{r,l}|k^*_{\mu,l,r} - \hat{k}_{\mu,l,r}| = O_P(1)$ under the three examples (IID), (MIX1), and (NED1). It is obvious that the examples of the previous paragraphs hold true if we replace the exact change-points in the mean by their estimates with this kind of estimation rate. Therefore, we consider a general estimation rate in the following. Set

$$\hat{\mu}^{(3)}_{l,i,n} = \overline{Z_l}^{\hat{k}_{\mu,l,r}}_{w,\hat{k}_{\mu,l,r-1}} = (\hat{k}_{\mu,l,r} - \hat{k}_{\mu,l,r-1})^{-1} \sum_{v=\hat{k}_{\mu,l,r-1}+1}^{\hat{k}_{\mu,l,r}} w_{v,\hat{k}_{\mu,l,r}-\hat{k}_{\mu,l,r-1}} Z_{l,v} \tag{4.3.5}$$

$\forall i \in (\hat{k}_{\mu,l,r-1},\hat{k}_{\mu,l,r}]$, $r = 1,\ldots,R_{l,n}+1$. Suppose the change-point estimates fulfill

$$\max_r \frac{|k^*_{\mu,l,r} - \hat{k}_{\mu,l,r}|}{c_{l,r,n}} = o_P(1) \qquad \text{with} \qquad c_{l,r,n} \vee c_{l,r-1,n} = O\left((k^*_{\mu,l,r} - k^*_{\mu,l,r-1})^{1/2}\right) \tag{4.3.6}$$

as $n \to \infty$, then, we get

$$\left|\hat{\mu}^{(3)}_{l,i,n} - \mu_{l,i}\right| = \begin{cases} O_P\left((k^*_{\mu,l,r} - k^*_{\mu,l,r-1})^{-1/2}\right), & \text{for all } i \in (k^*_{l,r-1,n},k^*_{l,r,n}] \cap (\hat{k}_{\mu,l,r-1},\hat{k}_{\mu,l,r}], \\ O_P(1), & \text{else.} \end{cases}$$

Hence, $\mu_{l,i}$ satisfies assumption (PEE3) with $\hat{\mu}^{(3)}_{l,i,n}$, $d_{l,i} \equiv 0$,

$$\hat{I}_{\mu,l,i} = \left\{(k^*_{l,j-1,n}, k^*_{l,j,n}] \cap (\hat{k}_{l,k-1,n}, \hat{k}_{l,k,n}],\ \ i = j + (k-1)\cdot(R_{l,n}+1),\ \ j,k = 1,\ldots,R_{l,n}+1, \right. \tag{4.3.7}$$

$$I_{\mu,l,i} = \begin{cases} (k^*_{l,j-1,n}, k^*_{l,j,n}], & i = j + (j-1)\cdot(R_{l,n}+1),\ \ j = 1,\ldots,R_{l,n}+1, \\ (k^*_{l,j,n}, c_{l,j,n} + k^*_{l,j,n}], & i = j + (j-2)\cdot(R_{l,n}+1),\ \ j = 2,\ldots,R_{l,n}+1, \\ (k^*_{l,j-1,n} - c_{l,j-1,n}, k^*_{l,j-1,n}], & i = j + j\cdot(R_{l,n}+1),\ \ j = 1,\ldots,R_{l,n}, \\ \emptyset & \text{else,} \end{cases} \tag{4.3.8}$$

and

$$b_{\mu,l,i} = \begin{cases} R^{1/2}_{l,n}(k^*_{\mu,l,r} - k^*_{\mu,l,r-1})^{-1/2}, & \text{for } i = j + (j-1)\cdot(R_{l,n}+1),\ \ j = 1,\ldots,R_{l,n}+1, \\ 1, & \text{else,} \end{cases} \tag{4.3.9}$$

where we use that

$$P\left(\bigcap_{j=1}^{(R_{l,n}+1)^2} \left\{\hat{I}_{\mu,l,i} \subset I_{\mu,l,i}\right\}\right) = 1 - P\left(\bigcup_{j=1}^{R_{l,n}} \left\{|k_{\mu,l,r} - \hat{k}_{\mu,l,r}| \geq c_{l,r,n}\right\}\right) = 1 + o(1).$$

We obtain that the rate assumptions of Theorem 4.1.10 are fulfilled if

$$R_{l,n} = o(n^{1/4}), \qquad \sum_{i=1}^{R_{l,n}} c_{l,i,n} = o(nR^{-2}_{l,n}), \qquad \text{and} \qquad \sum_{i_1=1}^{R_{1,n}}\sum_{i_2=1}^{R_{2,n}} c_{1,i_1,n} \wedge c_{2,i_2,n} = o(n^{1/2}). \tag{4.3.10}$$

Thus, if $c_n = \max_{l,r} c_{l,r,n} = o(n^{1/2})$, the rate assumptions of Theorem 4.1.10 and of Theorem 4.1.26 are fulfilled, which is the case for the three examples (IID), (MIX) and (NED) with $r'_l > 2$ and $p_l = 2$ for $l = 1,2$ and in the case of non-local change-points with $\min_{r,l}|k^*_{\mu,l,r} - k^*_{\mu,l,r-1}| \sim n$.

Now, we consider the second estimate type under the condition that the change-points in the means are unknown but can be estimated in the sense of $\max_{r,l}|k^*_{\mu,l,r} - \hat{k}_{\mu,l,r}| = O_P(1)$ which holds under a multitude of multiple change-point settings; cf. Theorem 2.1.22. We set

$$\begin{aligned}\hat{\mu}^{(4)}_{l,i,k,n} &= (k \wedge \hat{k}_{\mu,l,r} - \hat{k}_{\mu,l,r-1})^{-1} \sum_{k=\hat{k}_{\mu,l,r-1}+1}^{\hat{k}_{\mu,l,r}\wedge k} w_{i,k\wedge\hat{k}_{\mu,l,r}-\hat{k}_{\mu,l,r-1}} Z_{l,k} \\ &\quad \forall i \in (\hat{k}_{\mu,l,r-1},k \wedge \hat{k}_{\mu,l,r}],\ k = 2,\dots,n,\ r = 1,\dots,R+1.\end{aligned} \tag{4.3.11}$$

Then, Assumption (PEE4) is fulfilled with $d_{\mu,l,n} \equiv 0$, $\hat{I}_{1,\mu,l,i} = \hat{I}_{\mu,l,i}$, and $I_{1,\mu,l,i} = I_{\mu,l,i}$, which are displayed in (4.3.7) and (4.3.8) and fulfill

$$\begin{aligned}
\hat{I}_{\mu,l,2,2r+1} &= (\hat{k}_{\mu,l,r},\hat{k}_{\mu,l,r+1} \wedge (\hat{k}_{\mu,l,r} + c_{r,n})],\\
\hat{I}_{\mu,l,2,2r+2} &= (\hat{k}_{\mu,l,r+1} \wedge (\hat{k}_{\mu,l,r} + c_{r,n}), \hat{k}_{\mu,l,r+1}],\\
I_{\mu,l,2,2r+1} &= (k^*_{\mu,l,r} - c_n,(k^*_{\mu,l,r+1} + c_n) \wedge (k^*_{\mu,l,r} + c_{r,n} + c_n)],\\
I_{\mu,l,2,2r+2} &= (((k^*_{\mu,l,r+1} - c_n) \wedge (k^*_{\mu,l,r} + c_{r,n} - c_n), k^*_{\mu,l,r+1} + c_n],
\end{aligned}$$

and

$$\begin{aligned}
&(b_{\mu,l,j_1,j_2})_{\substack{j_1=1,\dots,(R+1)^2\\ j_2=1,\dots,2(R+1)}}\\
&= \begin{pmatrix}
1 & c_{0,n}^{-1/2} & (k^*_{\mu,l,1})^{-1/2} & \cdots & & & & \\
1 & 1 & 1 & \cdots & \cdots & \cdots & \cdots & 1\\
\vdots & \vdots & \vdots & \vdots & \vdots & \vdots & \vdots & \vdots\\
1 & 1 & 1 & \cdots & \cdots & \cdots & \cdots & 1\\
0 & 0 & 1 & 1 & 1 & \cdots & & \\
0 & 0 & 1 & c_{2,n}^{-1/2} & (k^*_{\mu,l,2} - k^*_{\mu,l,1})^{-1/2} & \cdots & & \\
\ddots & \ddots & \ddots & \ddots & \ddots & \ddots & \ddots & \ddots\\
0 & 0 & 0 & 0 & \cdots & 0 & 1 & c_{R+1,n}^{-1/2}
\end{pmatrix}.
\end{aligned}$$

Here we used $\max_{r,l}|k^*_{\mu,l,r} - \hat{k}_{\mu,l,r}| = O_P(1)$. Additionally, let $c_n,c_{0,n},\dots,c_{R,n}$ be sequences increasing towards infinity, e.g., $c_n = \log(n)$ and $c_{r,n} = (k^*_{\mu,l,r+1} - k^*_{\mu,l,r})^\delta$ for some $\delta \in (\frac{1}{4},\frac{1}{2})$.

If we just have $\max_{r,l}|k^*_{\mu,l,r} - \hat{k}_{\mu,l,r}| = o_P(c_n)$, we can replace $c_{r,n}$ and $(k^*_{\mu,l,r+1} - k^*_{\mu,l,r})$ by $c_{r,n} - 2c_n$ and $(k^*_{\mu,l,r+1} - k^*_{\mu,l,r} - 2c_n)$, respectively, which implies that c_n must not be too large. Now, it is easy to obtain that the rate assumptions of Theorem 4.1.13 are fulfilled so that the tests are available.

Sequential Procedures under Known Change-Points in the Means If we obtain the data one by one, we have to estimate the change-points of the mean sequentially. Therefore, we consider the simplified model characterized by:

$$R_{l,n} \equiv R_l \in \mathbb{N}_{>0},\quad k^*_{\mu,l,r} = [n\theta_{l,r}],\ \theta_{r,l} \in (0,1),\ r = 1,\dots,R-1, l = 1,2,\ \theta_{r,l} < \theta_{r+1,l}$$

and

$$k^*_{\mu,l,R} = [n\theta_{l,R}],\ \theta_{l,R} \in (1,1+m).$$

Firstly, we look at the procedure where we estimate the change-points of the training period one time and we estimate again and again the change-points in the means of the sequentially observed data. Hence, we use the previously presented estimate $\hat{\mu}^{(3)}_{l,i,n}$ for the training set, i.e., for $i \in \{1,\dots,n\}$. As in the above example, $\hat{\mu}^{(3)}_{l,i,n}$ estimates the means so that the estimation error does not influence the asymptotic behavior of $\hat{\rho}_{4,n}$. For the sequentially observed data, i.e., for $i \in \{n+1,\dots,n+nm\}$, we use an estimate-type such as $\hat{\mu}^{(4)}_{l,i,k,n}$. In this estimate we replace $\hat{k}_{\mu,l,R}$ by $\hat{k}_{\mu,l,R,k}$, since the last change-points $k^*_{\mu,l,R}$, $l=1,2$, are estimated sequentially. Thus,

$$\hat{\mu}^{(4')}_{l,i,k,n} = \begin{cases} \frac{1}{\hat{k}_{\mu,l,R,k}\wedge k}\sum_{i=n+1}^{n+(\hat{k}_{\mu,l,R,k}\wedge k)} Z_{l,i}, & i=n+1,\dots,n+(k\wedge \hat{k}_{\mu,l,R,k}) \\ \frac{1}{k-\hat{k}_{\mu,l,R,k}}\sum_{i=n+\hat{k}_{\mu,l,R,k}}^{n+k} Z_{l,i}, & i=n+\hat{k}_{\mu,l,R,k},\dots,n+k \end{cases} \tag{4.3.12}$$

where $\hat{k}_{\mu,l,R,k}$ can be an estimate based on the data $Z_{l,n+1},\dots,Z_{l,n+k}$ and is defined as k if there is no change detected. Using Theorem 2.2.1 or 2.2.3 and the following remarks we obtain that the change in the mean is asymptotically detected with probability one in the time-interval $(n,k^*_{l,R}+\sqrt{c_n n}]$ for any sequence $c_n\to\infty$. The number of early alarms of a stopping time, which controls a change in the mean, can be reduced if we replace the critical value by $c_n^{1/2}$. After this stopping time alarms, we assume a structural break in the mean and estimate the possible change-point by the least square estimate, Theorem 2.1.18. Henceforward, we can sequentially estimate the change-point in the correlations by using Theorem 2.1.20. Thus, we can apply Theorem 4.2.5 to sequentially test whether or not there is a change in the correlation if

$$\#(n,k^*_{1,R}+c_n\sqrt{n}]\cap(n,k^*_{2,R}+c_n\sqrt{n}]=o(n^{1/2}).$$

Now, we take a look at the second procedure where every change-point in the means is estimated sequentially, even the ones in the training period. Then, we apply Theorem 2.1.25 to estimate the numbers and the locations of structural breaks in the mean. The number will be asymptotically correct in a time-interval of length of $n^{1/2+\epsilon}$ after the latter appeared structural break. Additionally, each estimated change-point $k^*_{\mu.}$ is sufficiently close to the correct one so that we apply Theorem 4.2.7 in case for all $r_1,r_2=1,\dots,R+1$ it holds that

$$\#(k^*_{1,r_1}+n^{1/2+\epsilon}]\cap(n,k^*_{2,r_2}+n^{1/2+\epsilon}]=o(n^{1/2}).$$

General Mean Functions Until now, we have considered piecewise constant means. In this paragraph, we look at the following model:

$$\mu_{l,i}=\mu_{l,i,n}=m_{l,i}+g_{\mu,l}(i/n) \quad\text{and}\quad m_i=\sum_{j=1}^{R_l}\Delta_{\mu,l,j}\mathbb{1}_{\{i\le[n\theta^*_{\mu,l,j}]\}}, \tag{4.3.13}$$

where $g_{\mu,l}$ is a Lipschitz continuous function. We define the estimates

$$\hat{\mu}_{i,l,n}=\left\{\frac{1}{hn}\sum_{j\in I_{i,nh}} Z_{l,j} \quad\text{with}\quad I_{i,nh}=\begin{cases}(0,nh], & \text{if } 1\le i\le hn,\\ (i-nh,i], & \text{if } hn<i\le n,\end{cases}\right. \tag{4.3.14}$$

where $h=h_n\to 0$ but $nh\to\infty$, as $n\to\infty$.

Firstly, we assume $m_i\equiv 0$ and obtain that the estimate can be split in a deterministic and in a random part. Because of the Lipschitz continuity we get by

$$\Delta_{\mu,l,i,n}=g_{\mu,l}(i/n)-\frac{1}{nh}\sum_{j\in I_{i,nh}} g(j/n) \quad\text{that}\quad \max_i|\Delta_{\mu,i,n}|=O(h). \tag{4.3.15}$$

To illustrate that Theorem 2.1.1 holds true if we replace the exact means by the above estimates, it is sufficient to prove for $l=1,2$ that

$$\left\|\sum_{i=1}^{[n\cdot]}\frac{\epsilon_{3-l,i}}{\sigma_{l,i}}(\mu_{l,i}-\hat{\mu}_{i,l,n})\right\|=o_P(n^{1/2}) \quad\textit{and}\quad \left\|\sum_{i=1}^{[n\cdot]}\frac{(\mu_{1,i}-\hat{\mu}_{1,i,n})(\mu_{2,i}-\hat{\mu}_{2,i,n})}{\sigma_{1,i}\sigma_{2,i}}\right\|=o_P(n^{1/2}).$$

Theorem 4.3.3. *Under the (IID) assumption and the preceding mean model, the convergences of Theorem 2.1.1 hold true if* $n^{-1/2}\ll h\ll n^{-1/4}$.

Proof. The claim is established if the previous two equalities are fulfilled. Firstly, for a fixed $n \in \mathbb{N}$ we obtain that

$$\left(\sum_{i=1}^{m} \frac{\epsilon_{1,i}}{\sigma_{2,i}} \Delta_{\mu,2,i,n} \,; \sigma(\epsilon_{1,i}, i \leq m)\right) \quad \text{and} \quad \left(\sum_{i=1}^{m} \frac{\epsilon_{2,i}}{\sigma_{1,i}} \Delta_{\mu,1,i,n} \,; \sigma(\epsilon_{2,i}, i \leq m)\right)$$

are martingales. Thus, we can apply Doob's martingale inequality to obtain

$$P\left(\left\|\sum_{i=1}^{[n\cdot]} \frac{\epsilon_{l,i}}{\sigma_{3-l,i}} \Delta_{\mu,3-l,i,n}\right\| \geq \eta\sqrt{n}\right) \leq (\eta\sqrt{n})^{-r_l} \mathbb{E}\left[\left|\sum_{i=1}^{n} \frac{\epsilon_{l,i}}{\sigma_{3-l,i}} \Delta_{\mu,3-l,i,n}\right|^{r_l}\right]$$
$$\leq (\eta\sqrt{n})^{-r_l} \sum_{i=1}^{n} \mathbb{E}\left[|\frac{\epsilon_{l,i}}{\sigma_{3-l,i}} \Delta_{\mu,l,i,n}|^{r_l}\right] = O(n^{1-r_l/2} h^{r_l}),$$

where we use Bahr and Esseen (1964, Th. 2) for the last inequality. For $r_l = 2$, the right-hand side is of order $o(1)$ if $h \to 0$. Secondly, we get that

$$\max_{1\leq k\leq n}\left|\sum_{i=1}^{k} \frac{\epsilon_{l,i}}{\sigma_{3-l,i}} \frac{1}{nh} \sum_{j\in I_{i,nh}} \epsilon_{3-l,j}\right| \leq \max_{1\leq k\leq n}\left|\sum_{i=1}^{k} \frac{1}{nh} \sum_{j\in I_{i,nh}} \frac{\epsilon_{l,i}\epsilon_{3-l,j} - \mathbb{E}\left[\epsilon_{l,i}\epsilon_{3-l,j}\right]}{\sigma_{3-l,i}}\right|$$
$$+ \sum_{i=1}^{n}\left|\frac{1}{nh} \sum_{j\in I_{i,nh}} \frac{\mathbb{E}\left[\epsilon_{l,i}\epsilon_{3-l,j}\right]}{\sigma_{3-l,i}}\right|,$$

where the second term is equal to a rate of $O(h^{-1})$ which is $o(n^{1/2})$ if $h \gg n^{-1/2}$. For the first summand, we use the following decomposition

$$\sum_{i=1}^{k} \frac{\epsilon_{l,i}\epsilon_{3-l,i} - \mathbb{E}\left[\epsilon_{l,i}\epsilon_{3-l,i}\right]}{nh\sigma_{3-l,i}} + \sum_{i=1}^{k} \sum_{j=1\vee(i-nh)\wedge(n-nh)}^{i-1} \frac{\epsilon_{l,i}\epsilon_{3-l,j}}{nh\sigma_{3-l,i}}$$

and obtain that the two summands are martingales with respect to $\sigma\left(\epsilon_{1,i}, \epsilon_{2,i}, i \leq k\right)$. Hence, we can apply the σ-additivity and Doob's martingale inequality for each of the summands and obtain

$$P\left(\max_{1\leq k\leq n}\left|\sum_{i=1}^{k} \sum_{j\in I_{i,nh}} \frac{\epsilon_{l,i}\epsilon_{3-l,j} - \mathbb{E}\left[\epsilon_{l,i}\epsilon_{3-l,j}\right]}{nh\sigma_{3-l,i}}\right| \geq \eta\sqrt{n}\right)$$
$$\leq (\eta\sqrt{n})^{-r}\left[\mathbb{E}\left[\left|\sum_{i=1}^{n} \frac{\epsilon_{l,i}\epsilon_{3-l,i} - \mathbb{E}\left[\epsilon_{l,i}\epsilon_{3-l,i}\right]}{nh\sigma_{3-l,i}}\right|^{r}\right]\right.$$
$$\left. + \mathbb{E}\left[\left|\sum_{i=1}^{n-1} \sum_{j=1\vee(i-nh)\wedge(n-nh)}^{i-1} \frac{\epsilon_{l,i}\epsilon_{3-l,j}}{nh\sigma_{3-l,i}}\right|^{r}\right]\right]$$
$$\leq C_{\eta,r}(n^{3/2}h)^{-r} \sum_{i=1}^{n}\left[\mathbb{E}\left[|\epsilon_{1,i}\epsilon_{2,i} - \rho_i|^r\right] + \sum_{j=1\vee(i-nh)\wedge(n-nh)}^{i-1} \mathbb{E}\left[|\epsilon_{l,i}\epsilon_{3-l,j}|^r\right]\right]$$
$$= O\left(n^{2-3r/2} h^{1-r}\right)$$

by using Bahr and Esseen (1964, Th. 2) in the last inequality. For $r = 2$, it is $o(1)$ if $nh \to \infty$. Finally, we decompose

$$(\mu_{l,i} - \hat{\mu}_{i,1,n})(\mu_{l,i} - \hat{\mu}_{i,2,n}) = \Delta_{\mu,1,i,n}\Delta_{\mu,2,i,n} + \frac{\Delta_{\mu,1,i,n}}{nh} \sum_{j\in I_{i,nh}} \epsilon_{2,j} + \frac{\Delta_{\mu,2,i,n}}{nh} \sum_{j\in I_{i,nh}} \epsilon_{1,j}$$
$$+ \frac{1}{(nh)^2} \sum_{j_1\in I_i} \sum_{j_2\in I_i} (\epsilon_{1,j_1}\epsilon_{2,j_2} - \mathbb{E}\left[\epsilon_{1,j_1}\epsilon_{2,j_2}\right]) + \frac{1}{(nh)^2} \sum_{j_1\in I_i} \sum_{j_2\in I_i} \mathbb{E}\left[\epsilon_{1,j_1}\epsilon_{2,j_2}\right]$$

and get $\max_k \left| n^{-1/2} \sum_{i=1}^k \Delta_{\mu,1,i,n}\Delta_{\mu,2,i,n} \right| = O(\sqrt{n}h^2)$. Furthermore, we obtain by Markov's inequality that

$$\begin{aligned}
&P\left(\max_{1\leq k\leq n} \left| \sum_{i=1}^k \frac{\Delta_{\mu,1,i,n}}{nh} \sum_{j\in I_{i,nh}} \epsilon_{2,j} \right| \geq \eta\sqrt{n} \right) \\
&\leq \frac{1}{n(nh\eta)^2} \mathbb{E}\left[\max_{1\leq k\leq n} \left(\sum_{i=1}^k \Delta_{\mu,1,i,n} \sum_{j\in I_{i,nh}} \epsilon_{2,j} \right)^2 \right] \\
&\leq \frac{1}{n^3\eta^2} \sum_{i_1=1}^n \sum_{i_2=1}^n \mathbb{E}\left[\left| \sum_{j_1\in I_{i_1,nh}} \epsilon_{2,j_1} \sum_{j_2\in I_{i_1,nh}} \epsilon_{2,j_2} \right| \right] = O(h),
\end{aligned}$$

where we use in the last row that for each i_1 there are at most $2nh$ many i_2 such that there are at most nh many dependent and at most $(nh)^2$ independent summands in the double sum. Moreover, we use $\|X\|_1 \leq \|X\|_2$ for the rate of the sum of independent random variables. Similarly, we obtain that

$$\max_{1\leq k\leq n} \left| \sum_{i=1}^k \frac{1}{(nh)^2} \sum_{j_1\in I_i} \sum_{j_2\in I_i} \epsilon_{1,j_1}\epsilon_{2,j_2} \right| = O_P\left(\frac{1}{\sqrt{n}h} \right)$$

and

$$\max_{1\leq k\leq n} \left| \sum_{i=1}^k \frac{1}{(nh)^2} \sum_{j_1\in I_i} \sum_{j_2\in I_i} \mathbb{E}\left[\epsilon_{1,j_1}\epsilon_{2,j_2}\right] \right| = O_P\left(\frac{1}{\sqrt{n}h} \right).$$

Hence, if $\sqrt{n}h^2 = o(1)$ and $(\sqrt{n}h)^{-1} = o(1)$, the two necessary rates are fulfilled. □

Remark 4.3.4. *1. It is possible to weight the $\{Z_{l,n}\}$ by some deterministic numbers.*

2. *Similar results can be proven under assumptions such as (MIX1) or (NED1).*

3. *Suppose the preceding Lipschitz continuous change functions $g_{\mu,1}$ and $g_{\mu,2}$ are only piecewise Lipschitz continuous with finitely many bounded jumps. Then, we get the same result if each of the common abrupt changes $1 < k^*_{1,1} < \ldots < k^*_{1,m_1} < n$ and $1 < k^*_{2,1} < \ldots < k^*_{2,m_2} < n$ of $g_{\mu,1}$ and $g_{\mu,2}$ are isolated such that*

$$\sum_{i=1}^{m_1-1} \sum_{j=1}^{m_2-1} \#(k^*_{1,i} - nh, k^*_{1,i+1} + nh] \cap (k^*_{2,j} - nh, k^*_{2,j+1} + nh] \cap (0,n] = o(\sqrt{n}).$$

4. *In many types of structures of g_μ, we can apply research results to observe the behavior of $g_\mu - \hat{g}_\mu$ which we can use to prove the two sufficient rates. For instance, if X and Y follow processes with linear drift, i.e. $g_{\mu,l}(t) = a_{l,t}t$ with optionally piecewise constants $a_{l,t} \neq 0$, we refer Horváth and Steinebach (2000).*

5 Change-Point Analysis of the Correlation under Known Means and Unknown Variances

5.1 A Posteriori Analysis under a General Dependency Framework and General Variance Estimates

In this subsection, we consider a posteriori testing procedures and change-point estimates of structural breaks in the correlation under known expectations but unknown variances. We set for $k = 1, \dots, n$

$$\hat{\rho}_{k,4+\psi} = \frac{1}{k}\sum_{i=1}^{k} \frac{(X_i - \mu_{1,i})(Y_i - \mu_{2,i})}{\hat{\sigma}^{(\psi)}_{1,i,k,n}\hat{\sigma}^{(\psi)}_{2,i,k,n}} = \frac{1}{k}\sum_{i=1}^{k} Z^{(4+\psi)}_{i,k,n}, \tag{5.1.1}$$

where $\psi = 1, \dots, 4$ is a design index to distinguish between the four different variance estimate-types which we will present. Below, we will omit the index n of the estimates and of $Z^{(4+\psi)}_{i,k,n}$. For $\{Z^{(\psi)}_{i,k}\}_{1 \leq i \leq k \leq n \in \mathbb{N}}$ we use the decomposition

$$\begin{aligned}
Z^{(4+\psi)}_{i,\cdot} =& \frac{(X_i - \mu_{1,i})(Y_i - \mu_{2,i})}{\hat{\sigma}^{(\psi)}_{1,\cdot}\hat{\sigma}^{(\psi)}_{2,i,\cdot}} \\
=& Z^{(0)}_i + Z^{(0)}_i\left(\frac{\sigma_{1,i}}{\hat{\sigma}^{(\psi)}_{1,\cdot}} - 1\right) + Z^{(0)}_i\left(\frac{\sigma_{2,\cdot}}{\hat{\sigma}^{(\psi)}_{2,\cdot}} - 1\right) + Z^{(0)}_i\left(\frac{\sigma_{1,i}}{\hat{\sigma}^{(\psi)}_{1,\cdot}} - 1\right)\left(\frac{\sigma_{2,i}}{\hat{\sigma}^{(\psi)}_{2,\cdot}} - 1\right) \\
=& Z^{(0)}_i + R_{1 \cdot Z^{(j+4)}} + R_{2 \cdot Z^{(4+\psi)}} + R_{3 \cdot Z^{(4+\psi)}} = Z^{(0)}_i\left(1 + \sum_{l=1}^{3} r_{l \cdot Z^{(\psi+4)}}\right),
\end{aligned} \tag{5.1.2}$$

where the points stand for the index k and/or index i. Moreover, we will eliminate the unused indices.

5.1.1 Testing under a Functional Central Limit Theorem and General Variance Estimates

Nearly Constant Variances In this paragraph, we assume that the unknown variances are nearly constant. Firstly, we expect that the variances (or equivalently the standard deviations) satisfy Assumption (PEE1) with $\delta_l > 0$. Thus, the positive variance or rather the positive standard deviation can be estimated nearly consistently. Since the proof will be less technical if we postulate that the standard deviations satisfy Assumption (PEE1), we will do so without loss of generality.

Theorem 5.1.1. *Let the parameters σ_1 and σ_2 fulfill Assumption (PEE1) with $\delta_{\sigma,1}, \delta_{\sigma,2} > 0$, where $\sigma_{l,i} - d_{\sigma,l,i} = \sigma_l > 0$. Then, Theorem 2.1.1 holds true if we replace $B^{0,0,0}_n$ by $B^{5,0,0}_n$.*

Proof of Theorem 5.1.1. Firstly, we use the decomposition (5.1.2) and note that it is sufficient to show for $l = 1, 2, 3$

$$\max_{1 \leq k \leq n}\left|\frac{k}{\sqrt{n}}\left(\frac{1}{k}\sum_{i=1}^{k} Z^{(0)}_i r_{ilZ^{(5)}} - \frac{1}{n}\sum_{i=1}^{n} Z^{(0)}_i r_{ilZ^{(5)}}\right)\right| = o_P(1). \tag{5.1.3}$$

Furthermore, we obtain that the left–hand side can be bounded from above by

$$\max_{1 \leq k \leq n}\left|\frac{k}{\sqrt{n}}\left(\frac{1}{k}\sum_{i=1}^{k} \rho_i r_{ilZ^{(5)}} - \frac{1}{n}\sum_{i=1}^{n} \rho_i r_{ilZ^{(5)}}\right)\right| + 2\max_{1 \leq k \leq n}\left|\frac{1}{\sqrt{n}}\sum_{i=1}^{k}(Z^{(0)}_i - \rho_i) r_{ilZ^{(5)}}\right|. \tag{5.1.4}$$

Starting with the second term and $l = 1$, we obtain that

$$r_{i1Z^{(5)}} = \left(\frac{\sigma_{1,i}}{\hat{\sigma}_{1,n}} - 1\right) = o_P(1) + d_{1,\sigma,i}/\hat{\sigma}_{1,n},$$

where we use the consistence of the variance estimate. Hence, by the use of Lemma 4.1.1 it holds that

$$\max_{1\leq k\leq n}\left|\frac{1}{\sqrt{n}}\sum_{i=1}^{k}(Z_i^{(0)}-\rho_i)r_{ilZ^{(5)}}\right| = o_P(1)\max_{1\leq k\leq n}\left|\frac{1}{\sqrt{n}}\sum_{i=1}^{k}(Z_i^{(0)}-\rho_i)\right| + o_P(1) = o_P(1),$$

which similarly follows as in the proof of Theorem 4.1.2. In a similar way we get the rates for $l = 2,3$ so that it remains to consider the first summand of (5.1.4). We obtain that

$$\max_{1\leq k\leq n}\left|\frac{k}{\sqrt{n}}\left(\frac{1}{k}\sum_{i=1}^{k}\rho_i r_{ilZ^{(5)}} - \frac{1}{n}\sum_{i=1}^{n}\rho_i r_{ilZ^{(5)}}\right)\right| = o_P(1)\left(\max_{1\leq k\leq n}\left|\frac{k}{\sqrt{n}}\left(\frac{1}{k}\sum_{i=1}^{k}\rho_i - \frac{1}{n}\sum_{i=1}^{n}\rho_i\right)\right| + 1\right)$$

so that we just have to replace $R_{n,\rho}(\cdot)$ by $R_{n,\rho}(\cdot)(1+o_P(1))$ in the proof of Theorem 2.1.1 and the claimed convergences follow analogously. □

Corollary 5.1.2. *Under the assumptions of Theorem 5.1.1, let each* $\{d_n\} \in \{\{d_{\sigma,1,n}\},\{d_{\sigma,2,n}\},\{1\}\}$ *fulfill that*

$$\max_{1\leq k\leq [\epsilon n]}\left(\frac{n}{k}\right)^{\gamma}\frac{1}{\sqrt{n}}\left|\sum_{i=1}^{k}(Z_i^{(0)}-\rho_i)d_i\right| = o_P(1) \tag{5.1.5}$$

and

$$\max_{[(1-\epsilon)n]\leq k\leq n-1}\left(\frac{n}{n-k}\right)^{\gamma}\frac{1}{\sqrt{n}}\left|\sum_{i=k+1}^{n}(Z_i^{(0)}-\rho_i)d_i\right| = o_P(1), \tag{5.1.6}$$

as $n\to\infty$*, followed by* $\epsilon\to 0$*. Then, we can replace* $B_n^{0,0,\gamma}$ *by* $B_n^{5,0,\gamma}$ *in Theorem 2.1.4 and the convergences hold true.*

Proof. Firstly, we decompose

$$\begin{aligned} B_n^{5,0,\gamma}(\cdot) = B_n^{0,0,\gamma}(\cdot) - \sum_{l=1}^{3} w_\gamma([n\cdot]/n))&\left[\frac{n-[n\cdot]}{n}n^{-1/2}\sum_{i=1}^{[n\cdot]}\rho_i r_{i,n,l,z^{(5)}}\right.\\ &+\frac{n-[n\cdot]}{n}n^{-1/2}\sum_{i=1}^{[n\cdot]}(Z_i^{(0)}-\rho_i)r_{i,n,l,z^{(5)}}\\ &\left.-\frac{[n\cdot]}{n}n^{-1/2}\sum_{i=[\cdot n]}^{n}\rho_i r_{i,n,l,z^{(5)}} - \frac{[n\cdot]}{n}n^{-1/2}\sum_{i=[\cdot n]}^{n}(Z_i^{(0)}-\rho_i)r_{i,n,l,z^{(5)}}\right]. \end{aligned}$$

Then, the claim follows from the combination of the arguments of the proof of Theorem 5.1.1, (5.1.5), and (5.1.6). □

Remark 5.1.3. *Suppose the variances have non-local structural breaks in* $[[\epsilon n], n-[\epsilon n]]$, $\epsilon > 0$*. Then, the variance estimates* $\hat\sigma_{1,n}^2$ *and* $\hat\sigma_{2,n}^2$ *are not consistent. Assuming the estimates converge in probability towards a positive constant, then, it is still possible that the process* $B_n^{5,0,0}(\cdot)$ *converges in distribution towards a Gaussian process whose covariance structure depends on both variance functions. In particular, let us assume that* $\sigma_{1,i}^2 = g_{\sigma,1}(i/n)$ *and* $\sigma_{2,i}^2 = g_{\sigma,1}(i/n)$ *with bounded, positive, and piecewise continuous functions* $g_{\sigma,l}$*, the Assumption (IID) with* $\mathbb{E}\left[(\tilde\epsilon_{1,i}\tilde\epsilon_{2,i}-\rho_0)^2\right] \equiv 1$*, and the sample variances as estimates. Then, under* H_0 *we obtain that*

$$\begin{aligned} &w_\gamma\left(\frac{[n\cdot]}{n}\right)\frac{[n\cdot]}{\sqrt{n}}(\hat\rho_{5,[n\cdot]}-\hat\rho_{5,n})\\ &\xrightarrow{\mathcal{D}[0,1]} w_\gamma(\cdot)\frac{W(\cdot)\int_0^{\cdot} g_{\sigma,1}(x)g_{\sigma,2}(x)dx - (\cdot)W(1)\int_0^1 g_{\sigma,1}(x)g_{\sigma,2}(x)dx}{\sqrt{\int_0^1 g_{\sigma,1}(x)dx\int_0^1 g_{\sigma,2}(x)dx}}\\ &\qquad + \rho_0 w_\gamma(\cdot)\frac{\int_0^{\cdot} g_{\sigma,1}(x)g_{\sigma,2}(x)dx - (\cdot)\int_0^1 g_{\sigma,1}(x)g_{\sigma,2}(x)dx}{\sqrt{\int_0^1 g_{\sigma,1}(x)dx\int_0^1 g_{\sigma,2}(x)dx}}, \end{aligned}$$

which follows from an application of Davidson (1994, Corollary 29.11).

Now, we consider the estimates which are only calculated by the first k observations. In contrast to the estimates of the mean, where it was sufficient that the estimation error is small, it is not sufficient for the estimates of the variances. They have to fulfill a FCLT additionally.

Theorem 5.1.4. *Let the parameters σ_1 and σ_2 fulfill Assumption (PEE2), where $\sigma_{l,i} - d_{\sigma,l,i} = \sigma_l > 0$. Additionally, let $\max_{l=1,2,k\in\mathbb{N}} \hat{\sigma}_{l,k}^{-1} = O_P(1)$ for $l = 1,2$ and let*

$$\frac{1}{\sqrt{n}}\sum_{i=1}^{[n\cdot]}(Z_i^{(0)} - \rho_i) - \frac{[n\cdot]}{\sqrt{n}}(\hat{\sigma}_{1,[n\cdot]}^2 - \sigma_1^2)\frac{\overline{\rho}_{[n\cdot]}}{2\sigma_1^2} - \frac{[n\cdot]}{\sqrt{n}}(\hat{\sigma}_{2,[n\cdot]}^2 - \sigma_2^2)\frac{\overline{\rho}_{[n\cdot]}}{2\sigma_2^2} \xrightarrow{D[0,1]} D_{(6)}^{1/2}W(\cdot), \tag{5.1.7}$$

where $D_{(6)} > 0$. Then, the convergences of Theorem 2.1.1 hold if we replace each $B_n^{0,0}$ and D by $B_n^{6,0,0}$ and $D_{(6)}$.

Remark 5.1.5. *Suppose the variances are constant and the variance estimates are the sample variances with known means. Then,* (5.1.7) *fails for $\rho_i \equiv 1$ or $\rho_i \equiv -1$ since the term on the left-hand side equals zero.*

Proof of Theorem 5.1.4. Firstly, from the triangle inequality and from the combination of the FCTL from Theorem 2.1.1 and of (5.1.7) we obtain that

$$\left\| \frac{[n\cdot]}{\sqrt{n}}\frac{(\hat{\sigma}_{1,[n\cdot]}^2 - \sigma_1^2)}{2\sigma_1^2} + \frac{[n\cdot]}{\sqrt{n}}\frac{(\hat{\sigma}_{2,[n\cdot]}^2 - \sigma_2^2)}{2\sigma_2^2} \right\| = O_P(1).$$

Furthermore, we obtain that

$$B_n^{(6,0,0)}(k) = B_n^{(0,0,0)}(k) + \hat{D}_{0,n}^{-1/2}\sum_{l=1}^{3}\frac{k}{\sqrt{n}}\left(\frac{1}{k}\sum_{i=1}^{k} Z_i^{(0)} r_{ilZ^{(6)}} - \frac{1}{n}\sum_{i=1}^{n} Z_i^{(0)} r_{ilZ^{(6)}}\right).$$

Define $\sigma_l = \sigma_{l,i} - d_{\sigma,l,i}$. Now, we consider the summand with $l = 1$ and obtain that

$$\begin{aligned}
&\left(\frac{1}{[n\cdot]}\sum_{i=1}^{[n\cdot]} Z_i^{(0)} r_{i1Z^{(6)}} - \frac{1}{n}\sum_{i=1}^{n} Z_i^{(0)} r_{i1Z^{(6)}}\right)\\
&= \left(\frac{1}{[n\cdot]}\sum_{i=1}^{[n\cdot]} Z_i^{(0)}\frac{\sigma_1^2 - \hat{\sigma}_{1,[n\cdot]}^2}{\hat{\sigma}_{1,[n\cdot]}(\sigma_1 + \hat{\sigma}_{1,[n\cdot]})} - \frac{1}{n}\sum_{i=1}^{n} Z_i^{(0)}\frac{\sigma_1^2 - \hat{\sigma}_{1,n}^2}{\hat{\sigma}_{1,n}(\sigma_1 + \hat{\sigma}_{1,n})}\right)\\
&\quad + \left(\frac{1}{[n\cdot]}\sum_{i=1}^{[n\cdot]} Z_i^{(0)}\frac{d_{\sigma,1,i}}{\hat{\sigma}_{1,[n\cdot]}} - \frac{1}{n}\sum_{i=1}^{n} Z_i^{(0)}\frac{d_{\sigma,1,i}}{\hat{\sigma}_{1,n}}\right),
\end{aligned}$$

where the second summand weighted by $[n\cdot]/n^{1/2}$ uniformly vanishes in probability because of $\max_l \|\hat{\sigma}_{l,[n\cdot]}^{-1}\| = O_P(1)$ and the characteristic of the p.e.s. This follows analogously for the summand with $l = 2$. For the weighted first summand, temporarily defined as $S_{1,1,[n\cdot]}$ for $l = 1$ and $S_{2,1,[n\cdot]}$ for $l = 2$, we obtain by using $\max_l \|\hat{\sigma}_{l,[n\cdot]}^{-1}\| = O_P(1)$ and both estimation rates from Assumption (PEE2) that

$$\begin{aligned}
\frac{[n\cdot]}{\sqrt{n}}S_{1,1,[n\cdot]} &= \frac{[n\cdot]}{\sqrt{n}}(\sigma_1^2 - \hat{\sigma}_{1,[n\cdot]}^2)\left(\frac{[n\cdot]^{-1}\sum_{i=1}^{[n\cdot]}(Z_i^{(0)} - \rho_i)}{\hat{\sigma}_{1,[n\cdot]}(\sigma_1 + \hat{\sigma}_{1,[n\cdot]})} + \frac{[n\cdot]^{-1}\sum_{i=1}^{[n\cdot]}\rho_i}{\hat{\sigma}_{1,[n\cdot]}(\sigma_1 + \hat{\sigma}_{1,[n\cdot]})}\right)\\
&\quad - \sqrt{n}(\sigma_1^2 - \hat{\sigma}_{1,n}^2)\left(\frac{n^{-1}\sum_{i=1}^{n}(Z_i^{(0)} - \rho_i)}{\hat{\sigma}_{1,n}(\sigma_1 + \hat{\sigma}_{1,n})} + \frac{n^{-1}\sum_{i=1}^{n}\rho_i}{\hat{\sigma}_{1,n}(\sigma_1 + \hat{\sigma}_{1,n})}\right)\frac{[n\cdot]}{n}\\
&= \frac{[n\cdot]}{\sqrt{n}}(\sigma_1^2 - \hat{\sigma}_{1,[n\cdot]}^2)\frac{n^{-1}\sum_{i=1}^{[n\cdot]}\rho_i}{\hat{\sigma}_{1,[n\cdot]}(\sigma_1 + \hat{\sigma}_{1,[n\cdot]})} + o_P(1)\\
&\quad - \sqrt{n}(\sigma_1^2 - \hat{\sigma}_{1,n}^2)\frac{[n\cdot]}{n}\frac{n^{-1}\sum_{i=1}^{n}\rho_i}{\hat{\sigma}_{1,n}(\sigma_1 + \hat{\sigma}_{1,n})} + o_P(1)\\
&= \frac{[n\cdot]}{\sqrt{n}}\frac{(\sigma_1^2 - \hat{\sigma}_{1,[n\cdot]}^2)}{2\sigma_1^2}[n\cdot]^{-1}\sum_{i=1}^{[n\cdot]}\rho_i - \frac{\sqrt{n}(\sigma_1^2 - \hat{\sigma}_{1,n}^2)}{2\sigma_1^2}\frac{[n\cdot]}{n}n^{-1}\sum_{i=1}^{n}\rho_i + o_P(1)
\end{aligned}$$

as $n \to \infty$. Hence, we get

$$\frac{[n\cdot]}{\sqrt{n}}(S_{1,1,[n\cdot]} + S_{2,1,[n\cdot]}) = \frac{[n\cdot]}{\sqrt{n}}\left(\frac{(\sigma_1^2 - \hat{\sigma}_{1,[n\cdot]}^2)}{2\sigma_1^2} + \frac{(\sigma_2^2 - \hat{\sigma}_{2,[n\cdot]}^2)}{2\sigma_2^2}\right)\overline{\rho}_{[n\cdot]}$$
$$- \left(\frac{\sqrt{n}(\sigma_1^2 - \hat{\sigma}_{1,n}^2)}{2\sigma_1^2} + \frac{\sqrt{n}(\sigma_1^2 - \hat{\sigma}_{2,n}^2)}{2\sigma_2^2}\right)\frac{[n\cdot]}{n}\overline{\rho}_n + o_P(1)$$

as $n \to \infty$. Now, we consider the second summand of the first display with $l = 3$:

$$\left\|\frac{1}{[n\cdot]}\sum_{i=1}^{[n\cdot]} Z_i^{(0)} r_{i3Z^{(6)}} - \frac{1}{\sqrt{n}}\sum_{i=1}^{n} Z_i^{(0)} r_{i3Z^{(6)}}\right\| \leq c\left\|\sum_{i=1}^{[n\cdot]} Z_i^{(0)} \frac{\sigma_1 - \hat{\sigma}_{1,[n\cdot]}}{\hat{\sigma}_{1,[n\cdot]}}\frac{\sigma_2 - \hat{\sigma}_{2,[n\cdot]}}{\hat{\sigma}_{2,[n\cdot]}}\right\| + o_P(\sqrt{n})$$

as $n \to \infty$, where we use the characteristic of the p.e.s. and $\max_l \|\hat{\sigma}_{l,[n\cdot]}^{-1}\| = O_P(1)$ as well as $\max_l \|\hat{\sigma}_{l,[n\cdot]}\| = O_P(1)$. Furthermore, we obtain

$$\frac{2}{\sqrt{n}}\left\|\sum_{i=1}^{[n\cdot]} Z_i^{(0)} \frac{\sigma_1 - \hat{\sigma}_{1,[n\cdot]}}{\hat{\sigma}_{1,[n\cdot]}}\frac{\sigma_2 - \hat{\sigma}_{2,[n\cdot]}}{\hat{\sigma}_{2,[n\cdot]}}\right\| \leq \max_{N \leq k \leq n} \frac{2}{\sqrt{n}}\left|\sum_{i=1}^{k}(Z_i^{(0)} - \rho_i)\frac{\sigma_1 - \hat{\sigma}_{1,k}}{\hat{\sigma}_{1,k}}\frac{\sigma_2 - \hat{\sigma}_{2,k}}{\hat{\sigma}_{2,k}}\right|$$
$$+ \max_{N \leq k \leq n} \frac{2}{\sqrt{n}}\left|\sum_{i=1}^{k}\rho_i \frac{\sigma_1^2 - \hat{\sigma}_{1,k}^2}{\hat{\sigma}_{1,k}(\sigma_1 + \hat{\sigma}_{1,k})}\frac{\sigma_2 - \hat{\sigma}_{2,k}}{\hat{\sigma}_{2,k}}\right| + o_P(1)$$

as $n \to \infty$, where the first summand tends towards zero as $n \to \infty$, followed by $N \to \infty$, because of Assumption (PEE2)

For the second summand we use $\sum_{i=1}^{k} |\rho_i| \leq k$, $\|[n\cdot]/\sqrt{n}(\sigma_1^2 - \hat{\sigma}_{1,[\cdot n]}^2)\| = O_P(1)$ and $\max_{N \leq k \leq n} |\sigma_2^2 - \hat{\sigma}_{2,k}^2| = o_P(1)$ as $n \to \infty$, followed by $N \to \infty$ so that the second summand converges in probability towards zero, too.

Hence, under H_0 or $\mathbf{H_{LA}}$ we obtain

$$B_n^{(6,0,0)}([n\cdot]) = B_n^{(0,0,0)}([n\cdot]) + \hat{D}_{0,n}^{-1/2}\frac{[n\cdot]}{\sqrt{n}}\frac{(\overline{\rho}_n\hat{\sigma}_{1,n}^2 - \overline{\rho}_{[n\cdot]}\hat{\sigma}_{1,[n\cdot]}^2)}{2\sigma_1^2}$$
$$+ \hat{D}_{0,n}^{-1/2}\frac{[n\cdot]}{\sqrt{n}}\frac{(\overline{\rho}_n\hat{\sigma}_{2,n}^2 - \overline{\rho}_{[n\cdot]}\hat{\sigma}_{2,[n\cdot]}^2)}{2\sigma_2^2} + o_P(1)$$

which implies the convergences under H_0 and $\mathbf{H_{LA}}$ by the use of the CMT and the assumed FCLT.

Under Assumption $\mathbf{H_A}$ we obtain

$$B_n^{(6,0,0)}([n\cdot]) = B_n^{(0,0,0)}([n\cdot]) + \hat{D}_{0,n}^{-1/2} O_P(1),$$

where the dominating term is $B_n^{(0,0,0)}([n\cdot])$. Thus, under Assumption $\mathbf{H_A}$ the claimed convergence holds true. □

Remark 5.1.6. *1. If $\hat{\sigma}_n^2 \to \sigma^2$ a.s., it follows that $\max_{n \leq k} |\hat{\sigma}_k^2 - \sigma^2| = o_P(1)$.*

2. The influence of the variance estimates on the asymptotic behavior depends on the correlation. If $\rho_i \equiv 0$ under H_0, the asymptotic normality of $[n\cdot]/\sqrt{n}(\sigma_l - \hat{\sigma}_{l,[n\cdot]})$ can be reduced to $\hat{\sigma}_{l,n} \to \sigma_l$ a.s., for which the fourth moments of X and Y are not always necessary.

Corollary 5.1.7. *Let the assumptions in Theorem 5.1.4 hold true. Additionally, let both rate assumptions of Corollary 5.1.2 be fulfilled and let for $l = 1, 2$*

$$\max_{1 \leq k \leq [n\epsilon]} \left|w_\gamma(k/n)\frac{k}{\sqrt{n}}(\hat{\sigma}_{l,n}^2 - \hat{\sigma}_{l,k}^2)\right| = o_P(1) \quad \text{and} \quad \max_{n-[n\epsilon] \leq k \leq n} \left|w_\gamma(k/n)\frac{k}{\sqrt{n}}(\hat{\sigma}_{l,n}^2 - \hat{\sigma}_{l,k}^2)\right| = o_P(1)$$

as $n \to \infty$, followed by $\epsilon \to 0$. Then, Theorem 2.1.4 holds true if we replace $B_n^{0,0,\gamma}$ by $B_n^{6,0,\gamma}$.

Proof. Firstly, we obtain that

$$B_n^{6,0,\gamma}(\cdot) = B_n^{0,0,\gamma}(\cdot) - \hat{D}_{0,n}^{-1/2}\sum_{l=1}^{3} w_\gamma([n\cdot]/n)\Bigg[n^{-1/2}\sum_{i=1}^{[n\cdot]}\rho_i\left(r_{i,[n\cdot],l,z^{(6)}} - \frac{[n\cdot]}{n}r_{i,n,l,z^{(6)}}\right)$$
$$+ n^{-1/2}\sum_{i=1}^{[n\cdot]}(Z_i^{(0)} - \rho_i)\left(r_{i,[n\cdot],l,z^{(6)}} - \frac{[n\cdot]}{n}r_{i,n,l,z^{(6)}}\right)$$
$$- \frac{[n\cdot]}{n}n^{-1/2}\sum_{i=[\cdot n]+1}^{n}\rho_i r_{i,n,l,z^{(6)}} - \frac{[n\cdot]}{n}n^{-1/2}\sum_{i=[\cdot n]+1}^{n}(Z_i^{(0)} - \rho_i)r_{i,n,l,z^{(6)}})\Bigg].$$

Using the representation of $r_{i,[n\cdot],l,z^{(6)}}$, the rate assumptions imply the claim. □

Non-constant Variances In this paragraph, we consider settings where the variances satisfy Assumption (PEE4).

Remark 5.1.8. *1. Under the assumption that the parameters σ_1 and σ_2 fulfill Assumption (PEE3) with finitely many change-points and under some additional assumptions, we get under H_0 that*

$$B_n^{7,0,0}(\cdot) \xrightarrow{D[0,1]} B(\cdot)$$
$$- \rho_0\left(\sum_{j=1}^{m_1+1}\frac{((\cdot)\wedge\theta_{\sigma,1,j} - \theta_{\sigma,1,j-1}) - (\cdot)(\theta_{\sigma,1,j} - \theta_{\sigma,1,j-1})}{2\sigma_{1,j}^2}(W_{\sigma,1}(\theta_{\sigma,1,j}) - W_{\sigma,1}(\theta_{\sigma,1,j-1}))\right)$$
$$- \rho_0\left(\sum_{j=1}^{m_2+1}\frac{((\cdot)\wedge\theta_{\sigma,2,j} - \theta_{\sigma,2,j-1}) - (\cdot)(\theta_{\sigma,2,j} - \theta_{\sigma,2,j-1})}{2\sigma_{2,j}^2}(W_{\sigma,2}(\theta_{\sigma,2,j}) - W_{\sigma,2}(\theta_{\sigma,2,j-1}))\right),$$

where B is a Brownian bridge, $W_{\sigma,1}$ and $W_{\sigma,2}$ are Brownian motions, and $0 = \theta_{\sigma,l,0} < \ldots < \theta_{\sigma,l,m_l+1} = 1$ are the change-points of the variances. In order to get a test statistic, we apply a continuous function that maps in $\mathbb{R}$. But in general, this distribution depends on the unknown change-points $\theta_{\sigma,l,j}$ and therefore it is useless for testing if $\rho_0 \neq 0$. Hence, an assumption as (PEE3) is only suitable if we apply a continuous mapping into $\mathbb{R}$ so that the asymptotic distribution is independent of these change-points.

Theorem 5.1.9. *Let the parameters σ_1^2 and σ_2^2 fulfill Assumption (PEE4) and let the following conditions hold true:*

1. *$D_{(8)} > 0$ and*

$$\frac{1}{\sqrt{n}}\sum_{i=1}^{[n\cdot]}\left[(Z_i^{(0)} - \rho_i) - (\hat{\sigma}_{1,i,[n\cdot]}^2 - \sigma_{1,i}^2)\frac{\rho_i}{2\sigma_{1,i}^2} - (\hat{\sigma}_{2,i,[n\cdot]}^2 - \sigma_{2,i}^2)\frac{\rho_i}{2\sigma_{2,i}^2}\right] \xrightarrow{D[0,1]} D_{(8)}^{1/2}W(\cdot) \quad (5.1.8)$$

under H_0 and the left hand side is equal to $O_P(1)$ under the alternatives $\mathbf{H_A}$ or $\mathbf{H_{LA}}$;

2. *the sequence $\{Z_n^{(0)} - \rho_n\}$ fulfills $(\mathcal{K}_\mathbf{r}^{(\mathbf{1})})$ and $(\mathcal{K}_\mathbf{r}^{(\mathbf{2})})$ for $r_z > 1$;*

3. *the estimates fulfill $\max_{l=1,2,k\in\mathbb{N}}\hat{\sigma}_{l,k}^{-1} = O_P(1)$ for $l = 1, 2$;*

4. *the arrays $\{b_{\sigma,l,j_1,j_2}\}$, $\{I_{\sigma,l,1,j_1}\}$, and $\{I_{\sigma,l,2,j_2}\}$ of Assumption (PEE4) fulfill for $l = 1, 2$ and*

as $n \to \infty$

$$\sum_{j_2=1}^{m_{3-l,2}} \left(m_{3-l,1}^{r_z} \sum_{j_1=1}^{m_{3-l,1}} b_{3-l,j_1,j_2}^{r_z} \# I_{3-l,1,j_1} \cap (0, \max I_{3-l,2,j_2}] \right)^{1/r_z} = o(n^{1/2}), \tag{5.1.9}$$

$$\sum_{\substack{1 \le j_1 \le m_{2,1} \\ 1 \le j_2 \le m_{2,2}}} \left(\sum_{i_1=1}^{m_{1,1}} \sum_{i_2=1}^{m_{1,2}} \frac{\#(I_{1,1,i_1} \cap I_{2,1,i_1} \cap (0, \max I_{1,2,j_1} \cap I_{2,2,j_2}])(b_{1,i_1,j_1} b_{2,i_2,j_2})^{r_z}}{(m_{1,1} \vee m_{2,1})^{-1}} \right)^{1/r_z} \tag{5.1.10}$$

$$= o(\sqrt{n}),$$

$$\max_{\substack{1 \le j_1 \le m_{2,1} \\ 1 \le j_2 \le m_{2,2}}} \sum_{\substack{1 \le i_1 \le m_{1,1} \\ 1 \le i_2 \le m_{1,2}}} \#(I_{1,1,i_1} \cap I_{2,1,i_1} \cap (0, \max I_{1,2,j_1} \cap I_{2,2,j_2}]) b_{1,i_1,j_1} b_{2,i_2,j_2} = o(n^{1/2}), \tag{5.1.11}$$

$$\max_{1 \le j_2 \le m_{l,2}} \sum_{j_1=1}^{m_{l,1}} b_{l,j_1,j_2}^2 \# I_{l,1,j_1} \cap (0, \max I_{l,2,j_2}] = o_P(n^{1/2}), \tag{5.1.12}$$

and for $l_1, l_2 \in \{1,2\}$

$$\max_{1 \le j_2 \le m_{l_2,2}} \sum_{j_1=1}^{m_{l_1,1,n}} b_{l_1,j_1,j_2,n} \sum_{i \in I_{l_1,1,j_1} \cap (0, \max I_{l_1,2,j_2}]} |d_{l_2,i}| = o(\sqrt{n}). \tag{5.1.13}$$

Then, the convergences of Theorem 2.1.1 hold true if we replace $B_n^{0,0}$ *and* D *by* $B_n^{8,0,0}$ *and* $D_{(8)}$, *respectively.*

Proof. (To get more space we drop the index $[n\cdot]$ of $\hat{\sigma}_{l,i,[n\cdot]}$ but keep it in mind.) Firstly, we obtain that

$$\frac{1}{\sqrt{n}} \sum_{i=1}^{[n\cdot]} \rho_i r_{i1kZ^{(8)}} = \frac{1}{\sqrt{n}} \sum_{i=1}^{[n\cdot]} \rho_i \frac{\sigma_{1,i}^2 - \hat{\sigma}_{1,i}^2}{2\sigma_{1,i}^2} + \frac{1}{\sqrt{n}} \sum_{i=1}^{[n\cdot]} \rho_i \frac{(\sigma_{1,i}^2 - \hat{\sigma}_{1,i}^2)(2\sigma_{1,i}^2 - \hat{\sigma}_{1,i}(\hat{\sigma}_{1,i} + \sigma_{1,i}))}{2\sigma_{1,i}^2 \hat{\sigma}_{1,i}(\hat{\sigma}_{1,i} + \sigma_{1,i})},$$

where we note that the second sum can be estimated by Assumption (PEE4) in the following way

$$\left\| \frac{1}{\sqrt{n}} \sum_{i=1}^{[n\cdot]} \rho_i \frac{(\sigma_{1,i}^2 - \hat{\sigma}_{1,i}^2)(2\sigma_{1,i}^2 - \hat{\sigma}_{1,i}(\hat{\sigma}_{1,i} + \sigma_{1,i}))}{2\sigma_{1,i}^2 \hat{\sigma}_{1,i}(\hat{\sigma}_{1,i} + \sigma_{1,i})} \right\|$$

$$= O_P \left(n^{-1/2} \max_{1 \le j_2 \le m_{1,2}} \sum_{j_1=1}^{m_{1,1}} b_{1,j_1,j_2}^2 \# I_{1,1,j_1} \cap (0, \max I_{1,2,j_2}] \right) + o_P(1)$$

$$+ O_P \left(n^{-1/2} \max_{1 \le j_2 \le m_{1,2}} \sum_{j_1=1}^{m_{1,1,n}} b_{1,j_1,j_2,n} \sum_{i \in I_{1,1,j_1} \cap (0, \max I_{l,2,j_2}]} |d_{3-l,i}| \right).$$

Furthermore, with $\tilde{Z}_i^{(0)} = Z_i^{(0)} - \rho_i$ and the Kolmogorov-type inequalities we get that

$$\left\| \sum_{i=1}^{[n\cdot]} \tilde{Z}_i^{(0)} r_{i1[n\cdot]Z^{(8)}} \right\| = O_P \left(\sum_{j_2=1}^{m_{1,2}} \left(m_{1,1}^{r_1} \sum_{j_1=1}^{m_{1,1}} b_{2,j_1,j_2}^{r_z} \# I_{1,1,j_1} \cap (0, \max I_{1,2,j_2}] \right)^{1/r_z} \right).$$

Analogously, we obtain the rates for $r_{i2kZ^{(2)}}$. Secondly, we obtain with Assumption (PEE4) that

$$\left\| \sum_{i=1}^{[n\cdot]} \rho_i r_{i3kZ^{(8)}} \right\| = \left\| \frac{1}{\sqrt{n}} \sum_{i=1}^{[n\cdot]} \rho_i \frac{(\sigma_{1,i} - \hat{\sigma}_{1,i})(\sigma_{2,i} - \hat{\sigma}_{2,i})}{\hat{\sigma}_{1,i} \hat{\sigma}_{2,i}} \right\| = o_P(\sqrt{n})$$

$$+ O_P \left(\max_{\substack{1 \le j_1 \le m_{2,1} \\ 1 \le j_2 \le m_{2,2}}} \sum_{i_1=1}^{m_{1,1}} \sum_{i_2=1}^{m_{1,2}} \#(I_{1,1,i_1} \cap I_{2,1,i_1} \cap (0, \max I_{1,2,j_1} \cap I_{2,2,j_2}]) b_{1,i_1,j_1} b_{2,i_2,j_2} \right)$$

and

$$\left\|\sum_{i=1}^{[n\cdot]} \tilde{Z}_i^{(0)} r_{i3kZ^{(8)}}\right\| = o_P(\sqrt{n})$$

$$+ O_P\left(\sum_{\substack{1\leq j_1\leq m_{2,1}\\ 1\leq j_2\leq m_{2,2}}} \frac{\left(\sum_{i_1=1}^{m_{1,1}}\sum_{i_2=1}^{m_{1,2}} \#(I_{1,1,i_1}\cap I_{2,1,i_1}\cap (0,\max I_{1,2,j_1}\cap I_{2,2,j_2}])(b_{1,i_1,j_1}b_{2,i_2,j_2})^{r_z}\right)^{1/r_z}}{(m_{1,1}\vee m_{2,1})^{-1}}\right),$$

where we use (5.1.13). Now, we use the above rates and the equations (5.1.9), (5.1.11), (5.1.12), and (5.1.13) to get that

$$\begin{aligned}
B_n^{(8,0,0)}(\cdot) &= B_n^{(0,0,0)}(\cdot) + \hat{D}_{0,n}^{-1/2}\sum_{l=1}^{3}\frac{[\cdot n]}{\sqrt{n}}\left(\frac{1}{[\cdot n]}\sum_{i=1}^{[\cdot n]} Z_i^{(0)} r_{il[\cdot n]Z^{(8)}} - \frac{1}{n}\sum_{i=1}^{n} Z_i^{(0)} r_{ilnZ^{(8)}}\right)\\
&= B_n^{(0,0,0)}(\cdot) + \hat{D}_{0,n}^{-1/2}\frac{[\cdot n]}{\sqrt{n}}\left(\frac{1}{[\cdot n]}\sum_{i=1}^{[\cdot n]}\rho_i\frac{\sigma_{1,i}^2-\hat{\sigma}_{1,i}^2}{2\sigma_{1,i}^2} - \frac{1}{n}\sum_{i=1}^{n}\rho_i\frac{\sigma_{1,i}^2-\hat{\sigma}_{1,i}^2}{2\sigma_{1,i}^2}\right)\\
&\quad + \hat{D}_{0,n}^{-1/2}\frac{[\cdot n]}{\sqrt{n}}\left(\frac{1}{[\cdot n]}\sum_{i=1}^{[\cdot n]}\rho_i\frac{\sigma_{2,i}^2-\hat{\sigma}_{2,i}^2}{2\sigma_{2,i}^2} - \frac{1}{n}\sum_{i=1}^{n}\rho_i\frac{\sigma_{2,i}^2-\hat{\sigma}_{2,i}^2}{2\sigma_{2,i}^2}\right) + o_P(1).
\end{aligned}$$

Moreover, we obtain that the term on the right–hand side is a sum of $\hat{D}_{0,n}^{-1/2}\frac{[n\cdot]}{\sqrt{n}}\left(\overline{\rho}_{[n\cdot]}-\overline{\rho}_n\right)$ and of one continuous mapping of the left hand-side of (5.1.8). Hence, the convergences of Theorem 2.1.1 hold true.

□

Corollary 5.1.10. *Let the assumptions of Theorem 5.1.9 hold true and let* $d_{l,i}\equiv 0$ *be given. Additionally, let both rate assumptions of Corollary 5.1.7 be fulfilled and for* $l=1,2$ *let*

$$\sum_{j_2=1}^{m_{3-l,2}}\left(m_{3-l,1}^{r_z}\sum_{j_1=1}^{m_{3-l,1}}\sum_{\substack{i\in I_{3-l,1,j_1}\cap(0,[n\epsilon]\wedge\max I_{3-l,2,j_2}]\\ \cup I_{3-l,1,j_1}\cap(n-[n\epsilon],\max I_{3-l,2,j_2}]}}\frac{b_{3-l,j_1,j_2}^{r_z}}{i^{\gamma r_z}}\right)^{1/r_z} = o(n^{1/2-\gamma}), \tag{5.1.14}$$

$$\sup_{z\in(0,\epsilon]}\max_{\substack{1\leq j_1\leq m_{2,1}\\ 1\leq j_2\leq m_{2,2}}}\sum_{i_1=1}^{m_{1,1}}\sum_{i_2=1}^{m_{1,2}}\frac{\#(I_{1,1,i_1}\cap I_{2,1,i_1}\cap(0,[nz]\wedge\max I_{1,2,j_1}\cap I_{2,2,j_2}])}{(b_{1,i_1,j_1}b_{2,i_2,j_2})^{-1}[nz]^{\gamma}} = o(n^{1/2-\gamma}), \tag{5.1.15}$$

$$\sup_{z\in[1-\epsilon,1)}\max_{\substack{1\leq j_1\leq m_{2,1}\\ 1\leq j_2\leq m_{2,2}}}\sum_{i_1=1}^{m_{1,1}}\sum_{i_2=1}^{m_{1,2}}\frac{\#(I_{1,1,i_1}\cap I_{2,1,i_1}\cap([nz],\max I_{1,2,j_1}\cap I_{2,2,j_2}])}{(b_{1,i_1,j_1}b_{2,i_2,j_2})^{-1}(n-[nz])^{\gamma}} = o(n^{1/2-\gamma}), \tag{5.1.16}$$

$$\sup_{z\in(0,\epsilon]}\max_{1\leq j_2\leq m_{l,2}}\sum_{j_1=1}^{m_{l,1}}\frac{\#I_{l,1,j_1}\cap((0,[nz]\wedge\max I_{l,2,j_2}]\cup(n-[nz],\max I_{l,2,j_2}])}{b_{l,j_1,j_2}^{-2}[nz]^{\gamma}} = o_P(n^{1/2-\gamma}) \tag{5.1.17}$$

as $n\to\infty$ *and* $\epsilon\to 0$. *Then, we can replace* $B_n^{0,0,0}$ *by* $B_n^{8,0,\gamma}$ *and the convergences of Theorem 2.1.4 hold true.*

Proof. The proof strictly follows the proof of Theorem 5.1.9, where we just have to add the weighting function. Since the weighting function is continuous on $(\epsilon,1-\epsilon)$ for each $\epsilon >$, we just have to show that the test statistic vanishes in probability on $(0,\epsilon]\cup[1-\epsilon,1)$. Using the assumed rates after applying Assumption (PEE4) or both rate assumptions of Corollary 5.1.7 on the weighted error terms of the proof of Theorem 5.1.9 yields the claim. □

5.1.2 Change-Point Estimation under Unknown Variances

In this sub-subsection, we consider the change-point estimates of the correlation under the alternative and under unknown variances which are estimated. In doing so, we will focus on the general multiple change-point estimate. The special change-point estimate for epidemic change-points in the correlations can also be extended for the case of unknown variances, but we will not pursue these estimates.

Constant Variances

Theorem 5.1.11. *Define* $Q_n^{(5)}$ *as* $Q_n^{(0)}$ *with* $Z_i^{(5)}$ *instead of* $Z_i^{(0)}$. *Then, Theorem 2.1.22 holds true with* $\theta_n^{(5)}$ *instead of* $\theta_n^{(0)}$ *if the following conditions are additionally fulfilled:*

1. *the parameters* σ_1^2 *and* σ_2^2 *fulfill Assumption (PEE1) with* $d_{\sigma^2,l,i} \equiv 0$;
2. *for* $l = 1, 2$ *let* $P(\hat{\sigma}_l^{-1} > 0) \to 1$ *as* $n \to \infty$.

Proof. Since

$$Z_i^{(5)} = Z_i^{(0)} \left(1 + \sum_{l=1}^{3} r_{lZ^{(5)}}\right),$$

we get

$$Q_n^{(5)}(k) = \left(1 + \sum_{l=1}^{3} r_{lZ^{(5)}}\right)^2 \sum_{r=1}^{R+1} \sum_{i=k_{r-1}+1}^{k_r} (Z_i^{(0)} - \overline{Z^{(0)}}_{k_{r-1}}^{k_r})^2 = \left(1 + \sum_{l=1}^{3} r_{lZ^{(5)}}\right)^2 Q_n^{(0)}(k)$$

and

$$Q_n^{(5)}(k) - Q_n^{(5)}(k^*) = \left(1 + \sum_{l=1}^{3} r_{lZ^{(5)}}\right)^2 \left(Q_n^{(0)}(k) - Q_n^{(0)}(k^*)\right)$$
$$= \left(\frac{\sigma_1\sigma_2}{\hat{\sigma}_1\hat{\sigma}_2}\right)^2 \left(Q_n^{(0)}(k) - Q_n^{(0)}(k^*)\right).$$

Hence, we obtain

$$\begin{aligned} &P(a_n|\hat{\theta}_{r_0}^{(5)} - \theta_{r_0}| \geq N+1) \\ &\leq P\left(\left(\frac{\sigma_1\sigma_2}{\hat{\sigma}_1\hat{\sigma}_2}\right)^2 \min_{1<k_1<\ldots<k_R<n; |k_{r_0} - k_{r_0}^*| \geq a_{1,n,N}} Q_n^{(0)}(k) - Q_n^{(0)}(k^*) \leq 0\right) \\ &= o(1) + P\left(\min_{1<k_1<\ldots<k_R<n; |k_{r_0} - k_{r_0}^*| \geq a_{1,n,N}} Q_n^{(0)}(k) - Q_n^{(0)}(k^*) \leq 0\right) = o(1) \end{aligned}$$

as $n \to \infty$, followed by $N \to \infty$, where we use the arguments of the proof of Theorem 2.1.22. □

Remark 5.1.12. *The assumption* $P(\hat{\sigma}_l^{-1} > 0) \to 1$ *as* $n \to \infty$ *of the last theorem is weak and even allows that the estimates* $\hat{\sigma}_l$ *could tend to infinity with any arbitrary rate* $O_P(a_n)$, $a_n \to \infty$ *as* $n \to \infty$ *but* $a_n \neq \infty$ *for all* $n \in \mathbb{N}$.

Theorem 5.1.13. *Under the assumptions of Theorem 5.1.11 let sequences* $s_{1,n}$ *and* $s_{2,n}$ *exist so that*

$$(\hat{\sigma}_1^2\hat{\sigma}_2^2)^{-1} = O_P(s_{1,n}), \qquad \hat{\sigma}_1^2\hat{\sigma}_2^2 = O_P(s_{2,n}), \tag{5.1.18}$$
$$s_{1,n} n^{1/r + 1/r_z} \ll \beta_n^{(5)} \leq \frac{1}{4C^*} \min_{1 \leq i \leq m} \Delta_{\rho,i,n}^2 \underline{\Delta}_{k^*,n} s_{2,n}^{-1}. \tag{5.1.19}$$

Then, $\hat{R}^{(5)}$ *is a consistent estimate for the number of change-points* R^*.

Proof. Set $\beta_n = \beta_n^{(5)}$. Using the arguments of the proof of Theorem 2.1.22 combined with the ones of Theorem 5.1.11 yields that

$$P(\hat{R}^{(5)} < R^*) \leq o(1) + P\left(\min_{1\leq k\leq n; R<R^*} Q_n^{(0)}(k) - Q_n^{(0)}(k^*) - O_P(s_{2,n})\beta_n(R^*-1) < 0 \right) = o(1)$$

and

$$P(\hat{R}^{(5)} > R^*) \leq o(1) + P\left(O_P(s_{1,n}) \min_{1\leq k\leq n; R>R^*} Q_n^{(0)}(k) - Q_n^{(0)}(k^*) + \beta_n < 0 \right) = o(1).$$

This implies the claim. □

Non-constant Variances For the following theorem, we define for all $M \subset \mathbb{R}$ and $x \in \mathbb{R}$ $d(x,M) = \inf\{|x-y| \,:\, \forall y \in M\}$ and

$$d(x,M) = \inf\{|x-y| \,:\, \forall y \in M\} \quad \text{with} \quad d(x,\emptyset) = \infty.$$

Theorem 5.1.14. *Define $Q_n^{(7)}$ as $Q_n^{(0)}$ with $Z_i^{(7)}$ instead of $Z_i^{(0)}$. Then, Theorem 2.1.22 holds true with $\theta_n^{(7)}$ instead of $\theta_n^{(0)}$ if the following conditions are additionally fulfilled:*

1. *the parameters σ_1^2 and σ_2^2 fulfill Assumption (PEE3) with $d_{\sigma^2,l,i} \equiv 0$;*
2. *the estimates fulfill $\max_{l=1,2;k\in\mathbb{N}} \hat{\sigma}_{l,k}^{-1} = O_P(1)$;*
3. *let the following rates be satisfied*

$$a_{1,n} + a_{2,n} = o\left(\frac{\underline{\Delta}_{k^*,n}\underline{\Delta}_{\rho,n}^2}{\max_{1\leq i\leq R} |\Delta_{\rho,i,n}|} \right), \quad a_{1,n}^2 + a_{3,n}^2 = o\left(\underline{\Delta}_{k^*,n}\underline{\Delta}_{\rho,n}^2\right), \tag{5.1.20}$$

$$\sum_{l=1}^{2} \max_{1\leq r\leq R} m_{l,r,\epsilon,n} \left(\sum_{j=1}^{m_{l,n}} b_{l,j}^{r_z} \left(d(k_r^*,I_{l,j}) \vee Nn/a_n\right)^{-(r_z-1)} \right)^{1/r_z} = o\left(\frac{\underline{\Delta}_{\rho,n}^2}{\max_r |\Delta_{\rho,r}|} \right), \tag{5.1.21}$$

$$\max_{1\leq r\leq R} m_{1,r,\epsilon,n} \vee m_{2,r,\epsilon,n} \left(\sum_{j_1=1}^{m_{1,n}} \sum_{j_2=1}^{m_{2,n}} \frac{(b_{1,j_1}b_{2,j_2})^{r_z}}{(d(k_r^*,I_{1,j_1}\cap I_{2,j_2}) \vee Nn/a_n)^{r_z-1}} \right)^{1/r_z} = o\left(\frac{\underline{\Delta}_{\rho,n}^2}{\max_r |\Delta_{\rho,r}|} \right), \tag{5.1.22}$$

$$\sum_{l=1}^{2} \max_{1\leq r\leq R} \max_{1\leq |k_r-k_r^*|\leq \epsilon\underline{\Delta}_{k^*}} \sum_{j=1}^{m_{l,n}} b_{l,j} \frac{\#(k_r \wedge k_r^*, k_r \vee k_r^*] \cap I_{l,j}}{|k_r - k_r^*| \vee Nn/a_n} = o\left(\frac{\underline{\Delta}_{\rho,n}^2}{\max_r |\Delta_{\rho,r}|} \right), \tag{5.1.23}$$

$$\max_{1\leq r\leq R} \max_{1\leq |k_r-k_r^*|\leq \epsilon\underline{\Delta}_{k^*}} \sum_{j_1=1}^{m_{1,n}} \sum_{j_2}^{m_{2,n}} b_{1,j_1}b_{2,j_2} \frac{\#(k_r \wedge k_r^*, k_r \vee k_r^*] \cap I_{1,j_1} \cap I_{2,j_2}}{|k_r - k_r^*| \vee Nn/a_n} = o\left(\frac{\underline{\Delta}_{\rho,n}^2}{\max_r |\Delta_{\rho,r}|} \right) \tag{5.1.24}$$

as $n \to \infty$, followed by $N \to \infty$, where

$$M_{l,r,\epsilon,n} = \{1 \leq j \leq m_{l,n} \,:\, I_{l,j} \cap (k_r^* - \epsilon\underline{\Delta}_{k^*}, k_r^* + \epsilon\underline{\Delta}_{k^*}) \neq \emptyset\}, \quad m_{l,r,\epsilon,n} = \#M_{l,r,\epsilon,n}, \tag{5.1.25}$$

$$a_{1,n} = \sum_{l=1}^{2} m_{l,n} \left(\sum_{j=1}^{m_{l,n}} b_{l,j}^{r_z} \#I_{1,j_1} \right)^{\frac{1}{r_z}} + m_{1,n} \vee m_{2,n} \left(\sum_{j_1=1}^{m_{1,n}} \sum_{j_2=1}^{m_{2,n}} (b_{1,j_1}b_{2,j_2})^{r_z} \#(I_{1,j_1} \cap I_{2,j_2}) \right)^{\frac{1}{r_z}}, \tag{5.1.26}$$

$$a_{2,n} = \sum_{l=1}^{2} \sum_{j=1}^{m_{l,n}} b_{l,j} \#I_{l,j_1} + \sum_{j_1=1}^{m_{1,n}} \sum_{j_2=1}^{m_{2,n}} b_{1,j_1}b_{2,j_2} \#(I_{1,j_1} \cap I_{2,j_2}), \tag{5.1.27}$$

$$a_{3,n} = \max_{k_1\leq k_2} \sum_{l=1}^{2} \sum_{j=1}^{m_{l,n}} b_{l,j} \frac{\#I_{l,j_1} \cap [k_1,k_2]}{\sqrt{k_2-k_1+1}} + \max_{k_1\leq k_2} \sum_{j_1=1}^{m_{1,n}} \sum_{j_2=1}^{m_{2,n}} b_{1,j_1}b_{2,j_2} \frac{\#(I_{1,j_1} \cap I_{2,j_2} \cap [k_1,k_2])}{\sqrt{k_2-k_1+1}}. \tag{5.1.28}$$

Proof of Theorem 5.1.14. We define

$$\tilde{Z}_i^{(7)} = Z_i^{(7)} - \rho_i = \tilde{Z}_i^{(0)} + \tilde{Z}_i^{(0)} \sum_{l=1}^{3} r_{lin} + \rho_i \sum_{l=1}^{3} r_{lin}$$

and use $(\mathcal{K}_{\mathbf{r}}^{(\mathbf{2})})$, Assumption (PEE3), and the triangle inequality to obtain the following estimations:

$$\max_{1 \le k_1 \le k_2 \le n} \left| \sum_{i=k_1}^{k_2} \tilde{Z}_i^{(0)} \sum_{l=1}^{3} r_{lin} \right| = O_P(a_{1,n}), \qquad \max_{1 \le k_1 \le k_2 \le n} \left| \sum_{i=k_1}^{k_2} \rho_i \sum_{l=1}^{3} r_{lin} \right| = O_P(a_{2,n}),$$

and

$$\max_{1 \le k_1 \le k_2 \le n} \frac{\left| \sum_{i=k_1}^{k_2} \rho_i \sum_{l=1}^{3} r_{lin} \right|}{\sqrt{k_2 - k_1 + 1}} = O_P(a_{3,n}).$$

Now, we follow the proof of Theorem 2.1.22, and obtain

$$\begin{aligned} Q_n^{(7)}(k) - Q^{(7)}(k^*) &= \sum_{r=1}^{R+1} \sum_{i=k_{r-1}+1}^{k_r} \left[2\left(\rho_i - \overline{\rho}(k_{r-1},k_r)\right) \tilde{Z}_i^{(7)} + \left(\rho_i - \overline{\rho}(k_{r-1},k_r)\right)^2 \right] \\ &+ \sum_{r=1}^{R+1} \left[(k_r^* - k_{r-1}^*) \left(\overline{\tilde{Z}^{(7)}}(k_{r-1}^*,k_r^*) \right)^2 - (k_r - k_{r-1}) \left(\overline{\tilde{Z}^{(7)}}(k_{r-1},k_r) \right)^2 \right]. \end{aligned}$$

If there are $r^* \in \{1,\dots,R\}$ and $\epsilon > 0$ so that $|k_{r^*} - k_{r^*}^*| \ge \epsilon \underline{\Delta}_{k^*,n} \vee a_{1,n,N}$, then, it is clear that

$$\max_k \left| \sum_{r=1}^{R+1} \sum_{i=k_{r-1}+1}^{k_r} \left(\rho_i - \overline{\rho}(k_{r-1},k_r)\right) \tilde{Z}_i^{(7)} \right| = O_P\big(\max_{1 \le i \le R} |\Delta_{\rho,i,n}| [n^{1/r_z} + a_{1,n} + a_{2,n}]\big)$$

as $n \to \infty$. Furthermore, using the estimation of Theorem 2.1.22 we observe that

$$\max_k (k_r - k_{r-1}) \left(\overline{\tilde{Z}^{(7)}}(k_{r-1},k_r) \right)^2 = O_P\left(n^{1/r + 1/r_z} + a_{1,n}^2 + a_{3,n}^2 \right)$$

as $n \to \infty$, which is a quite rough estimation but reduce the complexity. Additionally, we define the rate of the right–hand side by $O_P(b_n^{(7)})$. Using the (2.1.28) from Theorem 2.1.22, the previous rates, and (5.1.20) yields that

$$P\left(\min_{1 < k_1 < \ldots < k_R < n; \|k - k^*\| \ge \epsilon \underline{\Delta}_{k^*,n}} Q_n^{(7)}(k) - Q_n^{(7)}(k^*) \le 0 \right) = o(1)$$

as $n \to \infty$. Now, we consider the case where we minimize over each k_r which is inside an $(\epsilon \underline{\Delta}_{k^*,n})$–neighborhood of k_r^*. Then, we get that

$$\begin{aligned} &Q_n^{(7)}(k) - Q_n^{(7)}(k^*) \\ &\ge c \|k^* - k\| \underline{\Delta}_\rho^2 \Bigg(1 - o_P(1) - O_P\left(\frac{\max_{1 \le r \le R} |\Delta_{r,\rho}| (a_{1,n} + a_{2,n})}{\underline{\Delta}_{k^*,n} \underline{\Delta}_\rho^2} \right) - O_P\left(\frac{a_{1,n}^2 + a_{3,n}^2}{\underline{\Delta}_{k^*,n} \underline{\Delta}_\rho^2} \right) \\ &\quad - C(R+1) \frac{\max_{1 \le i \le R} |\Delta_{\rho,i,n}|}{\underline{\Delta}_\rho^2} \max_{1 \le r \le R} \max_{1 \le |k_r - k_r^*| \le \epsilon \underline{\Delta}_{k^*}} \frac{\left| \sum_{i=k_r \wedge k_r^* + 1}^{k_r \vee k_r^*} \tilde{Z}_i^{(7)} \right|}{|k_r - k_r^*| \vee (Nn/a_n)} \Bigg) \end{aligned}$$

as $n \to \infty$, followed by $N \to \infty$. Furthermore, we obtain that

$$\max_{1 \le r \le R} \max_{1 \le |k_r - k_r^*| \le \epsilon \underline{\Delta}_{k^*}} \frac{\left| \sum_{i=k_r \wedge k_r^* + 1}^{k_r \vee k_r^*} \tilde{Z}_i^{(7)} \right|}{|k_r - k_r^*| \vee (Nn/a_n)} = O_P((Nn/a_n)^{-(r_z - 1)/r_z})$$

$$+\sum_{r=1}^{R}\sum_{l=1}^{2} O_P\left(\#M_{l,r,\epsilon,n}\left(\sum_{j=1}^{m_{l,n}} b_{l,j}^{r_z}\left(d(k_r^*,I_{l,j})\vee Nn/a_n\right)^{-(r_z-1)}\right)^{1/r_z}\right)$$

$$+\sum_{r=1}^{R} O_P\left(\#M_{1,r,\epsilon,n}\vee\#M_{2,r,\epsilon,n}\left(\sum_{j_1=1}^{m_{1,n}}\sum_{j_2=1}^{m_{2,n}}\frac{(b_{1,j_1}b_{2,j_2})^{r_z}}{(d(k_r^*,I_{1,j_1}\cap I_{2,j_2})\vee Nn/a_n)^{r_z-1}}\right)^{1/r_z}\right)$$

$$+\sum_{l=1}^{2} O_P\left(\max_{1\le r\le R}\max_{1\le|k_r-k_r^*|\le\epsilon\Delta_{k^*}}\sum_{j=1}^{m_{l,n}} b_{l,j}\frac{\#(k_r\wedge k_r^*,k_r\vee k_r^*]\cap I_{l,j}}{|k_r-k_r^*|\vee Nn/a_n}\right)$$

$$+O_P\left(\max_{1\le r\le R}\max_{1\le|k_r-k_r^*|\le\epsilon\Delta_{k^*}}\sum_{j_1=1}^{m_{1,n}}\sum_{j_2}^{m_{2,n}} b_{1,j_1}b_{2,j_2}\frac{\#(k_r\wedge k_r^*,k_r\vee k_r^*]\cap I_{1,j_1}\cap I_{2,j_2}}{|k_r-k_r^*|\vee Nn/a_n}\right).$$

Due to the combination of (2.1.29) and (5.1.21)–(5.1.24), each of the five $O_P(\cdot)$-terms is equal to $o_P(1)$. This finally implies the claim. □

Theorem 5.1.15. *Under the assumptions of Theorem 5.1.14 let*

$$d_n^{(7)} \ll \beta_n^{(7)} \le \frac{1}{4C^*}\min_{1\le i\le m}\Delta_{\rho,i,n}^2\Delta_{k^*,n} \tag{5.1.29}$$

with

$$d_n^{(7)} = b_n^{(7)} + a_{1,n} + a_{2,n} \quad \text{and} \quad b_n^{(7)} = n^{1/r+1/r_z} + a_{1,n}^2 + a_{3,n}^2, \tag{5.1.30}$$

where $a_{1,n}$, $a_{2,n}$, and $a_{3,n}$ are defined as in Theorem 5.1.14. Then, $\hat{R}^{(7)}$ is a consistent estimate for the number of change-points R^*.

Proof. Set $\beta_n = \beta_n^{(7)}$. This proof follows by the arguments of the proof of Theorem 2.1.25. Hence, it is sufficient to show that the sets $\{\hat{R}^{(7)} < R^*\}$ and $\{\hat{R}^{(7)} > R^*\}$ are asymptotically empty. The asymptotic behavior of $\{\hat{R}^{(7)} < R^*\}$ follows in the same way as that of $\{\hat{R} < R^*\}$ in the proof of Theorem 2.1.25, while here we use the arguments of Theorem 5.1.14 instead of Theorem 2.1.22.

Now, we consider $\{\hat{R}^{(7)} > R^*\}$ and obtain the lower bounds by using the same arguments as in the proof of Theorem 2.1.25:

$$Q_n^{(7)}(k) - Q_n^{(7)}(k^*) = \sum_{r=1}^{R+1}\sum_{i=k_{r-1}+1}^{k_r}\left[2\left(\rho_i-\overline{\rho}(k_{r-1},k_r)\right)\tilde{Z}_i^{(7)} + \left(\rho_i-\overline{\rho}(k_{r-1},k_r)\right)^2\right] - O_P(b_n^{(7)})$$

$$\ge \sum_{r=1}^{R+1}\sum_{i=k_{r-1}+1}^{k_r}\left(\rho_i-\overline{\rho}(k_{r-1},k_r)\right)^2 - O_P\left(\max_{1\le i\le R}|\Delta_{\rho,i,n}|[n^{(2-r_z)/r_z}+a_{1,n}+a_{2,n}]\right) - O_P(b_n^{(7)}),$$

which is dominated by β_n. This implies the claim. □

5.1.3 Long-run Variance Estimation under Unknown Variances

In this sub-subsection, we present some LRV estimates corresponding to the LRVs used in the Theorems 5.1.1, 5.1.4, and 5.1.9.

Constant Variances The following theorem yields a consistent estimate which can be used for the test statistics presented in Theorem 5.1.1.

Theorem 5.1.16. *Let the assumptions of Theorem 2.1.33 and the following assumptions hold true:*

1. *The parameters σ_1 and σ_2 fulfill Assumption (PEE1) with $d_{\sigma,l,i}\equiv 0$;*
2. *$\frac{\hat{\sigma}_{1,n}\hat{\sigma}_{2,n}}{\sigma_{1,j}\sigma_{2,j}}\tilde{\rho}_j^{(5)}$ fulfills the same condition as $\tilde{\rho}_n(\psi)$ in Theorem 2.1.33.*

Then, it holds that

$$\hat{D}_{5,n} = D + O_P(n^{-(\delta_{\sigma,1}\wedge\delta_{\sigma,2})}) + O_P(n^{-(\delta_{\sigma,1}\wedge\delta_{\sigma,2})} \vee 1)\hat{R}_n^{(0)},$$

where $\hat{R}_n^{(0)}$ *is defined as in Theorem 2.1.33 and*

$$\hat{D}_{5,n} = \frac{1}{n}\sum_{i=1}^{n}\sum_{j=1}^{n}\mathfrak{f}\left(\frac{i-j}{q_n}\right)(Z_i^{(5)} - \tilde{\rho}_i^{(5)})(Z_j^{(5)} - \tilde{\rho}_j^{(5)}).$$

Proof. Firstly, we obtain that

$$\begin{aligned}
&\frac{1}{n}\sum_{i=1}^{n}\sum_{j=1}^{n}\mathfrak{f}\left(\frac{i-j}{q_n}\right)(Z_i^{(5)} - \tilde{\rho}_i^{(5)})(Z_j^{(5)} - \tilde{\rho}_j^{(5)}) \\
&= \frac{\sigma_1\sigma_2\sigma_1\sigma_2}{\hat{\sigma}_{1,n}^2\hat{\sigma}_{2,n}^2}\frac{1}{n}\sum_{i=1}^{n}\sum_{j=1}^{n}\mathfrak{f}\left(\frac{i-j}{q_n}\right)(Z_i^{(0)} - \frac{\hat{\sigma}_{1,n}\hat{\sigma}_{2,n}}{\sigma_1\sigma_2}\tilde{\rho}_i^{(5)})(Z_j^{(0)} - \frac{\hat{\sigma}_{1,n}\hat{\sigma}_{2,n}}{\sigma_1\sigma_2}\tilde{\rho}_j^{(5)}) \\
&= \hat{D}_{0,n}\left(1 + O_P(n^{-(\delta_{\sigma,1}\wedge\delta_{\sigma,2})})\right)
\end{aligned}$$

since $|\frac{\sigma_{1,i}\sigma_{2,i}\sigma_{1,j}\sigma_{2,j}}{\hat{\sigma}_{1,n}^2\hat{\sigma}_{2,n}^2} - 1| = O_P(n^{-(\delta_{\sigma,1}\wedge\delta_{\sigma,2})})$. □

Lemma 5.1.17. *If for each* $W_{1n}, W_{2,n} \in \{\overline{Z^{(0)}}_n, \hat{\sigma}_{1,n}^2, \hat{\sigma}_{2,n}^2\}$ *there exists a* $\sigma \in \mathbb{R}$ *so that*

$$n\mathrm{Cov}(W_{1,n}, W_{2,n}) \to \sigma,$$

it holds that there exists a $D \geq 0$ *such that*

$$\mathrm{Var}\left[\frac{1}{\sqrt{n}}\sum_{i=1}^{n}\left[(Z_i^{(0)} - \rho_i) - \sqrt{n}(\hat{\sigma}_{1,n}^2 - \sigma_1^2)\frac{\rho_i}{2\sigma_1^2} - \sqrt{n}(\hat{\sigma}_{2,n}^2 - \sigma_2^2)\frac{\rho_i}{2\sigma_2^2}\right]\right] \to D. \tag{5.1.31}$$

Proof. Using the bilinear property of the covariance we get the claim. □

Now, we present a LRV estimate which can be used in the setting of Theorem 5.1.4. To this goal, we consider the special type of variance estimates:

$$\hat{\sigma}_{[n\cdot]}^2 = \frac{1}{[\cdot n]}\sum_{i=1}^{[n\cdot]}(Z_i - \mathbb{E}[Z_i])^2$$

which yields the following FCLT in Theorem 5.1.4

$$\frac{1}{\sqrt{n}}\sum_{i=1}^{[n\cdot]}\left(Z_i^{(0)} - \rho_i - [(X_i - \mu_{1,i})^2 - \sigma_1^2]\frac{\rho_i}{2\sigma_1^2} - [(Y_i - \mu_{2,i})^2 - \sigma_2^2]\frac{\rho_i}{2\sigma_2^2}\right) \xrightarrow{D[0,1]} D_{(6)}^{1/2}W(\cdot). \tag{5.1.32}$$

This provides a test statistic which is similar to the one of Wied et al. (2012), presented in Subsection 1.3. Here we use the exact mean instead of the sequence of sample means. However, with our approach we get another LRV estimate. In particular, the parameters σ_1^2 and σ_2^2 in (5.1.32) can be estimated in many different ways in which the sample variance is the natural estimate.

Theorem 5.1.18. *Let the parameters* σ_1^2 *and* σ_2^2 *fulfill Assumption (PEE1) with* $\delta_{l,i} \equiv 0$. *Additionally, let the following conditions hold true:*

1. $\mathfrak{f}$ *is a kernel with a bandwidth* $q_n \to \infty$*;*
2. *the sequence* $\{Z_n^{(0)} - \rho_n\}$ *fulfills* $(\mathcal{K}_{\mathbf{r}}^{(1)})$ *for* $r_z > 1$*;*
3. *the sequences* $\{\epsilon_{1,n}^2 - 1\}$ *and* $\{\epsilon_{1,n}^2 - 1\}$ *fulfill* $(\mathcal{K}_{\mathbf{r}}^{(1)})$ *for* $r_{x^2}, r_{y^2} > 1$*;*

4. *set*

$$\{(W_{1,n},W_{2,n},W_{3,n})\} = \{(Z_n^{(0)} - \rho_n,\rho_n/2 - (X_n - \mu_{1,n})^2\rho_n/(2\sigma_1^2),\\ \rho_n/2 - (Y_n - \mu_{2,n})^2\rho_n/(2\sigma_2^2))\}$$

and let for each $j_1,j_2 \in \{1,2,3\}$ *hold true that* $\sup_{n_1,n_2\in\mathbb{N}} \mathbb{E}[|W_{j_1,n_1}W_{j_2,n_2}|] < \infty$ *and that there exist constants* $D_{j_1,j_2} \in \mathbb{R}$ *so that*

$$n^{-1}\sum_{i=1}^{n}\sum_{j=1}^{n}\operatorname{Cov}(W_{j_1,i},W_{j_2,j}) \to D_{j_1,j_2} \tag{5.1.33}$$

and

$$\hat{D}_{j_1,j_2} = n^{-1}\sum_{i=1}^{n}\sum_{j=1}^{n}\mathfrak{f}\left(\frac{i-j}{q_n}\right)W_{j_1,i}W_{j_2,j} \xrightarrow{P} D_{j_1,j_2} \tag{5.1.34}$$

with $D = \sum_{i,j=1}^{3} D_{i,j} > 0$;

5. *the estimate* $\hat{\rho}_i = \hat{\rho}_n(i)$ *fulfills the 4th condition of Theorem 2.1.33.*

Then, it holds with

$$\hat{D}_0 = \sum_{i,j=1}^{3}\hat{D}_{i,j},$$

$$\tilde{Z}_i^{(6)} = \left(Z_i^{(5)} - \hat{\rho}_n - \frac{(X_i-\mu_{1,i})^2 - \hat{\sigma}_{1,n}^2}{2\hat{\sigma}_{1,n}^2}\hat{\rho}_n - \frac{(Y_i-\mu_{2,i})^2 - \hat{\sigma}_{2,n}^2}{2\hat{\sigma}_{2,n}^2}\hat{\rho}_n\right),$$

$$\hat{D}_{6,n} = n^{-1}\sum_{i=1}^{n}\sum_{j=1}^{n}\mathfrak{f}\left(\frac{i-j}{q_n}\right)\tilde{Z}_i^{(6)}\tilde{Z}_j^{(6)}$$

that $\hat{D}_{6,n} = \hat{D}_0 + \hat{R}_n^{(6)}$ *with*

$$\begin{aligned}\hat{R}_n^{(6)} &= O_P(q_n n^{-(\delta_{\sigma,1}\wedge\delta_{\sigma,2})})\\ &+ \begin{cases} O_P(q_n n^{-\delta_1-(\delta_{\sigma,1}\wedge\delta_{\sigma,2})}) + O_P(q_n n^{-2\delta_1}) \\ \quad +O_P(n^{-\delta_1}) + O_P(q_n n^{-1-(\delta_{\sigma,1}\wedge\delta_{\sigma,2}\wedge\delta_1)+1/(r_z\wedge r_{x^2}\wedge r_{y^2})}), & (A),\\ O_P(\max_{1\le j_1,j_2\le m}[(\#C_{j_1}\vee\#C_{j_2})\wedge q_n](\#C_{j_1}\wedge\#C_{j_2})\,n^{-1-\delta_{j_1}-\delta_{j_2}}) \\ \quad +O_P(q_n\max_{1\le j\le m}\#C_j n^{-1-\delta_j}) & (E),\\ \quad +O_P(q_n\max_{1\le j\le m} n^{-1-(\delta_j\delta_{\sigma,1}\wedge\delta_{\sigma,2})+1/(r_z\wedge r_{x^2}\wedge r_{y^2})}), \end{cases}\end{aligned}$$

plus $o_P(q_n n^{-1/2})$, $O_P(q_n n^{-1/2})$, *or* $O_P(q_n)$ *in cases of (B), (C), or (D) and in cases of (F), (G), or (H), respectively.*

Proof. Firstly, we obtain that

$$\begin{aligned}&\left(Z_i^{(5)} - \hat{\rho}_i - \frac{(X_i-\mu_{1,i})^2 - \hat{\sigma}_{1,n}^2}{2\hat{\sigma}_{1,n}^2}\hat{\rho}_i - \frac{(Y_i-\mu_{2,i})^2 - \hat{\sigma}_{2,n}^2}{2\hat{\sigma}_{2,n}^2}\hat{\rho}_i\right)\\ &= (Z_i^{(0)} - \rho_i) - \frac{1}{2}\rho_i(\epsilon_{1,i}^2 - 1) - \frac{1}{2}\rho_i(\epsilon_{2,i}^2 - 1) - R_i = \tilde{W}_i - R_i\end{aligned}$$

with

$$\begin{aligned}R_i &= \left(1 - \frac{\sigma_1\sigma_2}{\hat{\sigma}_{1,n}\hat{\sigma}_{2,n}}\right)(Z_i^{(0)} - \rho_i) + \left(\frac{1}{2} - \frac{\sigma_1^2}{2\hat{\sigma}_{1,n}^2}\right)\rho_i(\epsilon_{1,i}^2 - 1) - \left(\frac{1}{2} - \frac{\sigma_2^2}{2\hat{\sigma}_{2,n}^2}\right)\rho_i(\epsilon_{2,i}^2 - 1)\\ &+ \frac{\sigma_1^2}{2\hat{\sigma}_{1,n}^2}(\hat{\rho}_i - \rho_i)(\epsilon_{1,i}^2 - 1) + \frac{\sigma_2^2}{2\hat{\sigma}_{2,n}^2}(\hat{\rho}_i - \rho_i)(\epsilon_{2,i}^2 - 1)\\ &+ \frac{\sigma_1^2}{2\hat{\sigma}_{1,n}^2}(\hat{\rho}_i - \rho_i) + \frac{\sigma_2^2}{2\hat{\sigma}_{2,n}^2}(\hat{\rho}_i - \rho_i) + \rho_i\left(\frac{\sigma_1^2}{2\hat{\sigma}_{1,n}^2} + \frac{\sigma_2^2}{2\hat{\sigma}_{2,n}^2} - 1\right).\end{aligned}$$

Thus, we get

$$\hat{D}_{6,n} = \hat{D}_{0,n} + n^{-1}\sum_{i=1}^{n}\sum_{j=1}^{n} \mathfrak{f}\left(\frac{i-j}{q_n}\right)(R_iR_j - R_i\tilde{W}_j - \tilde{W}_iR_j).$$

Furthermore, we obtain by Slutsky's Theorem that it is sufficient to consider the second summand. Using Assumption (PEE1) and (5.1.34) we get that

$$n^{-1}\sum_{i=1}^{n}\sum_{j=1}^{n} \mathfrak{f}\left(\frac{i-j}{q_n}\right) R_iR_j = O_P(q_n n^{-2(\delta_{\sigma,1}\wedge\delta_{\sigma,2})})$$
$$+\begin{cases} O_P(q_n n^{-\delta_1-(\delta_{\sigma,1}\wedge\delta_{\sigma,2})}) + O_P(q_n n^{-2\delta_1}), & \text{under (A)},\\ O_P(\sum_{j_1=1}^{m}\sum_{j_2=1}^{m}[(\#C_{j_1}\vee\#C_{j_2})\wedge q_n](\#C_{j_1}\wedge\#C_{j_2})n^{-1-\delta_{j_1}-\delta_{j_2}}) & \text{under (E)},\\ \quad +O_P(q_n\sum_{j=1}^{m}\#C_j n^{-1-\delta_j-(\delta_{\sigma,1}\wedge\delta_{\sigma,2})}), & \end{cases}$$

and

$$n^{-1}\sum_{i=1}^{n}\sum_{j=1}^{n} \mathfrak{f}\left(\frac{i-j}{q_n}\right) \tilde{W}_iR_j = O_P(n^{-(\delta_{\sigma,1}\wedge\delta_{\sigma,2})})$$
$$+\begin{cases} O_P(n^{-\delta_1}) + O_P(q_n n^{-1-(\delta_{\sigma,1}\wedge\delta_{\sigma,2}\wedge\delta_1)+1/(r_z\wedge r_{x^2}\wedge r_{y^2})}), & \text{(A)},\\ O_P(q_n\sum_{j=1}^{m}\#C_j n^{-1-\delta_j}) + O_P(q_n \max_{1\le j\le m} n^{-1-\delta_j+1/(r_z\wedge r_{x^2}\wedge r_{y^2})}) & \text{(E)},\\ \quad +O_P(q_n n^{-1-(\delta_{\sigma,1}\wedge\delta_{\sigma,2})+1/(r_z\wedge r_{x^2}\wedge r_{y^2})}), & \end{cases}$$

where we use the triangle inequality, Markov's inequality, and the Kolmogorov-type inequalities. In the cases of (B), (C), and (D) we add the rates $o_P(q_n n^{-1/2})$, $O_P(q_n n^{-1/2})$, and $O_P(q_n)$, respectively. The same rates are also added in the cases of (F) , (G), and (H). Thus, if we combine all above rates and use the finiteness of m, then, the claim finally follows.

□

Remark 5.1.19. *It is possible to use different bandwidths* q_{n,j_1,j_2} *instead of one single bandwidth* q_n *for all* $j_1, j_2 = 1, 2, 3$ *in* (5.1.33).

Non-constant Variances In this paragraph, we present a consistent LRV estimate for $D_{(8)}$ (cf. Theorem 5.1.9). Therefore, we specify the test statistic and choose the following type of variance estimate:

$$\hat{\sigma}^2_{l,i,[n\cdot]} = \frac{1}{\#(\hat{I}_{l,j}\cap[1,[n\cdot]])}\sum_{i\in(\hat{I}_{l,j}\cap[1,[n\cdot]])}(Z_{l,i}-\mu_{l,i})^2 \qquad \forall i\in\hat{I}_{l,j}, \tag{5.1.35}$$

where $\{(Z_{1,n}, Z_{2,n})\} = \{(X_n, Y_n)\}$ and $\hat{I}_{l,j}$ is an estimate for the exact change set, i.e., where the variances are non-constant. To be precise, this estimate yields the following FCLT under the assumptions of Theorem 5.1.9:

$$\frac{1}{\sqrt{n}}\sum_{i=1}^{[n\cdot]}\left(Z_i^{(0)} - \rho_i - [(X_i-\mu_{1,i})^2 - \sigma^2_{1,i}]\frac{\rho_i}{2\sigma^2_{1,i}} - [(Y_i-\mu_{2,i})^2-\sigma^2_{2,i}]\frac{\rho_i}{2\sigma^2_{2,i}}\right) \overset{D[0,1]}{\longrightarrow} D^{1/2}W(\cdot), \tag{5.1.36}$$

where $D > 0$. Again, to estimate the LRV, we use general variance estimates for the remaining parameters.

Theorem 5.1.20. *Let the parameters* σ_1^2 *and* σ_2^2 *fulfill Assumption (PEE3) with* $d_{\sigma,l,i}\equiv 0$ *and let the following conditions hold true:*

1. $\mathfrak{f}$ *is an absolutely integrable kernel with a bandwidth* $q_n\to\infty$*;*

2. *the sequence* $\{Z_n^{(0)} - \rho_n\}$ *fulfills* $(\mathcal{K}_{\mathbf{r}}^{(\mathbf{1})})$ *and* $(\mathcal{K}_{\mathbf{r}}^{(\mathbf{2})})$ *for some* $r_z > 1$;
3. *the sequences* $\{\epsilon_{1,n}^2 - 1\}$ *and* $\{\epsilon_{2,n}^2 - 1\}$ *fulfill* $(\mathcal{K}_{\mathbf{r}}^{(\mathbf{2})})$ *for* $r_{x^2}, r_{y^2} > 1$;
4. *set*

$$\{(W_{1,n},W_{2,n},W_{3,n})\} = \{(Z_n^{(0)}-\rho_n,\rho_n/2-(X_n-\mu_{1,n})^2\rho_n/(2\sigma_{1,n}^2),\rho_n/2-(Y_n-\mu_{2,n})^2\rho_n/(2\sigma_{2,n}^2))\}$$

and let for each $j_1,j_2 \in \{1,2,3\}$ *hold true that* $\sup_{n_1,n_2\in\mathbb{N}} I\!E[|W_{j_1,n_1}W_{j_2,n_2}|] < \infty$ *and that there exist constants* $D_{j_1,j_2} \in \mathbb{R}$ *so that*

$$n^{-1}\sum_{i=1}^{n}\sum_{j=1}^{n} \operatorname{Cov}(W_{j_1,i},W_{j_2,j}) \to D_{j_1,j_2} \quad \textit{and} \tag{5.1.37}$$

$$\hat{D}_{j_1,j_2} = n^{-1}\sum_{i=1}^{n}\sum_{j=1}^{n} \mathfrak{f}\left(\frac{i-j}{q_n}\right) W_{j_1,i}W_{j_2,j} \xrightarrow{P} D_{j_1,j_2} \tag{5.1.38}$$

as $n \to \infty$, *where* $D = \sum_{i,j=1}^{3} D_{i,j} > 0$ *and* $\lim_{n\to\infty} \overline{\rho}_n = \rho_0 \in [-1,1]$;
5. *the estimate* $\hat{\rho}_i = \hat{\rho}_n(i)$ *fulfills the 4th condition of Theorem 2.1.33.*

Then, it holds with

$$\hat{D}_0 = \sum_{i,j=1}^{3} \hat{D}_{i,j}, \quad \hat{D}_8 = n^{-1}\sum_{i=1}^{n}\sum_{j=1}^{n} \mathfrak{f}\left(\frac{i-j}{q_n}\right) \tilde{Z}_i^{(8)}\tilde{Z}_j^{(8)},$$
$$\tilde{Z}_i^{(8)} = \left(Z_i^{(7)} - \hat{\rho}_n - \frac{(X_i-\mu_{1,i})^2 - \hat{\sigma}_{1,i,n}^2}{2\hat{\sigma}_{1,i,n}^2}\hat{\rho}_n - \frac{(Y_i-\mu_{2,i})^2 - \hat{\sigma}_{2,i,n}^2}{2\hat{\sigma}_{2,i,n}^2}\hat{\rho}_n\right),$$

that $\hat{D}_{8,n} = \hat{D}_{0,n} + \hat{R}_n^{(8)}$ *with*

$$\hat{R}_n^{(8)} = O_P\Bigg(n^{-1}\sum_{i_1,j_1=1}^{m_{\sigma,1}}\sum_{i_2,j_2=1}^{m_{\sigma,1}} [(\#(I_{\sigma,1,i_1}\cap I_{\sigma,2,i_2}) \vee \#(I_{\sigma,1,j_1}\cap I_{\sigma,2,j_2})) \wedge q_n]$$
$$\cdot (\#(I_{\sigma,1,i_1}\cap I_{\sigma,2,i_2}) \wedge \#(I_{\sigma,1,j_1}\cap I_{\sigma,2,j_2}))\,(b_{\sigma,1,i_1}\vee b_{\sigma,2,i_2})(b_{\sigma,1,j_1}\vee b_{\sigma,2,j_2})\Bigg)$$
$$+ O_P\left(n^{-1}q_n\sum_{i_1=1}^{m_{\sigma,1}}\sum_{i_2=1}^{m_{\sigma,1}} \#(I_{\sigma,1,i_1}\cap I_{\sigma,2,i_2})(b_{\sigma,1,i_1}\vee b_{\sigma,2,i_2})\right)$$
$$+\begin{cases} O_P(q_n n^{-2\delta_1}) + O_P(q_n n^{-1-\delta_1+1/(r_z\wedge r_{x^2}\wedge r_{y^2})}), & (A),\\ O_P(\max_{1\le j_1,j_2\le m}[(\#C_{j_1}\vee\#C_{j_2})\wedge q_n]\,(\#C_{j_1}\wedge\#C_{j_2})\,n^{-1-\delta_{j_1}-\delta_{j_2}}) & \\ \quad +O_P\Big(n^{-1}\sum_{j=1}^{m}\sum_{i_1=1}^{m_{\sigma,1}}\sum_{i_2=1}^{m_{\sigma,1}}[(\#(I_{\sigma,1,i_1}\cap I_{\sigma,2,i_2})\vee\#C_j)\wedge q_n] & \\ \quad \cdot(\#(I_{\sigma,1,i_1}\cap I_{\sigma,2,i_2})\wedge\#C_j)\,(b_{\sigma,1,i_1}\vee b_{\sigma,1,i_2})n^{-\delta_j}\Big) & (E),\\ \quad +O_P(q_n\sum_{j=1}^{m}\#C_j n^{-1-\delta_j}), & \end{cases}$$

plus $o_P(q_n n^{-1/2})$, $O_P(q_n n^{-1/2})$, *or* $O_P(q_n)$ *in cases of (B), (C), or (D) and in cases of (F), (G), or (H), respectively.*

Proof. Firstly, we obtain that

$$\left(Z_i^{(7)} - \hat{\rho}_i - \frac{(X_i-\mu_{1,i})^2 - \hat{\sigma}_{1,i,n}^2}{2\hat{\sigma}_{1,i,n}^2}\hat{\rho}_i - \frac{(Y_i-\mu_{2,i})^2 - \hat{\sigma}_{2,i,n}^2}{2\hat{\sigma}_{2,i,n}^2}\hat{\rho}_i\right)$$
$$= (Z_i^{(0)} - \rho_i) - \frac{1}{2}\rho_i(\epsilon_{1,i}^2 - 1) - \frac{1}{2}\rho_i(\epsilon_{2,i}^2 - 1) - R_i = \tilde{W}_i - R_i$$

with

$$
\begin{aligned}
R_i =&(1 - \frac{\sigma_{1,i}\sigma_{2,i}}{\hat{\sigma}_{1,i,n}\hat{\sigma}_{2,i,n}})(Z_i^{(0)} - \rho_i) + (\frac{1}{2} - \frac{\sigma_{1,i}^2}{2\hat{\sigma}_{1,i,n}^2})\rho_i(\epsilon_{1,i}^2 - 1) - (\frac{1}{2} - \frac{\sigma_{2,i}^2}{2\hat{\sigma}_{2,i,n}^2})\rho_i(\epsilon_{2,i}^2 - 1) \\
&+ \frac{\sigma_{1,i}^2}{2\hat{\sigma}_{1,i,n}^2}(\hat{\rho}_i - \rho_i)(\epsilon_{1,i}^2 - 1) + \frac{\sigma_{2,i}^2}{2\hat{\sigma}_{2,i,n}^2}(\hat{\rho}_i - \rho_i)(\epsilon_{2,i}^2 - 1) \\
&+ \frac{\sigma_{1,i}^2}{2\hat{\sigma}_{1,i,n}^2}(\hat{\rho}_i - \rho_i) + \frac{\sigma_{2,i}^2}{2\hat{\sigma}_{2,i,n}^2}(\hat{\rho}_i - \rho_i) + \rho_i\left(\frac{\sigma_{1,i}^2}{2\hat{\sigma}_{1,n}^2} + \frac{\sigma_{2,i}^2}{2\hat{\sigma}_{2,i,n}^2} - 1\right).
\end{aligned}
$$

Thus, we get

$$
\hat{D}_{8,n} = \hat{D}_{0,n} + n^{-1}\sum_{i=1}^{n}\sum_{j=1}^{n}\mathfrak{f}\left(\frac{i-j}{q_n}\right)(R_iR_j - R_i\tilde{W}_j - \tilde{W}_iR_j).
$$

Furthermore, we obtain with Slutsky's Theorem that it is sufficient to consider the second summand. Using Assumption (PEE3), we get that

$$
\begin{aligned}
&n^{-1}\sum_{i=1}^{n}\sum_{j=1}^{n}\mathfrak{f}\left(\frac{i-j}{q_n}\right)R_iR_j \\
&= O_P\Bigg(n^{-1}\sum_{i_1,j_1=1}^{m_{\sigma,1}}\sum_{i_2,j_2=1}^{m_{\sigma,1}}[(\#(I_{\sigma,1,i_1}\cap I_{\sigma,2,i_2}) \vee \#(I_{\sigma,1,j_1}\cap I_{\sigma,2,j_2})) \wedge q_n] \\
&\quad \cdot (\#(I_{\sigma,1,i_1}\cap I_{\sigma,2,i_2}) \wedge \#(I_{\sigma,1,j_1}\cap I_{\sigma,2,j_2}))(b_{\sigma,1,i_1}\vee b_{\sigma,2,i_2})(b_{\sigma,1,j_1}\vee b_{\sigma,2,j_2})\Bigg) \\
&+ \begin{cases}
O_P\left(q_n n^{-1-\delta_1}\sum_{i_1=1}^{m_{\sigma,1}}\sum_{i_2=1}^{m_{\sigma,1}}\#(I_{\sigma,1,i_1}\cap I_{\sigma,2,i_2})(b_{\sigma,1,i_1}\vee b_{\sigma,1,i_2})\right) \\
\quad +O_P(q_n n^{-2\delta_1}), & \text{(A)}, \\
O_P(\sum_{j_1=1}^{m}\sum_{j_2=1}^{m}[(\#C_{j_1}\vee\#C_{j_2})\wedge q_n](\#C_{j_1}\wedge\#C_{j_2})n^{-1-\delta_{j_1}-\delta_{j_2}}) \\
\quad +O_P\Big(n^{-1}\sum_{j=1}^{m}\sum_{i_1=1}^{m_{\sigma,1}}\sum_{i_2=1}^{m_{\sigma,1}}[(\#(I_{\sigma,1,i_1}\cap I_{\sigma,2,i_2})\vee\#C_j)\wedge q_n] & \text{(E)}, \\
\quad \cdot(\#(I_{\sigma,1,i_1}\cap I_{\sigma,2,i_2})\wedge\#C_j)(b_{\sigma,1,i_1}\vee b_{\sigma,1,i_2})n^{-\delta_j}\Big),
\end{cases}
\end{aligned}
$$

and

$$
\begin{aligned}
&n^{-1}\sum_{i=1}^{n}\sum_{j=1}^{n}\mathfrak{f}\left(\frac{i-j}{q_n}\right)\tilde{W}_iR_j \\
&= O_P\left(n^{-1}q_n\sum_{i_1=1}^{m_{\sigma,1}}\sum_{i_2=1}^{m_{\sigma,1}}\#(I_{\sigma,1,i_1}\cap I_{\sigma,2,i_2})(b_{\sigma,1,i_1}\vee b_{\sigma,2,i_2})\right) \\
&+ \begin{cases}
O_P(q_n n^{-1-\delta_1+1/(r_z\wedge r_{x^2}\wedge r_{y^2})}), & \text{under (A)}, \\
O_P(q_n\sum_{j=1}^{m}\#C_j n^{-1-\delta_j}), & \text{under (E)},
\end{cases}
\end{aligned}
$$

where we use the triangle inequality, Markov's inequality, and the Kolmogorov-type inequalities. In the cases of (B) , (C), and (D) we add the rates $o_P(q_n n^{-1/2})$, $O_P(q_n n^{-1/2})$, and $O_P(q_n)$, respectively. The same rates are also added in the cases of (F) , (G), and (H). Hence, the claim follows by combining the above rates.

□

5.2 Sequential Analysis under a General Dependency Framework and General Variance Estimates

In this subsection, we consider the asymptotic behavior of the stopping time, where the variances are unknown. We define for $k = 1, \dots$

$$\hat{\rho}^n_{\psi,k,1} = \frac{1}{n}\sum_{i=1}^{n} Z^{(\psi)}_{i,k,n} \qquad \text{and} \qquad \hat{\rho}^k_{\psi,k,n+1} = \frac{1}{k}\sum_{i=n+1}^{n+k} Z^{(\psi)}_{i,k,n}, \tag{5.2.1}$$

where $\psi = 5, \dots$ is a design index for different variance estimate types and

$$Z^{(\psi)}_{i,k,n} = \frac{(X_i - \mu_{1,i})(Y_i - \mu_{2,i})}{\hat{\sigma}^{(\psi)}_{1,i,k,n}\hat{\sigma}^{(\psi)}_{2,i,k,n}}.$$

We distinguish between the variance estimate types as we did for the mean estimates in Subsection 4.2. One estimate type uses the whole observation, i.e., from 1 to $n+k$, to estimate the unknown variances. The other estimate, on the one hand, uses the observations from 1 to n and, on the other hand, the observations from $n+1$ until $n+k$.

5.2.1 Closed-end Procedure under Unknown Variances

Nearly Constant Variances In the following theorem we sequentially use the whole sample to estimate the variances. Since the proofs for the results of the weighted and unweighted testing procedures are similar, we just present the convergence of the weighted stopping times. In addition, it is obvious that we can replace the assumptions of Theorem 2.2.3 by the slightly weaker ones of Theorem 2.2.1 in order to get the result for the unweighted stopping times.

Theorem 5.2.1. *Let the parameters σ_1 and σ_2 fulfill Assumption (PEE5) with $d_{l,n} \equiv 0$ and $\delta_{\sigma,1}, \delta_{\sigma,2} > 0$. Then, Theorem 2.2.3 holds true if we replace $\tau^{(c)}_{n,\iota,0,0,\gamma}$ by $\tau^{(c)}_{n,\iota,5,0,\gamma}$.*

Proof. Firstly, we obtain that $\max_{l,k} \hat{\sigma}^{-1}_{l,k,n} = O_P(1)$ holds true since $\delta_{\sigma,1} \wedge \delta_{\sigma,2} > 0$. Moreover, we observe that

$$\tilde{B}^{5,0,\gamma}_n([n\cdot]) = \tilde{B}^{0,0,\gamma}_n([n\cdot]) + \left(\frac{\sigma_1\sigma_2}{\hat{\sigma}_{1,[n\cdot],n}\hat{\sigma}_{2,[n\cdot],n}} - 1\right)\tilde{B}^{0,0,\gamma}_n([n\cdot]). \tag{5.2.2}$$

Due to Assumption (PEE5) and $\max_{l,k} \hat{\sigma}^{-1}_{l,k,n} = O_P(1)$, it follows that

$$\max_{1\le k\le [nm]}\left|\frac{\sigma_1\sigma_2}{\hat{\sigma}_{1,k,n}\hat{\sigma}_{2,k,n}} - 1\right| = O_P(n^{-(\delta_{\sigma,1}\wedge\delta_{\sigma,2})}).$$

Hence, it holds that

$$\tilde{B}^{5,0,\gamma}_n([n\cdot]) = (1 + O_P(n^{-(\delta_{\sigma,1}\wedge\delta_{\sigma,2})}))\tilde{B}^{0,0,\gamma}_n([n\cdot]),$$

which implies that the convergences in Theorem 2.2.3 hold true. □

Remark 5.2.2. *Like in Remark 5.1.3, it is still possible that the test statistic converges towards a Gaussian process if there are non-local structural breaks in the variances. Note that the covariance structure now depends on these structural breaks. Let us assume, for example, that $\sigma^2_{1,i} = g_{\sigma,1}(i/n)$ and $\sigma^2_{2,i} = g_{\sigma,1}(i/n)$ with bounded, positive, and piecewise continuous functions $g_{\sigma,l}$, that $\hat{\sigma}^2_{1,i} \equiv \overline{(X-\mu_1)^2}_n$ and $\hat{\sigma}^2_{2,i} \equiv \overline{(Y-\mu_2)^2}_n$, and that the Assumption (IID) is fulfilled with*

$I\!E\left[(\tilde{\epsilon}_{1,i}\tilde{\epsilon}_{2,i}-\rho_0)^2\right]\equiv 1$. *Then, under* $H_0^{(2)}$ *we obtain that*

$$\begin{aligned}&w_\gamma\left(\frac{[n\cdot]}{n}\right)\frac{[n\cdot]}{\sqrt{n}}(\hat{\rho}_{5,n,n+1}^{n+[n\cdot]}-\hat{\rho}_{5,n,1})\\&\overset{\mathcal{D}[0,m]}{\longrightarrow} w_\gamma(\cdot)\frac{W(1+\cdot)\int_1^{1+\cdot}g_{\sigma,1}(x)g_{\sigma,2}(x)dx-(\cdot+1)W(1)\int_0^1 g_{\sigma,1}(x)g_{\sigma,2}(x)dx}{\sqrt{\int_0^1 g_{\sigma,1}(x)dx\int_0^1 g_{\sigma,2}(x)dx}}\\&\qquad+\rho_0 w_\gamma(\cdot)\frac{\int_1^{1+\cdot}g_{\sigma,1}(x)g_{\sigma,2}(x)dx-(\cdot+1)\int_0^1 g_{\sigma,1}(x)g_{\sigma,2}(x)dx}{\sqrt{\int_0^1 g_{\sigma,1}(x)dx\int_0^1 g_{\sigma,2}(x)dx}},\end{aligned}$$

which follows from an application from Davidson (1994, Corollary 29.11).

Now, we consider the stopping times where the variances are estimated piecewise from 1 to n and from $n+1$ until $n+k$, $k=1,2,\ldots$.

Theorem 5.2.3. *Let the FCLT (as displayed in* (5.1.7)*) be fulfilled on* $\mathcal{D}[0,1+m]$. *Additionally, let the variance estimates hold for all* $m_1,m_2\in\mathbb{N}$, $m_2>m_1$

$$m_2\hat{\sigma}_{l,1}^{2,m_2}-m_1\hat{\sigma}_{l,1}^{2,m_1}=(m_2-m_1)\hat{\sigma}_{l,m_1+1}^{2,m_2}\quad\textit{and}\quad\left\|\frac{[n\cdot]+n}{\sqrt{n}}(\hat{\sigma}_{l,1}^{2,n+[n\cdot]}-\sigma_l^2)\right\|_{[0,m]}=O_P(1).\tag{5.2.3}$$

Then, Theorem 2.2.1 holds true if we replace $\tau_{n,\iota,0,0}^{(c)}$ *by* $\tau_{n,\iota,6,0}^{(c)}$.

Proof. Firstly, we obtain that

$$\begin{aligned}\tilde{B}_n^{6,0,0}([n\cdot])=&\tilde{B}_n^{0,0,\gamma}([n\cdot])+\left(\frac{\sigma_1\sigma_2}{\hat{\sigma}_{1,n+1}^{n+[n\cdot]}\hat{\sigma}_{2,n+1}^{n+[n\cdot]}}-1\right)\hat{D}^{-1/2}\frac{n}{n+[n\cdot]}\frac{1}{\sqrt{n}}\sum_{i=n+1}^{n+[n\cdot]}(Z_i^{(0)}-\rho_i)\\&+\left(\frac{\sigma_1\sigma_2}{\hat{\sigma}_{1,1}^{n}\hat{\sigma}_{2,1}^{n}}-1\right)\hat{D}^{-1/2}\frac{n}{n+[n\cdot]}\frac{[n\cdot]}{n}\frac{1}{\sqrt{n}}\sum_{i=1}^{n}(Z_i^{(0)}-\rho_i)\\&+\hat{D}^{-1/2}\frac{n}{n+[n\cdot]}\frac{[n\cdot]}{\sqrt{n}}\left(\left(\frac{\sigma_1\sigma_2}{\hat{\sigma}_{1,n+1}^{n+[n\cdot]}\hat{\sigma}_{2,n+1}^{n+[n\cdot]}}-1\right)\overline{\rho}_{n+1}^{n+[n\cdot]}-\left(\frac{\sigma_1\sigma_2}{\hat{\sigma}_{1,1}^{n}\hat{\sigma}_{2,1}^{n}}-1\right)\overline{\rho}_1^{n}\right).\end{aligned}$$

On the right–hand side we obtain that the first and second summand are equal to $O_P(\hat{D}^{-1/2}n^{-1/2})$ due to (5.2.3). Hence, it remains to consider the last term, where the last bracket can be rewritten in the following way:

$$\begin{aligned}&\left(\frac{\sigma_1^2-\hat{\sigma}_{1,n+1}^{2,n+k}}{\hat{\sigma}_{1,n+1}^{n+k}(\sigma_1+\hat{\sigma}_{1,n+1}^{2,n+k})}\overline{\rho}_{n+1}^{n+k}-\frac{\sigma_1^2-\hat{\sigma}_{1,1}^{2,n}}{\hat{\sigma}_{1,1}^{n}(\sigma_1+\hat{\sigma}_{1,1}^{2,n})}\overline{\rho}_1^{n}\right)\\&+\left(\frac{\sigma_2^2-\hat{\sigma}_{2,n+1}^{2,n+k}}{\hat{\sigma}_{2,n+1}^{n+k}(\sigma_2+\hat{\sigma}_{2,n+1}^{2,n+k})}\overline{\rho}_{n+1}^{n+k}-\frac{\sigma_2^2-\hat{\sigma}_{2,1}^{2,n}}{\hat{\sigma}_{2,1}^{n}(\sigma_2+\hat{\sigma}_{2,1}^{2,n})}\overline{\rho}_1^{n}\right)\\&+\left(\frac{\sigma_1^2-\hat{\sigma}_{1,n+1}^{2,n+k}}{\hat{\sigma}_{1,n+1}^{n+k}(\sigma_1+\hat{\sigma}_{1,n+1}^{2,n+k})}\frac{\sigma_2^2-\hat{\sigma}_{2,n+1}^{2,n+k}}{\hat{\sigma}_{2,n+1}^{n+k}(\sigma_2+\hat{\sigma}_{2,n+1}^{2,n+k})}\overline{\rho}_{n+1}^{n+k}-\frac{\sigma_1^2-\hat{\sigma}_{1,1}^{2,n}}{\hat{\sigma}_{1,1}^{n}(\sigma_1+\hat{\sigma}_{1,1}^{2,n})}\frac{\sigma_2^2-\hat{\sigma}_{2,1}^{2,n}}{\hat{\sigma}_{2,1}^{n}(\sigma_2+\hat{\sigma}_{2,1}^{2,n})})\overline{\rho}_1^{n}\right).\end{aligned}$$

Using the same arguments as before and

$$(\sigma_l^2-\hat{\sigma}_{l,n+1}^{2,n+k})=\frac{1}{k}\left((n+k)(\hat{\sigma}_{l,1}^{2,n+k}-\sigma_l^2)-n(\hat{\sigma}_{l,1}^{2,n}-\sigma_l^2)\right)$$

we obtain that

$$\tilde{B}_n^{6,0,0}([n\cdot])=\tilde{B}_n^{0,0,\gamma}([n\cdot])+\hat{D}^{-1/2}\frac{n}{n+[n\cdot]}\frac{[n\cdot]}{\sqrt{n}}\left((\sigma_1^2-\hat{\sigma}_{1,n+1}^{2,n+[n\cdot]})\frac{\overline{\rho}_{n+1}^{n+[n\cdot]}}{2\sigma_1^2}+(\sigma_2^2-\hat{\sigma}_{2,n+1}^{2,n+[n\cdot]})\frac{\overline{\rho}_{n+1}^{n+[n\cdot]}}{2\sigma_2^2}\right.$$

$$- (\sigma_1^2 - \hat{\sigma}_{1,1}^{2,n})\frac{\overline{\rho}_1^n}{2\sigma_1^2} + (\sigma_2^2 - \hat{\sigma}_{2,1}^{2,n})\frac{\overline{\rho}_1^n}{2\sigma_2^2}\Bigg) + o_P(\hat{D}^{-1/2}),$$

where $\overline{\rho}_1^n \to \rho_0$ and even $\overline{\rho}_{n+1}^{n+[n\cdot]} \to \rho_0$ uniformly under H_0 and $\mathbf{H_{LA}}$. In contrast to this, under $\mathbf{H_A}$, $\overline{\rho}_{n+1}^{n+[n\cdot]}$ is just uniformly bounded. Hence, we get

$$\begin{aligned}\tilde{B}_n^{6,0,0}([n\cdot]) &= \tilde{B}_n^{0,0,\gamma}([n\cdot]) \\ &+ \hat{D}^{-1/2}\frac{n}{n+[n\cdot]}\frac{[n\cdot]}{\sqrt{n}}\left((\hat{\sigma}_{1,1}^{2,n} - \hat{\sigma}_{1,n+1}^{2,n+[n\cdot]})\frac{\rho_0}{2\sigma_1^2} + (\hat{\sigma}_{2,1}^{2,n} - \hat{\sigma}_{2,n+1}^{2,n+[n\cdot]})\frac{\rho_0}{2\sigma_1^2}\right) \\ &\cdot (1+o(1)) + o_P(\hat{D}_0^{-1/2}),\end{aligned}$$

under H_0 and $\mathbf{H_{LA}}$, whereas $o(1)$ is replaced by $O(1)$ if $\mathbf{H_A}$ is assumed instead. Since

$$\frac{[n\cdot]}{\sqrt{n}}(\hat{\sigma}_{l,1}^{2,n} - \hat{\sigma}_{l,n+1}^{2,n+[n\cdot]}) = \frac{[n\cdot]+n}{\sqrt{n}}(\sigma_l^2 - \hat{\sigma}_{l,1}^{2,n+[n\cdot]}) - \left(1+\frac{[n\cdot]}{n}\right)\frac{n}{\sqrt{n}}(\sigma_l^2 - \hat{\sigma}_{l,1}^{2,n})$$

by using the assumed property of the variance estimates, we obtain that

$$\begin{aligned}\tilde{B}_n^{6,0,0}([\cdot]) =& \hat{D}^{-1/2} f\Bigg(\frac{1}{\sqrt{n}}\sum_{i=1}^{[n\cdot]}(Z_i^{(0)} - \rho_i) - \frac{[n\cdot]}{\sqrt{n}}(\hat{\sigma}_{1,[n\cdot]}^2 - \sigma_1^2)\frac{\rho_0}{2\sigma_1^2} \\ &- \frac{[n\cdot]}{\sqrt{n}}(\hat{\sigma}_{2,[n\cdot]}^2 - \sigma_2^2)\frac{\rho_0}{2\sigma_2^2}, \frac{[n\cdot]}{n}, \frac{n}{n+[n\cdot]}\Bigg),\end{aligned}$$

where $f: \mathcal{D}[0,1+m]^3 \to \mathcal{D}[0,m]$ with $f(x,y,z) = z(\cdot)(x(1+\cdot) - x(1) - y(\cdot)x(1))$. Since f is continuous, the first and second convergences of Theorem 2.2.1 hold true due the CMT. The third convergence towards $\tilde{B}_n^{0,0,\gamma}([n\cdot])$ analogously follows since $|\tilde{B}_n^{0,0,\gamma}([n\cdot]) - \tilde{B}_n^{6,0,\gamma}([n\cdot])| = O_P(\hat{D}^{-1/2})$. □

Corollary 5.2.4. *Under the assumptions of Theorem 2.2.3 and Theorem 5.2.3 let additionally*

$$\left\|\left(\frac{[n\cdot]}{n}\right)^{-\gamma}\frac{[n\cdot]}{\sqrt{n}}(\hat{\sigma}_{l,n}^{2,n+[n\cdot]} - \sigma_l^2)\right\|_{[0,\epsilon]} = o_P(1) \tag{5.2.4}$$

as $n \to \infty$, *followed by* $\epsilon \to 0$. *Then, Theorem 2.2.3 holds true if we replace* $\tau_{n,\iota,0,0,\gamma}^{(c)}$ *by* $\tau_{n,\iota,6,0,\gamma}^{(c)}$.

Proof. We just have to multiply each term by the weighting function $w_\gamma([n\cdot]/n)$ and additionally use (5.2.4). □

Non-constant Variances

Theorem 5.2.5. *Let the parameters* σ_1 *and* σ_2 *fulfill Assumption (PEE7) with* $d_{l,n} \equiv 0$ *and let the convergence in (5.1.8) be satisfied on* $\mathcal{D}[0,1+m]$. *Additionally, let* $\{Z_n^{(0)} - \rho_n\}$ *fulfill* $(\mathcal{K}_{\mathbf{r}}^{(2)})$ *for* $r_z > 1$ *and the sequences* $b_{l,j,n} = b_{\sigma_l,j,n}$ $(l = 1,2,\ j = 1,\ldots,m_{l,n},\ n = 1,\ldots)$ *fulfill*

$$\left(\sum_{j_1=1}^{m_{1,n}}\sum_{j_2=1}^{m_{2,n}}(b_{1,j_1} \vee b_{2,j_2})^{r_z}\#(I_{1,j_1} \cap I_{2,j_2})\right)^{1/r_z} = o((m_{1,n} \vee m_{2,n})^{-1}n^{1/2}), \tag{5.2.5}$$

$$\left(\sum_{j_1=1}^{m_{1,n}}\sum_{j_2=1}^{m_{2,n}}(b_{1,2,j_1} \vee b_{2,2,j_2})^{r_z}\#(I_{1,2,j_1} \cap I_{2,2,j_2})\right)^{1/r_z} = o((m_{1,n} \vee m_{2,n})^{-1}n^{1/2}), \tag{5.2.6}$$

$$\sum_{i=1}^{m_{1,n}}\sum_{j=1}^{m_{2,n}}\#(I_{\sigma,1,i} \cap I_{\sigma,2,j})b_{\sigma,1,i}b_{\sigma,2,j} = o(n^{1/2}), \tag{5.2.7}$$

$$\sum_{i=1}^{m_{1,n}}\sum_{j=1}^{m_{2,n}}\#(I_{\sigma,1,2,i} \cap I_{\sigma,2,2,j})b_{2,\sigma,1,2,i}b_{\sigma,2,2,j} = o(n^{1/2}). \tag{5.2.8}$$

Then, Theorem 2.2.1 holds true if we replace $\tau_{n,\iota,0,0}^{(c)}$ *by* $\tau_{n,\iota,7,0}^{(c)}$.

Proof. Firstly, we obtain that

$$\tilde{B}_n^{7,0,0}(k) = \tilde{B}_n^{0,0,\gamma}(k) + \hat{D}^{-1/2}\frac{n}{n+k}\frac{1}{\sqrt{n}}\sum_{i=n+1}^{n+k}\left(\frac{\sigma_{1,i}\sigma_{2,i}}{\hat{\sigma}_{1,i,n+1}^{n+k}\hat{\sigma}_{2,i,n+1}^{n+k}} - 1\right)(Z_i^{(0)} - \rho_i)$$
$$+ \hat{D}^{-1/2}\frac{n}{n+k}\frac{k}{n}\frac{1}{\sqrt{n}}\sum_{i=1}^{n}\left(\frac{\sigma_{1,i}\sigma_{2,i}}{\hat{\sigma}_{1,i,1}^{n}\hat{\sigma}_{2,i,1}^{n}} - 1\right)(Z_i^{(0)} - \rho_i)$$
$$+ \hat{D}^{-1/2}\frac{n}{n+k}\frac{1}{\sqrt{n}}\left(\sum_{i=n+1}^{n+k}\left(\frac{\sigma_{1,i}\sigma_{2,i}}{\hat{\sigma}_{1,i,n+1}^{n+k}\hat{\sigma}_{2,i,n+1}^{n+k}} - 1\right)\rho_i - \frac{k}{n}\sum_{i=1}^{n}\left(\frac{\sigma_{1,i}\sigma_{2,i}}{\hat{\sigma}_{1,i,1}^{n}\hat{\sigma}_{2,i,1}^{n}} - 1\right)\rho_i\right).$$

Using Assumption (PEE7) and the Kolmogorov-type inequality we get that the third summand is equal to

$$O_P\left((m_{1,n} \vee m_{2,n})n^{-1/2}\left(\sum_{j_1=1}^{m_{1,n}}\sum_{j_2=1}^{m_{2,n}}(b_{1,j_1} \vee b_{2,j_2})^{r_z}\#(I_{\sigma,1,j_1} \cap I_{\sigma,2,j_2})\right)^{1/r_z}\right) + o_P(1).$$

Analogously, we get for the second summand the preceding rate plus

$$O_P\left((m_{1,n} \vee m_{2,n})n^{-1/2}\left(\sum_{j_1=1}^{m_{1,n}}\sum_{j_2=1}^{m_{2,n}}(b_{1,2,j_1} \vee b_{2,2,j_2})^{r_z}\#(I_{\sigma,1,2,j_1} \cap I_{\sigma,2,2,j_2})\right)^{1/r_z}\right) + o_P(1).$$

Using that ρ_i is constant, constant plus $O(n^{-1/2})$, or constant on at most three subsets of $[1,n]$ under H_0, $\mathbf{H_{LA}}$, or $\mathbf{H_A}$, we get with Assumption (PEE7) that

$$\left\|\sum_{i=n+1}^{n+[n\cdot]}\frac{\sigma_{1,i}^2 - \hat{\sigma}_{1,i,n+1}^{2,n+[n\cdot]}}{\hat{\sigma}_{1,i,n+1}^{n+[n\cdot]}(\sigma_{1,i} + \hat{\sigma}_{1,i,n+1}^{2,n+[n\cdot]})}\frac{\sigma_{2,i}^2 - \hat{\sigma}_{2,i,n+1}^{2,n+[n\cdot]}}{\hat{\sigma}_{2,i,n+1}^{n+[n\cdot]}(\sigma_{2,i} + \hat{\sigma}_{2,i,n+1}^{2,n+[n\cdot]})}\rho_i\right\|_{[0,m]}$$
$$= O_P\left(\sum_{i=1}^{m_{1,n}}\sum_{j=1}^{m_{2,n}}\#(I_{\sigma,1,2,i} \cap I_{\sigma,2,2,j})b_{\sigma,1,2,i}b_{\sigma,2,2,j}\right)$$

and

$$\left|\sum_{i=1}^{n}\frac{\sigma_{1,i}^2 - \hat{\sigma}_{1,i,1}^{2,n}}{\hat{\sigma}_{1,i,1}^{n}(\sigma_{1,i} + \hat{\sigma}_{1,i,1}^{2,n})}\frac{\sigma_{2,i}^2 - \hat{\sigma}_{2,i,1}^{2,n}}{\hat{\sigma}_{2,i,1}^{n}(\sigma_{2,i} + \hat{\sigma}_{2,i,1}^{2,n})})\rho_i\right| = O_P\left(\sum_{i=1}^{m_{1,n}}\sum_{j=1}^{m_{2,n}}\#(I_{\sigma,1,i} \cap I_{\sigma,2,j})b_{\sigma,1,i}b_{\sigma,2,j}\right).$$

Furthermore, we obtain that

$$\tilde{B}_n^{7,0,0}([n\cdot]) = \tilde{B}_n^{0,0,\gamma}([n\cdot]) + \hat{D}^{-1/2}\frac{n}{n+[n\cdot]}\frac{1}{\sqrt{n}}\left[\sum_{i=1+n}^{n+[n\cdot]}\rho_i\left(\frac{\sigma_{1,i}^2 - \hat{\sigma}_{1,i,n+1}^{2,n+[n\cdot]}}{2\sigma_{1,i}^2} + \frac{\sigma_{2,i}^2 - \hat{\sigma}_{2,i,n+1}^{2,n+[n\cdot]}}{2\sigma_{2,i}^2}\right)\right.$$
$$\left. - \frac{[n\cdot]}{n}\sum_{i=1}^{n}\rho_i\left(\frac{\sigma_{1,i}^2 - \hat{\sigma}_{1,i,1}^{2,n}}{2\sigma_{1,i}^2} + \frac{\sigma_{2,i}^2 - \hat{\sigma}_{2,i,1}^{2,n}}{2\sigma_{2,i}^2}\right)\right] + o_P(\hat{D}^{-1/2}),$$

where under $\mathbf{H_{LA}^{(c)}}$ and $\mathbf{H_A^{(c)}}$ the summand $\tilde{B}_n^{0,0,\gamma}([n\cdot])$ is the dominating one and we have to add $O_P(\hat{D}^{-1/2})$. Finally, we obtain that the right–hand side is a continuous mapping of the modified display (5.1.8). Hence, the claim follows. □

Corollary 5.2.6. *Under the assumptions of Theorem 2.2.3 and Theorem 5.2.5 let additionally*

$$\left\|\left(\frac{[n\cdot]}{n}\right)^{\gamma}\sum_{i=n+1}^{n+[n\cdot]}(\hat{\sigma}_{l,i,,n}^{2,n+[n\cdot]} - \sigma_{l,i}^2)\right\|_{[0,\epsilon]} = o_P(\sqrt{n}) \tag{5.2.9}$$

as $n \to \infty$*, followed by* $\epsilon \to 0$*. Then, Theorem 2.2.3 holds true if we replace* $\tau_{n,\iota,0,0,\gamma}^{(c)}$ *by* $\tau_{n,\iota,7,0,\gamma}^{(c)}$.

Theorem 5.2.7. *Under the assumption of Theorem 5.2.5 let Assumption (PEE7) be replaced by Assumption (PEE8). Additionally, let the rates as displayed in* (5.2.5) *and* (5.2.7) *hold true. Then, Theorem 2.2.1 holds true if we replace* $\tau^{(c)}_{n,\iota,0,0}$ *by* $\tau^{(c)}_{n,\iota,8,0}$.

Proof. The proof follows in a similar way as the proof of Theorem 5.2.7. □

Corollary 5.2.8. *Under the assumptions of Theorem 2.2.3 and Theorem 5.2.7 let additionally*

$$\left\| \left(\frac{[n\cdot]}{n}\right)^{\gamma} \sum_{i=n+1}^{n+[\cdot]} (\hat\sigma^2_{l,i,[n\cdot],n} - \sigma^2_{l,i}) \right\|_{[0,\epsilon]} = o_P(\sqrt{n}) \tag{5.2.10}$$

as $n \to \infty$, *followed by* $\epsilon \to 0$. *Then, Theorem 2.2.3 holds true if we replace* $\tau^{(c)}_{n,\iota,0,0,\gamma}$ *by* $\tau^{(c)}_{n,\iota,8,0,\gamma}$.

5.2.2 Open-end Procedure under Unknown Variances

Constant Variances

Theorem 5.2.9. *Let the parameters* σ_1 *and* σ_2 *fulfill Assumption (PEE5) with* $\delta_{\sigma,1}, \delta_{\sigma,2} > 0$ *and* $d_{l,i} \equiv 0$. *Then, Theorem 2.2.5 holds true if we replace* $\tau^{(o)}_{n,\iota,0,0,\gamma}$ *by* $\tau^{(o)}_{n,\iota,5,0,\gamma}$.

Proof. The proof follows in a similar way to the proof of Theorem 5.2.1 where we use Theorem 2.2.5 instead of Theorem 2.2.3. □

Theorem 5.2.10. *Under the assumptions of Corollary 5.2.4 let for* $l = 1, 2$ *and some* $\lambda' > 0$

$$\max_{nm \le k} |\sigma_l^2 - \hat\sigma^{2,n+k}_{l,1}| = O_P(\sqrt{n}m^{-\lambda'})$$

as $n \to \infty$, *followed by* $m \to \infty$ *be given. Then, if we replace* $\tau^{(o)}_{n,\iota,0,0,\gamma}$ *by* $\tau^{(o)}_{n,\iota,6,0,\gamma}$, *Theorem 2.2.5 holds true.*

Proof. Following the proof of Theorem 2.2.5 we use Theorem 5.2.3 instead of Theorem 2.2.3 to obtain the claim under Assumption $\mathbf{H}^{(\mathbf{o})}_{\mathbf{A}}$. Under $H_0^{(o)}$ we get that

$$\begin{aligned} P(\tau^{(o)}_{n,\iota,6,\gamma} < \infty) =& P\left(\sup_{z\in[0,1]} u_\iota((\cdot)^{-\gamma}W(\cdot))(z) \ge c_\alpha \right) + o(1) \\ &+ P\left(\max_{1\le k\le mn} u_\iota(\tilde B_n^{6,0,\gamma})(k) < c_\alpha, \max_{k>nm} u_\iota(\tilde B_n^{6,0,\gamma})(k) \ge c_\alpha \right). \end{aligned}$$

Under Assumption $\mathbf{H}^{(\mathbf{o})}_{\mathbf{LA}}$ we just have to add $(1-\cdot)\int_1^{1/(1-\cdot)} g_\rho(x)dx$ to $(\cdot)^{-\gamma}W(\cdot)$ in the first line. Furthermore, with the same arguments used in the proof of Theorem 2.2.5 combined with the above rate assumption we obtain that the last row is equal to

$$\begin{aligned} P\Bigg(&\sup_{x>m} u_\iota\left(w_\gamma G \mathbb{1}_{[0,m]}(\cdot) - [w_\gamma(\cdot)\frac{\cdot}{\cdot+1}W(1) + \delta N m^{-\lambda}]\mathbb{1}_{\{\cdot\ge m\}} \right)(x) \\ &- \sup_{0\le x\le m} u_\iota(w_\gamma G)(x) \ge \epsilon \Bigg) + o(1) \end{aligned}$$

as $n \to \infty$, followed by $m \to \infty$, $N \to \infty$, and $\epsilon \to 0$, where $G(\cdot) = \frac{1}{1+\cdot}(W(1+\cdot) - (1+\cdot)W(1))$, $\delta = \mathrm{sign}(W(1))$, and $\lambda \in (0, \lambda' \wedge \frac{1}{2}]$. □

Non-constant Variances

Theorem 5.2.11. *Under the assumptions of Corollary 5.2.6 let additionally hold that*

$$\sum_{j_1=1}^{m_{1,n}}\sum_{j_2=1}^{m_{2,n}}\sum_{i=0}^{\infty}(b_{2,\sigma,1,j_1}\vee b_{2,\sigma,2,j_2})^{r_z}\frac{\#(I_{2,\sigma,1,j_1}\cap I_{2,\sigma,2,j_2}\cap[1,n+nm2^{i+1}])}{(2^{r_z})^i} = o\left(\frac{m^{r_z}n^{r_z/2}}{(m_{1,n}\vee m_{2,n})^{r_z}}\right), \tag{5.2.11}$$

$$\sum_{j_1=1}^{m_{1,n}}\sum_{j_2=1}^{m_{1,n}} b_{2,\sigma,1,j_1}b_{2,\sigma,2,j_2}\max_{k\geq mn}\frac{\#I_{2,\sigma,1,j_1}\cap I_{2,\sigma,2,j_2}\cap[n,n+k]}{k} = o(1), \tag{5.2.12}$$

$$\max_{nm\leq k}\frac{1}{n+k}\left|\sum_{i=n+1}^{n+k}(\sigma_{l,i}^2-\hat{\sigma}_{l,i,n+1}^{2,n+k})/\sigma_{l,i}^2\right| = o_P(\sqrt{n}) \tag{5.2.13}$$

as $n\to\infty$*, followed by* $m\to\infty$*. Then, Theorem 2.2.5 holds true if we replace* $\tau_{n,\iota,0,0,\gamma}^{(o)}$ *by* $\tau_{n,\iota,7,0,\gamma}^{(o)}$*.*

Proof. Combining the arguments of the proofs of Theorem 5.2.10, Theorem 5.2.5, Theorem 4.2.11 and the assumed rates (5.2.11), (5.2.12) implies

$$\max_{nm\leq k}\frac{n}{n+k}\frac{1}{\sqrt{n}}\left|\sum_{i=n+1}^{n+k}\left(\frac{\sigma_{1,i}\sigma_{2,i}}{\hat{\sigma}_{1,i,n+1}^{n+k}\hat{\sigma}_{2,i,n+1}^{n+k}}-1\right)(Z_i^{(0)}-\rho_i)\right| = o_P(1)$$

and

$$\max_{nm\leq k}\frac{n}{n+k}\frac{1}{\sqrt{n}}\left|\sum_{i=n+1}^{n+k}\frac{\sigma_{1,i}^2-\hat{\sigma}_{1,i,n+1}^{2,n+k}}{\hat{\sigma}_{1,i,n+1}^{n+k}(\sigma_{1,i}+\hat{\sigma}_{1,i,n+1}^{2,n+k})}\frac{\sigma_{2,i}^2-\hat{\sigma}_{2,i,n+1}^{2,n+k}}{\hat{\sigma}_{2,i,n+1}^{n+k}(\sigma_{2,i}+\hat{\sigma}_{2,i,n+1}^{2,n+k})}\rho_i\right| = o_P(1)$$

as $n\to\infty$, followed by $m\to\infty$. Now, we can follow Theorem 2.2.5, use the assumed rates, and claim is proven. □

Theorem 5.2.12. *Under the assumptions of Corollary 5.2.8 let the equations* (5.2.11), (5.2.12), *and*

$$\max_{nm\leq k}\frac{1}{n+k}\left|\sum_{i=n+1}^{n+k}(\sigma_{l,i}^2-\hat{\sigma}_{l,i,k,n}^2)/\sigma_{l,i}^2\right| = o_P(\sqrt{n}),$$

as $n\to\infty$*, followed by* $m\to\infty$*, be fulfilled. Then, Theorem 2.2.5 holds true if we replace* $\tau_{n,\iota,0,0,\gamma}^{(o)}$ *by* $\tau_{n,\iota,8,0,\gamma}^{(o)}$*.*

Proof. The proof follows in a similar way as the proof of Theorems 5.2.11. □

5.3 Examples

In this subsection, we continue with the three examples (IID), (MIX1), and (NED1), where we restrict the assumptions depending on the testing procedures and estimates. We focus on scenarios where we apply the different main results of this Section 5. Therefore, we set

$$\sigma_{l,i}^2 = \sigma_{l0}^2 + \sum_{j=1}^{R_{l,n}} \Delta_{\sigma,l,j} \mathbb{1}_{\{i \le k^*_{\mu,l,j}\}}, \tag{5.3.1}$$

where $R_{l,n} \geq 0$, $\Delta_{\sigma,l,j} \neq 0$, and $0 = k^*_{\mu,l,0} < k^*_{\mu,l,1} < \ldots < k^*_{\mu,l,R_{l,n}} < k^*_{\mu,l,R_{l,n}+1} = N_n$ with $N_n \in \{n, n(1+m), \infty\}$ depending on the considered procedure: a posteriori, closed-end, or open-end. In addition, we assume that $k^*_{\sigma,l,i+1} - k^*_{\sigma,l,i} \to \infty$ as $n \to \infty$.

5.3.1 Constant Variances

Firstly, we consider the special case of $R_{l,n} \equiv 0$. Then, we use the variance estimates based on the whole sample, i.e.,

$$(\hat{\sigma}_{l,n}^{(1)})^2 = n^{-1} \sum_{i=1}^{n} w_{i,n}(Z_{l,i} - \mu_{l,i})^2, \tag{5.3.2}$$

where the deterministic, positive, uniform bounded weights fulfill $\sum_{i=1}^n w_{i,n} = n$ for all $n \in \mathbb{N}$. Then, we get under Assumption (IID)

$$|(\hat{\sigma}_{l,n}^{(1)})^2 - \sigma_l^2| = O_P\left(n^{-(\frac{r_l'-2}{2} \wedge \frac{1}{2})}\right)$$

which implies that the variance estimate is consistent for $r_l' > 2$.

Under Assumption (MIX1) we obtain by Davidson (1994, Th. 14.1) that $\{Z_{l,n}^2\}$ is α-mixing of the same size as $\{Z_{l,n}\}$. Hence, $\{Z_{l,n}^2, \mathcal{F}_n\}$ is an L_{p_l}-mixingale, $p_l \leq r_l'/2$, with $\mathcal{F}_n := \bigvee_{i=1}^n \bigvee_{j=1}^\infty \sigma(\epsilon_{j,i})$ and sequence $\xi_n = \tilde{\alpha}(n)^{1/p_l - 2/r_l'}$, which directly follows from Davidson (1994, Th. 14.2). This implies that

$$P\left(|(\hat{\sigma}_{l,n}^{(1)})^2 - \sigma_l^2| \geq \eta\right) \leq (n\eta)^{-p_l} \mathbb{E}\left[\left(\sum_{i=1}^{n}(Z_{l,i}^2 - \sigma_l^2)\right)^{p_l}\right]$$
$$\leq \begin{cases} \eta^{-p_l} n^{-1} C\left(\sum_{m=0}^\infty \left(\sum_{k=0}^m \xi_k^{-2}\right)^{-1/2}\right)^2, & \text{if } p_l = 2, \\ \eta^{-p_l} n^{1-p_l} C\left(\sum_{k=0}^\infty \xi_k\right)^{p_l}, & \text{if } p_l < 2. \end{cases}$$

Thus, if $\sum_{m=0}^\infty \left(\sum_{k=0}^m \xi_k^{-2}\right)^{-1/2} < \infty$, we choose $p_l = 2$ and we get $|(\hat{\sigma}_{l,n}^{(1)})^2 - \sigma_l^2| = O_P\left(n^{-1/2}\right)$. Otherwise, we choose $p_l = r'/2 \wedge (2-\epsilon)$, $\epsilon > 0$, and $r' \in (2, r]$ from assumption (MIX) to obtain that the sum is finite, since $1/p_l - 2/r_l' \geq 2(1/r' - 1/r)$. This implies that $|(\hat{\sigma}_{l,n}^{(1)})^2 - \sigma_l^2| = O_P\left(n^{-(p_l-1)/p_l}\right)$.

Under Assumption (NED) with $p_l = 2$ and $r_l' > 4$, we obtain by Davidson (1994, Th. 17.17) that $\{Z_{l,n}^2\}$ is L_2-NED of size $-a_l(r_l'-4)/(2r_l'-2)$. By Lemma 2.3.7 we obtain that there are more combinations of p_l and r_l' that are suitable. However, if we additionally assume under (NED) that $a_l \geq (2r_l'-2)/(2r_l'-8)$ and $a_V \geq r_l'/(r_l'-4)$, we obtain that $\{Z_{l,n}^2\}$ is a L_2-mixingal of size $-\frac{1}{2}$. This implies that $\{Z_{l,n}^2 - \sigma_{l,n}^2\}$ fulfills the Kolmogorov-type inequalities for $r_l = 2$ and, especially, $(\hat{\sigma}_{l,n}^{(1)})^2 - \sigma_l^2 = O_P(n^{-1/2})$.

Assumption (NED2). *Let (NED) be fulfilled with* $a_l \geq (2r_l'-2)/(2r_l'-8)$ *and* $a_V \geq r_l'/(r_l'-4)$.

Thus, in all three previous examples the variance estimates are consistent and we can apply Theorem 5.1.1 and Corollary 5.1.2 to test whether there is a change in the correlation or not. Therefore, the LRV estimates from Theorem 5.1.16 are available. The estimation of change-points in the correlation works with even less assumptions. It is sufficient that $(\hat{\sigma}_{l,n}^{(1)})^2 = O_P(1)$. The sequential procedures, Theorem 5.2.1 and 5.2.9, can be applied by using $\hat{\sigma}_{l,k,n}^{(1)} = \hat{\sigma}_{l,n}^{(1)}$ for all $k \in \mathbb{N}$.

Remark 5.3.1. *The test statistic* $\phi_n^{\iota,5,5,\gamma}$ *and the stopping time* $\tau_{n,\iota,5,5,\gamma}$ *are equal to* $\phi_n^{\iota,0,0,\gamma}$ *and* $\tau_{n,\iota,0,0,\gamma}$*, respectively, if* $\hat{D}_{5,n}$ *uses* $\tilde{\rho}_j^{(5)} = \frac{\sigma_1\sigma_2}{\hat{\sigma}_{1,n}\hat{\sigma}_{2,n}}\tilde{\rho}_n(j)$*, where* $\tilde{\rho}_n(j)$ *is as in Theorem 2.1.33 and the variance estimates are the same as used for* $Z_n^{(5)}$*. This implies, that in this special case the assumptions on Section 2.3 are sufficient for the convergence, even if the variance estimates are not consistent.*

Assumption (IID2). *Let Assumption (IID) be fulfilled for* $r_l' > 4$*.*

Assumption (LRV2). *Let Assumption (LRV) be fulfilled with*

$$Z_i^0 - \rho_i\left((X_i - \mu_{1,i})^2/(2\sigma_{1,i}^2) + (Y_i - \mu_{2,i})^2/(2\sigma_{2,i}^2)\right) \quad \textit{instead of } Z_i^0.$$

Now, we consider the cumulative sample variances as estimates, i.e.

$$\hat{\sigma}_{l,k}^{(2)} = (\hat{\sigma}_{l,k}^{(1)})^2, \quad k = 1,2,\ldots,n. \tag{5.3.3}$$

Suppose Assumption (IID2) or Assumption (NED2) combined with Assumption (LRV2) hold true. Then, we can apply Theorem 5.1.4 to test for a constant correlation. Furthermore, we can use the weighted test statistic of Corollary 5.1.7.

5.3.2 Non-constant Variances

In this sub-subsection, we consider two examples for Theorem 5.1.9. Additionally, we assume that there are non-local changes and that the distance between the change-points increases with rate n.

Firstly, we postulate that we already know the change-points of the variance. Then, we set for all $i \in (k_{\sigma,j-1}^*, k \wedge k_{\sigma,j}^*]$, $j = 1,\ldots,R_l$, $k = 1,2,\ldots$

$$(\hat{\sigma}_{l,i,k}^{(4')})^2 = \frac{1}{k \wedge k_{\sigma,j}^* - k_{\sigma,j-1}^*} \sum_{v=k_{\sigma,j-1}^*+1}^{k \wedge k_{\sigma,j}^*} (Z_{l,v} - \mu_{l,v})^2. \tag{5.3.4}$$

We observe that

$$\sum_{i=1}^{[n\cdot]}((\hat{\sigma}_{l,i,[n\cdot]}^{(4')})^2 - \sigma_{l,i}^2)\sigma_{l,i}^{-2} = \sum_{i=1}^{[n\cdot]}((Z_{l,v} - \mu_{l,v})^2 - \sigma_{l,i}^2)\sigma_{l,i}^{-2}.$$

Hence, under Assumption (IID2) or Assumption (NED2) combined with Assumption (LRV2) we obtain that under H_0

$$\frac{1}{\sqrt{n}}\sum_{i=1}^{[n\cdot]}\left[(Z_i^{(0)} - \rho_i) - (\hat{\sigma}_{1,i,[n\cdot]}^2 - \sigma_{1,i}^2)\frac{\rho_i}{2\sigma_{1,i}^2} - (\hat{\sigma}_{2,i,[n\cdot]}^2 - \sigma_{2,i}^2)\frac{\rho_i}{2\sigma_{2,i}^2}\right] \overset{D[0,1]}{\longrightarrow} D_{(8)}^{1/2}W(\cdot).$$

Thus, the assumptions of Theorem 5.1.9 are fulfilled by using the same arguments as used for the mean estimation in Sub-Subsection 4.3.2.

Remark 5.3.2. *For the sake of completeness, we define for all* $i = 1,\ldots,n$

$$(\hat{\sigma}_{l,i}^{(3)})^2 = \sum_{j=1}^{R_l+1} \mathbb{1}_{\{i \in (\hat{k}_{\sigma,j-1}, \hat{k}_{\sigma,j}]\}} \frac{1}{\hat{k}_{\sigma,j} - \hat{k}_{\sigma,j-1}} \sum_{v=\hat{k}_{\sigma,j-1}+1}^{\hat{k}_{\sigma,j}} (Z_{l,v} - \mu_{l,v})^2 \tag{5.3.5}$$

but we bear in mind that this estimate is useless for change-point analysis of the correlations.

Now, we consider the second example where we drop the assumptions on known variance change-points. Therefore, we set for all $i \in (\hat{k}_{\sigma,j-1}, k \wedge \hat{k}_{\sigma,j}]$, $j = 1,\ldots,R_l + 1$, $k = 1,2,\ldots$

$$(\hat{\sigma}_{l,i,k}^{(4)})^2 = \frac{1}{k \wedge \hat{k}_{\sigma,j} - \hat{k}_{\sigma,j-1}} \sum_{v=\hat{k}_{\sigma,j-1}+1}^{k \wedge \hat{k}_{\sigma,j}} (Z_{l,v} - \mu_{l,v})^2. \tag{5.3.6}$$

Suppose that there exist sequences $a_{l,j,n}$ such that

$$\hat{k}_{\sigma,l,j} - k^*_{\sigma,l,j} = O_P(a_{\sigma,l,j,n}) \tag{5.3.7}$$

and

$$a_{\sigma,l,j,n} = o\left((k^*_{\sigma,l,j} - k^*_{\sigma,l,j-1}) \wedge (k^*_{\sigma,l,j+1} - k^*_{\sigma,l,j})\right) \tag{5.3.8}$$

for $j = 1, \ldots, R_l$ and as $n \to \infty$. Those two conditions imply that

$$P\left(\#(\hat{k}_{\sigma,l,j-1}, \hat{k}_{\sigma,l,j}] \cap (k^*_{\sigma,l,j-1\pm 2}, k^*_{\sigma,l,j\pm 2}] = 0\right) \to 1.$$

Thus, we obtain with $J_{l,j,[n\cdot]} = (k_{\sigma,l,j-1}, k_{\sigma,l,j}] \cap (0,[n\cdot]]$ and $\hat{J}_{l,j,[n\cdot]} = (\hat{k}_{\sigma,l,j-1}, \hat{k}_{\sigma,l,j}] \cap (0,[n\cdot]]$ that

$$\begin{aligned}
\sum_{i=1}^{[n\cdot]} (\hat{\sigma}^2_{l,i,[n\cdot]} - \sigma^2_{l,i})\sigma^{-2}_{l,i} &= \sum_{i=1}^{[n\cdot]} ((Z_{l,i} - \mu_{l,i})^2 - \sigma^2_{l,i})\sigma^{-2}_{l,i} \\
&+ O_P\left(\left\| \sum_{j=1}^{R_l+1} \frac{\#\hat{J}_{l,j,[n\cdot]} \cap J^c_{l,j,[n\cdot]}}{\#\hat{J}_{l,j,[n\cdot]}} \sum_{i\in \hat{J}_{l,j,[n\cdot]}} (Z_{l,i} - \mu_{l,i})^2 \right\|\right) \\
&+ O_P\left(\left\| \sum_{j=2}^{R_l+1} \frac{\#\hat{J}_{l,j-1,[n\cdot]} \cap J_{l,j,[n\cdot]}}{\#\hat{J}_{l,j-1,[n\cdot]}} \sum_{i\in \hat{J}_{l,j-1,[n\cdot]}} (Z_{l,i} - \mu_{l,i})^2 \right\|\right) \\
&+ O_P\left(\left\| \sum_{j=1}^{R_l} \frac{\#\hat{J}_{l,j+1,[n\cdot]} \cap J_{l,j,[n\cdot]}}{\#\hat{J}_{l,j+1,[n\cdot]}} \sum_{i\in \hat{J}_{l,j+1,[n\cdot]}} (Z_{l,i} - \mu_{l,i})^2 \right\|\right).
\end{aligned}$$

The first summand on the right–hand side weighted by $n^{-1/2}$ fulfills the FCLT under suitable assumptions such as in the case of known change-points in the variance. Hence, it remains to show that the three rates are equal to $o_P(n^{-1/2})$. Since each of the three rates can similarly be treated, we just consider the first one (w.l.o.g. let $\mu_{l,i} \equiv 0$):

$$\left\| \sum_{j=1}^{R_l+1} \frac{\#\hat{J}_{l,j,[n\cdot]} \cap J^c_{l,j,[n\cdot]}}{\#\hat{J}_{l,j,[n\cdot]}} \sum_{i\in \hat{J}_{l,j,[n\cdot]}} Z^2_{l,i} \right\| \le \sum_{j=1}^{R_l+1} O_P\left(\left\| \frac{a_{\sigma,l,j-1} \vee a_{\sigma,l,j}}{\#\hat{J}_{l,j,[n\cdot]}} \sum_{i\in \hat{J}_{l,j,[n\cdot]}} (Z^2_{l,i} - 1) \right\|\right) + O_P(a_{\sigma,l,j}).$$

Furthermore, we get that

$$\begin{aligned}
&P\left(\left\| \frac{a_{\sigma,l,j-1} \vee a_{\sigma,l,j}}{\#\hat{J}_{l,j,[n\cdot]}} \sum_{i\in \hat{J}_{l,j,[n\cdot]}} (Z^2_{l,i} - 1) \right\| \ge \eta\sqrt{n}\right) \\
&\le P\left(\max_{\substack{k_1 \in (k^*_{\sigma,l,j-1} - a_{\sigma,j-1}, k^*_{\sigma,l,j-1} + a_{\sigma,j-1}] \\ k_2 \in (k^*_{\sigma,l,j-1} - a_{\sigma,j-1}, k^*_{\sigma,l,j+1} + a_{\sigma,j+1}], k_1 < k_2}} \left| \frac{a_{\sigma,l,j-1} \vee a_{\sigma,l,j}}{k_2 - k_1} \sum_{i=k_1+1}^{k_2} (Z^2_{l,i} - 1) \right| \ge \eta\sqrt{n} \right) + o(1) \\
&\le \sum_{k_1 = k^*_{\sigma,l,j-1} - a_{\sigma,j-1}}^{k^*_{\sigma,l,j-1} + a_{\sigma,j-1}} P\left(\max_{k_2 \in (k_1, k^*_{\sigma,l,j+1} + a_{\sigma,j+1}]} \left| \frac{a_{\sigma,l,j-1} \vee a_{\sigma,l,j}}{k_2 - k_1} \sum_{i=k_1+1}^{k_2} (Z^2_{l,i} - 1) \right| \ge \eta\sqrt{n} \right) + o(1) \\
&\le (a_{\sigma,l,j-1} \vee a_{\sigma,l,j}/(\eta\sqrt{n}))^2 C \sum_{k_1 = k^*_{\sigma,l,j-1} - a_{\sigma,j-1}}^{k^*_{\sigma,l,j-1} + a_{\sigma,j-1}} \sum_{k_2 = k_1+1}^{k^*_{\sigma,l,j+1} + a_{\sigma,j+1}]} \left(\frac{1}{k_2 - k_1}\right)^{p_l} + o(1) \\
&= O\left(n^{-1}(a_{\sigma,l,j-1} \vee a_{\sigma,l,j})^2 a_{\sigma,l,j-1}\right) + o(1),
\end{aligned}$$

where we use (5.3.7), σ-additivity, and a Kolmogorov-type inequality. Thus, we obtain

$$\sum_{i=1}^{[n\cdot]}(\hat{\sigma}^2_{l,i,[n\cdot]} - \sigma^2_{l,i})\sigma^{-2}_{l,i} = \sum_{i=1}^{[n\cdot]}(Z^2_{l,i} - \sigma^2_{l,i})\sigma^{-2}_{l,i} + o_P(1)$$

if $\max_j a_{\sigma,l,j,n} = o(n^{1/3})$. Suppose that these restrictions combined with Assumption (IID2) or Assumption (NED2) and Assumption (LRV2) are fulfilled. Then, we get the assumed FCLT (5.1.8) from Theorem 5.1.9. Furthermore, under these assumptions $\{(Z_{l,n} - \mu_{l,n})^2 - \sigma^2_{l,n}\}$ fulfills the Kolmogorov-type inequalities.

Now, the technical rate assumptions on several results remain to be considered. For that purpose, we can use the examples of the non-constant means of Sub-subsection 4.3.2. There, we have seen that the model of the structure breaks and the Kolmogorov-type inequalities for $\{Z_{l,n} - \mu_{l,n}\}$ was sufficient. If we transfer the conditions of these models to models of changes in the variances and assume that $\{(Z_{l,n} - \mu_{l,n})^2 - \sigma^2_{l,n}\}$ fulfills the Kolmogorov-type inequalities for $r_l = 2$, as in the previously presented assumptions, we can apply the results of Subsection 5.1 and Subsection 5.2.

6 Change-Point Analysis of the Correlation under Unknown Means and Variances

In this section, we consider the setting where the four parameters $\mu_{1,i}$, $\mu_{2,i}$, $\sigma^2_{1,i}$, and $\sigma^2_{2,i}$ are unknown and estimated. Naturally, the variance estimates depend on the mean estimates. But first, we neglect this feature and will only come back to it at the end of each subsection.

6.1 A Posteriori Analysis under a General Dependency Framework, General Mean and Variance Estimates

In this subsection, we set

$$\hat{\rho}_{k,4+\psi_1+4\psi_2} = \frac{1}{k}\sum_{i=1}^{k} \frac{(X_i - \hat{\mu}^{(\psi_1)}_{1,i,k,n})(Y_i - \hat{\mu}^{(\psi_1)}_{2,i,k,n})}{\hat{\sigma}^{(\psi_2)}_{1,i,k,n}\hat{\sigma}^{(\psi_2)}_{2,i,k,n}} = \frac{1}{k}\sum_{i=1}^{k} Z^{(4+\psi_1+4\psi_2)}_{i,k,n}, \tag{6.1.1}$$

where $\psi_1, \psi_2 = 1,\ldots,4$ are design indices to distinguish between the parameter estimates. Furthermore, in each estimate and $Z^{(4+\psi_1+4\psi_2)}_{i,k,n}$, we will drop the index n and any other index which will not be used in the following.

For $\{Z^{(4+\psi_1+4\psi_2)}_{i,k,n}\}_{1\le i\le k\le n\in\mathbb{N}}$ we will frequently use the decomposition

$$\begin{aligned} Z^{(4+\psi_1+4\psi_2)}_{i,\cdot} &= \frac{(X_i - \hat{\mu}^{(\psi_1)}_{1,\cdot})(Y_i - \hat{\mu}^{(\psi_1)}_{2,\cdot})}{\hat{\sigma}^{(\psi_2)}_{1,\cdot}\hat{\sigma}^{(\psi_2)}_{2,\cdot}} \\ &= Z^{(0)}_i + Z^{(0)}_i\left(\frac{\sigma_{1,i}\sigma_{2,i}}{\hat{\sigma}^{(\psi_2)}_{1,\cdot}\hat{\sigma}^{(\psi_2)}_{2,\cdot}} - 1\right) + \sum_{l=1}^{3} R_{liknZ^{(\psi_1)}} + \sum_{l=1}^{3} R_{liknZ^{(\psi_1)}}\left(\frac{\sigma_{1,i}\sigma_{2,i}}{\hat{\sigma}^{(\psi_2)}_{1,\cdot}\hat{\sigma}^{(\psi_2)}_{2,\cdot}} - 1\right), \end{aligned} \tag{6.1.2}$$

where $R_{liknZ^{(\psi_1)}}$, $l = 1, 2, 3$, $\psi_1, \psi_2 = 1,\ldots,4$ are defined in Section 4, see (4.1.2). They are the error terms of the mean estimation. The dot stands for the index k which will only appear if it is in use. Thus, if the mean and variance estimates fulfill their corresponding assumptions on Section 4 and Section 5, only the last summand is new. Furthermore, in this section we will write $Z^{(8+\psi)}_{i,\cdot}$ or just $Z^{(\psi)}_{i,\cdot}$ instead of $Z^{(4+\psi_1+4\psi_2)}_{i,\cdot}$ for a suitable ψ. If ψ_1 and ψ_2 are equal to 1, 2, 3, or 4 the corresponding parameters fulfill Assumption (PEE1), (PEE2), (PEE3), or (PEE4), respectively. To avoid repetitions of similar results, we use the following notation for $j, l \in \{1, 2\}$

$$\begin{aligned} m_{l,n} &= m_{\mu,l,1,n} \vee m_{\sigma,1,1,n} \vee m_{\sigma,2,1,n}, \quad M^{(j)}_{n,l} = (0,m_{\mu,l,j,n}] \times (0,m_{\sigma,1,j,n}] \times (0,m_{\sigma,2,j,n}], \\ M^{(j)}_{n,0} &= (0,m_{\mu,1,j,n}] \times (0,m_{\mu,2,j,n}] \times (0,m_{\sigma,1,j,n}] \times (0,m_{\sigma,2,j,n}], \\ I^{(l,j)}_{i_1,i_2,i_3} &= I_{\mu,l,j,i_1,n} \cap I_{\sigma,1,j,i_2} \cap I_{\sigma,2,j,i_3}, \quad I^{(0,j)}_{i_1,i_2,i_3,i_4} = I_{\mu,1,j,i_1,n} \cap I_{\mu,2,j,i_2,n} \cap I_{\sigma,1,j,i_3} \cap I_{\sigma,2,j,i_4}, \\ b^{(l)}_{\mathbf{i},\mathbf{j}} &= b_{\mu,l,i_1,j_1,n}(b_{\sigma,1,i_2,j_2} \vee b_{\sigma,2,i_3,j_3}), \quad \mathbf{j} \in M^{(2)}_{n,l}, \mathbf{i} \in M^{(1)}_{n,l}, \\ b^{(0)}_{\mathbf{i},\mathbf{j}} &= b_{\mu,1,i_1,j_1,n} b_{\mu,2,i_2,j_2,n}(b_{\sigma,1,i_3,j_3} \vee b_{\sigma,2,i_4,j_4}), \quad \mathbf{j} \in M^{(2)}_{n,0}, \mathbf{i} \in M^{(1)}_{n,0}, \end{aligned}$$

where we set

$$\begin{aligned} m_{\mu,l,j} &= 1, \quad I_{\mu,l,j,1} = (0,n], \quad \text{if } \mu_1,\ \mu_2 \text{ fulfill (PEE1) or (PEE2)}, \\ m_{\sigma,l,j} &= 1, \quad I_{\sigma,l,j,1} = (0,n], \quad \text{if } \sigma_1,\ \sigma_2 \text{ fulfill (PEE1) or (PEE2)}, \\ m_{\mu,l,2} &= 1, \quad I_{\mu,l,2,1} = (0,n], \qquad \text{if } \mu_1,\ \mu_2 \text{ fulfill (PEE3)}, \\ m_{\sigma,l,2} &= 1, \quad I_{\sigma,l,2,1} = (0,n], \qquad \text{if } \sigma_1,\ \sigma_2 \text{ fulfill (PEE3)}. \end{aligned}$$

6.1.1 Testing under a Functional Central Limit Theorem, General Mean and Variance Estimates

Theorem 6.1.1. *Let one of the assumptions of the Theorems 4.1.2, 4.1.6, 4.1.10, or 4.1.13 and one of the assumptions of the Theorems 5.1.1, 5.1.4, or 5.1.9 be fulfilled, where we assume that*

$d_{\mu,l,i} = d_{\sigma,l,i} \equiv 0$. *Additionally, let the following conditions be fulfilled as* $n \to \infty$ *and for* $l = 1,2$

$$\sum_{\mathbf{j}\in M^{(1)}_{n,3-l}} \left((m_{n,3-l})^{r_l} \sum_{\mathbf{i}\in M^{(2)}_{n,3-l}} (b^{(3-l)}_{\mathbf{i},\mathbf{j}})^{r_l} \#[I^{(3-l,1)}_{\mathbf{i}} \cap (0, \max I^{(3-l,2)}_{\mathbf{j}})] \right)^{1/r_l} = o(n^{1/2}) \tag{6.1.3}$$

and

$$\max_{\mathbf{j}\in M^{(2)}_{n,0}} \sum_{\mathbf{i}\in M^{(1)}_{n,0}} \#(I^{(0,1)}_{\mathbf{i}} \cap (0, \max I^{(0,2)}_{\mathbf{j}}])b^{(0)}_{\mathbf{i},\mathbf{j}} = o(n^{1/2}). \tag{6.1.4}$$

Then, Theorem 2.1.1 holds true if we replace $B^{0,0}_n$ *by* $B^{4+j_1+4j_2,0,0}_n$.

Remark 6.1.2. *The sequences* $\{b^{(l)}_{\mathbf{i},\mathbf{j}}\}$ *and* $I^{(l,j)}_{\mathbf{i}}$ *are allowed to depend on an integer* N*, which tends towards infinity after* n *does, cf. Assumption (PEE2). Then, in the case of Assumption (PEE2) we can set for* $l = 1,2$

$$I^{(3-l,2)}_{\mathbf{j}} = \begin{cases} (0,N], & \text{if } \mathbf{j} = (1,1,1), \\ (N,n], & \text{if } \mathbf{j} = (2,2,2), \\ \emptyset, & \text{else} \end{cases} \quad \text{and} \quad b^{(3-l)}_{1,\mathbf{j}} = \begin{cases} 1, & \text{if } \mathbf{j} = (1,1,1), \\ N^{-1/2}, & \text{if } \mathbf{j} = (2,2,2), \\ 1, & \text{else,} \end{cases}$$

and the rate displayed in (6.1.3) *is directly fulfilled as* $n \to \infty$, *followed by* $N \to \infty$.

Proof of Theorem 6.1.1. Firstly, we obtain with (6.1.2) that

$$B^{j,0,0}_n(\cdot) = \hat{D}_0^{-1/2} \left(B_n(\cdot) + \frac{[n\cdot]}{\sqrt{n}} \left(\frac{1}{[n\cdot]} \sum_{i=1}^{[n\cdot]} \sum_{l=1}^{3} R_{li[n\cdot]nZ^{(\theta)}} \left(\frac{\sigma_{1,i}\sigma_{2,i}}{\hat{\sigma}^{(\psi)}_{1,i,[n\cdot]} \hat{\sigma}^{(\psi)}_{2,i,[n\cdot]}} - 1 \right) - \frac{1}{n} \sum_{i=1}^{n} \sum_{l=1}^{3} R_{linnZ^{(j_1)}} \left(\frac{\sigma_{1,i}\sigma_{2,i}}{\hat{\sigma}^{(\psi)}_{1,i,n} \hat{\sigma}^{(\psi)}_{2,i,n}} - 1 \right) \right) \right),$$

where the asymptotic behavior of $\hat{D}_0^{-1/2} B_n(\cdot)$ is known under each combination of the theorems of Section 4 and 5. Hence, it remains to consider the last summand. Here, we just consider the most general case, i.e., that the assumptions of Theorem 4.1.13 and Theorem 5.1.9 are fulfilled. Using the triangle inequality we individually consider each summand for $l = 1, 2, 3$. We start with $l = 1$ and obtain by Kolmogorov's inequality that

$$\left\| \sum_{i=1}^{[n\cdot]} R_{1i[n\cdot]nZ^{(\theta)}} \left(\frac{\sigma_{1,i}\sigma_{2,i}}{\hat{\sigma}^{(\psi)}_{1,i,[n\cdot]} \hat{\sigma}^{(\psi)}_{2,i,[n\cdot]}} - 1 \right) \right\|$$
$$= O_P \left(\sum_{(j_1,j_2,j_3)\in M^{(1)}_{n,2}} \left((m_{n,2})^{r_1} \sum_{(i_1,i_2,i_3)\in M^{(2)}_{n,2}} (b^{(2)}_{\mathbf{i},\mathbf{j}})^{r_1} \#[I^{(1,2)}_{i_1,i_2,i_3} \cap (0, \max I^{(2,2)}_{j_1,j_2,j_3})] \right)^{1/r_1} \right).$$

Hence, this is equal to $o_P(n^{-1/2})$ by (6.1.3). The same rate can be observed for the second term of the difference. Quite similarly we can treat the term for $l = 2$ so that it remains to consider the term for $l = 3$:

$$\left\| \sum_{i=1}^{[n\cdot]} R_{3i[n\cdot]nZ^{(\theta)}} \left(\frac{\sigma_{1,i}\sigma_{2,i}}{\hat{\sigma}^{(\psi)}_{1,i,[n\cdot]} \hat{\sigma}^{(\psi)}_{2,i,[n\cdot]}} - 1 \right) \right\|$$
$$= O_P \left(\max_{\mathbf{j}\in M^{(2)}_{n,0}} \sum_{\mathbf{i}\in M^{(1)}_{n,0}} \#(I^{(0,1)}_{\mathbf{i}} \cap (0, \max I^{(0,2)}_{\mathbf{j}}])b^{(0)}_{\mathbf{i},\mathbf{j}} \right).$$

Hence, using (6.1.4) it follows that $B^{j,0,0}_n(\cdot) = \hat{D}_0^{-1/2}(B_n(\cdot) + o_P(1))$ such that the claim finally follows. □

Influence of the Mean Estimates on the Variance Estimates Now, we consider the influence of the mean estimates on the variance estimates. Let m_2 denote the second moment. Then, we can decompose a variance parameter σ^2 in m_2 and μ^2, i.e., $\sigma^2 = m_2 - \mu^2$. Thus, we get

$$\sigma^2 - \hat{\sigma}_n^2 = m_2 - \hat{m}_{2,n} + \hat{\mu}_n^2 - \mu^2 = O_P(n^{-(\delta_m \wedge \delta_\mu)})$$

if

$$m_2 - \hat{m}_{2,n} = O_P(n^{-\delta_m}) \quad \text{and} \quad \mu - \hat{\mu}_n = O_P\left(n^{-\delta_\mu}\right),$$

which is relevant under the assumption of Theorem 4.1.2 and Theorem 5.1.1. Under the assumptions of Theorem 5.1.4 the variance estimates $\hat{\sigma}^2_{[n\cdot]}$ can be decomposed similarly. Then, there are at least three possible cases

$$\hat{\sigma}^2_{[n\cdot]} = \hat{m}_{2,[n\cdot]} - \hat{\mu}^2_{[n\cdot]}, \quad \hat{\sigma}^2_{[n\cdot]} = \hat{m}_{2,[n\cdot]} - \hat{\mu}^2_n, \quad \text{or} \quad \hat{\sigma}^2_{[n\cdot]} = \hat{m}_{2,n} - \hat{\mu}^2_{[n\cdot]}.$$

In the last two cases, we can reduce the convergence assumption of Theorem 5.1.4 displayed in (5.1.7) by replacing $\hat{\sigma}^2_{l,[n\cdot]} - \sigma_l^2$ by $\hat{m}_{2,l,[n\cdot]} - m_{2,l}$ or $\mu_l^2 - \hat{\mu}^2_{[n\cdot]}$, respectively, since in the statistic $B_n^{6,0,\gamma}$ the other terms are canceled out.

6.1.2 Change-Point Estimation under Unknown Means and Variances

In this sub-subsection, we consider change-point estimates for the structural breaks in the correlation under the setting where the means and the variances are unknown. For the estimation of the parameters we always use the whole sample. More precisely, we consider the change-point estimates under the assumption that each parameter μ_1, μ_2, σ_1, and σ_2 fulfills Assumption (PEE1) and/or (PEE3). This implies that the notations introduced at the beginning of this subsection will only be used for $l = 1$. Hence, we will drop this index and will use $M_{n,l}$, $I_{\mathbf{i}}^{(l)}$, $b_{\mathbf{i}}^{(l)}$, for $l = 0,1,2$.

Theorem 6.1.3. *Let one of the assumptions on the Theorems 4.1.18 or 4.1.22 and one of the assumptions on the Theorems 5.1.11 or 5.1.14 be fulfilled with $d_{l,i} \equiv 0$. Additionally, let for $l = 1,2$*

$$a_{1,n} + a_{2,n} = o\left(\frac{\underline{\Delta}_{k^*,n}\underline{\Delta}^2_{\rho,n}}{\max_{1\le i\le R}|\Delta_{\rho,i,n}|}\right), \quad a_{1,n}^2 + a_{3,n}^2 = o\left(\underline{\Delta}_{k^*,n}\underline{\Delta}^2_{\rho,n}\right), \tag{6.1.5}$$

$$\max_{1\le r\le R} a_{l,r,n}^{r_{3-l}} \sum_{j\in A_{\mu,l,r,n}} \sum_{i_1\in A_{\sigma,1,r,n}} \sum_{i_2\in A_{\sigma,2,r,n}} (b^{(l)}_{j,i_1,i_2})^{r_{3-l}} (Nn/a_n)^{-(r_{3-l}-1)} = o\left(\frac{\underline{\Delta}^2_{\rho,n}}{\max_r|\Delta_{\rho,r}|}\right), \tag{6.1.6}$$

$$\max_{1\le r\le R} \sum_{i_1\in A_{\sigma,1,r,n}} \sum_{i_2\in A_{\sigma,2,r,n}} \sum_{j_1\in A_{\mu,1,r,n}} \sum_{j_2\in A_{\mu,2,r,n}} b^{(0)}_{j_1,j_2,i_1,i_2} \cdot \frac{\#((k_r^* - \epsilon\underline{\Delta}_{k^*,n}, k_r^* + \epsilon\underline{\Delta}_{k^*,n}] \cap I^{(0)}_{j_1,j_2,i_1,i_2})}{\#((k_r^* - \epsilon\underline{\Delta}_{k^*,n}, k_r^* + \epsilon\underline{\Delta}_{k^*,n}] \cap I^{(0)}_{j_1,j_2,i_1,i_2}) \vee (Nn/a_n)} = o\left(\frac{\underline{\Delta}^2_{\rho,n}}{\max_r|\Delta_{\rho,r}|}\right), \tag{6.1.7}$$

as $n\to\infty$, followed by $N\to\infty$, where for $\xi \in \{\mu,\sigma\}$ and for $\epsilon > 0$

$$a_{1,n} = \sum_{l=1}^{2} m_{l,n}\left(\sum_{\mathbf{j}\in M_{n,l}} (b_{\mathbf{j}}^{(l)})^{r_{3-l}} \# I_{\mathbf{j}}^{(l)}\right)^{\frac{1}{r_{3-l}}}, \quad a_{2,n} = \sum_{\mathbf{j}\in M_{n,0}} b_{\mathbf{j}} \# I_{\mathbf{j}}^{(0)}, \tag{6.1.8}$$

$$a_{3,n} = \max_{k_1\le k_2} \sum_{\mathbf{j}\in M_{n,0}} b_{\mathbf{j}} \frac{\#(I_{\mathbf{j}}^{(0)} \cap [k_1,k_2])}{\sqrt{k_2-k_1+1}}, \tag{6.1.9}$$

$$A_{\xi,l,r,n} = \left\{1 \le j \le m_{\xi,l,n} \ : \ I_{\xi,l,j} \cap (k_r^* - \epsilon\underline{\Delta}_{k^*,n}, k_r^* + \epsilon\underline{\Delta}_{k^*,n}] \neq \emptyset\right\}, \tag{6.1.10}$$

$$a_{l,r,n} = \#A_{\mu,l,r,n} \vee \#A_{\sigma,1,r,n} \vee \#A_{\sigma,2,r,n}, \tag{6.1.11}$$

and

$$m_{\mu,l,n} = 1, \quad I_{\mu,l,1} = (0,n] \quad \text{under the assumptions of Theorem 4.1.18,} \tag{6.1.12}$$

$$m_{\sigma,l,n} = 1, \quad I_{\sigma,l,1} = (0,n] \quad \text{under the assumptions of Theorem 5.1.11.} \tag{6.1.13}$$

Then, it holds

$$\max_{1\le i\le R} a_n|\theta_{\rho,i} - \hat{\theta}_{\rho,i}^{(8+\psi)}| = O_P(1),$$

where $\theta_{\rho,i} = \lim_{n\to\infty} k_i^*/n$ *and* $\hat{\theta}_i^{(8+\psi)} = \hat{k}_i^{(8+\psi)}/n$ *with*

$$(\hat{k}_1^{(8+\psi)},\ldots,\hat{k}_R^{(8+\psi)}) \in \arg\max\left\{Q_n^{(8+\psi)}(k_1,\ldots,k_R)\ :\ 1 = k_0 < \ldots < k_R < k_{R+1} = n\right\}, \tag{6.1.14}$$

$$Q_n^{(8+\psi)}(k_1,\ldots,k_R) = \sum_{r=1}^{R+1}\sum_{i=k_{r-1}+1}^{k_r} (Z_i^{(8+\psi)} - \overline{Z^{(8+\psi)}}_{k_{r-1}}^{k_r})^2. \tag{6.1.15}$$

Proof. Since the settings of Theorems 4.1.22 and 5.1.14 are more general than the other ones, we just consider this case and note that the other combinations can similarly be proven.

Define $\tilde{Z}_i^{(8+\psi)} = Z_i^{(8+\psi)} - \rho_i$ and $a_{1,n,N} = Nn/a_n$. Following the proof of Theorem 2.1.22 yields

$$\begin{aligned}
&\min_{\|k-k^*\|\ge a_{1,n,N}} \frac{Q_n^{(8+\psi)}(k) - Q(k^*)^{(8+\psi)}}{\frac{1}{2}\min_i \Delta_{\rho,i,n}^2 \|k-k^*\| \wedge \underline{\Delta}_{k^*,n}}\\
&\qquad\ge 1 - \max_{\|k-k^*\|\ge a_{1,n,N}} \frac{\sum_{r=1}^{R+1}\sum_{i=k_{r-1}+1}^{k_r}\left[2\left(\rho_i - \overline{\rho}(k_{r-1},k_r)\right)\tilde{Z}_i^{(8+\psi)}\right]}{\frac{1}{2}\min_i \Delta_{\rho,i,n}^2 \|k-k^*\| \wedge \underline{\Delta}_{k^*,n}}\\
&\qquad\quad - \max_{\|k-k^*\|\ge a_{1,n,N}} \frac{\sum_{r=1}^{R+1}(k_r - k_{r-1})\left(\overline{\tilde{Z}^{(8+\psi)}}(k_{r-1},k_r)\right)^2}{\frac{1}{2}\min_i \Delta_{\rho,i,n}^2 \|k-k^*\| \wedge \underline{\Delta}_{k^*,n}}.
\end{aligned}$$

Since $\tilde{Z}_i^{(8+\psi)}$ is a sum of terms considered in the proofs of the Theorems 4.1.22, 5.1.14 plus

$$\tilde{r}_i = \sum_{l=1}^{3} R_{liknZ^{(\theta)}}\left(\frac{\sigma_{1,i}\sigma_{2,i}}{\hat{\sigma}_{1,i}^{(\psi)}\hat{\sigma}_{2,i}^{(\psi)}} - 1\right),$$

it remains to consider both maxima of the partial sums of $\tilde{r}_i$ instead of $\tilde{Z}_i^{(8+\psi)}$ in the first display. Using similar arguments as in proof of Theorem 4.1.22 yields that

$$\max_{\|k-k^*\|\ge a_{1,n,N}} \frac{\sum_{r=1}^{R+1}\sum_{i=k_{r-1}+1}^{k_r}\left[2\left(\rho_i - \overline{\rho}(k_{r-1},k_r)\right)\tilde{r}_i\right]}{0.5\min_i \Delta_{\rho,i,n}^2 \|k-k^*\| \wedge \underline{\Delta}_{k^*,n}} = o_P(1)$$

$$\text{and}\quad \max_{\|k-k^*\|\ge a_{1,n,N}} \frac{\sum_{r=1}^{R+1}(k_r - k_{r-1})\left(\overline{\tilde{r}}(k_{r-1},k_r)\right)^2}{0.5\min_i \Delta_{\rho,i,n}^2 \|k-k^*\| \wedge \underline{\Delta}_{k^*,n}} = o_P(1)$$

as $n\to\infty$ and $N\to\infty$ by using (6.1.5)–(6.1.7).

□

Theorem 6.1.4. *Under the assumptions of Theorem 6.1.3 let*

$$d_n^{(8+\psi)} \ll \beta_n^{(8+\psi)} \le \frac{1}{4C^*}\min_{1\le i\le m} \Delta_{\rho,i,n}^2 \underline{\Delta}_{k^*,n} \tag{6.1.16}$$

with

$$d_n^{(8+\psi)} = d_n^{(\psi_1)} + d_n^{(4+\psi_2)} + a_{1,n}^2 + a_{3,n}^2 + a_{1,n} + a_{2,n}, \tag{6.1.17}$$

where $a_{1,n}$, $a_{2,n}$, *and* $a_{3,n}$ *are defined as in* (6.1.8) *and* (6.1.9). *Then,* $\hat{R}^{(8+\psi)}$ *is a consistent estimate for the number of change-points* R^*.

Proof. Set $\beta_n = \beta_n^{(8+\psi)}$. This proof follows the arguments of the one of Theorem 2.1.25. Hence, it is sufficient to show that the sets $\{\hat{R}^{(8+\psi)} < R^*\}$ and $\{\hat{R}^{(8+\psi)} > R^*\}$ are asymptotically empty. The asymptotic behavior of $\{\hat{R}^{(8+\psi)} < R^*\}$ follows in the same way as the one for $\{\hat{R} < R^*\}$ in the proof of Theorem 2.1.25 with the arguments of Theorem 6.1.3 instead of those of Theorem 2.1.22.

Therefore, we consider $\{\hat{R}^{(8+\psi)} > R^*\}$ and by using the same arguments as in the proof of Theorem 6.1.3 combined with the change-point estimate results of Section 4 and Section 5 we obtain that the lower boundary rate:

$$Q_n^{(8+\psi)}(k) - Q_n^{(8+\psi)}(k^*) \geq -O_P(d_n^{(\psi_1)} + d_n^{(4+\psi_2)} + a_{1,n}^2 + a_{3,n}^2 + a_{1,n} + a_{2,n}).$$

□

Similar to the previous section we can split the variance estimates in $\hat{m}_{2,l,i}$ and $\hat{\mu}_{l,i}^2$. Then, we obtain that $\{\hat{\sigma}_{l,n}^2\} = \{\hat{m}_{2,l,n} - \hat{\mu}_{l,n}^2\}$ fulfills the conditions of Assumptions (PEE3) if $\{\hat{m}_{2,l,n}\}$ and $\{\hat{\mu}_{l,n}^2\}$ fulfill them with $d_{m_l,n} \equiv 0$ and $d_{\mu_l^2,n} \equiv 0$, too.

6.1.3 Long-run Variance Estimation under Unknown Means and Variances

In this sub-subsection, we present some LRV estimates corresponding to the ones in Theorem 6.1.1. Therefore, we have to separate the LRV estimates under the assumptions of Theorem 5.1.1 and under the assumptions of Theorem 5.1.4 or 5.1.9.

Theorem 6.1.5. *Let one of the assumptions of Theorem 4.1.25 or 4.1.26 and the assumptions of Theorem 5.1.16 be fulfilled. Then, it holds for $l = 1, 2$ that*

$$\hat{D}_{8+l,n} = D + O_P(n^{-(\delta_{\sigma,1,n}\wedge\delta_{\sigma,2,n})}) + O_P(n^{-(\delta_{\sigma,1,n}\wedge\delta_{\sigma,2,n})} \vee 1)(\hat{R}_n^{(0)} + \mathbb{1}_{\{l=1\}}\hat{R}_n^{(1)} + \mathbb{1}_{\{l=2\}}o_P(1)),$$

where $\hat{R}_n^{(0)}$ and $\hat{R}_n^{(1)}$ are the corresponding estimation errors of the Theorems 2.1.33 and 4.1.25, as well as

$$\hat{D}_{8+l,n} = \frac{1}{n}\sum_{i=1}^{n}\sum_{j=1}^{n} \mathfrak{f}\left(\frac{i-j}{q_n}\right)(Z_i^{(8+l)} - \tilde{\rho}_i^{(5)})(Z_j^{(8+l)} - \tilde{\rho}_j^{(5)}).$$

Proof. Firstly, for $l = 1, 2$ we obtain that

$$\begin{aligned}
&\frac{1}{n}\sum_{i=1}^{n}\sum_{j=1}^{n} \mathfrak{f}\left(\frac{i-j}{q_n}\right)(Z_i^{(8+l)} - \tilde{\rho}_i^{(5)})(Z_j^{(8+l)} - \tilde{\rho}_j^{(5)})\\
&= \frac{1}{n}\sum_{i=1}^{n}\sum_{j=1}^{n} \frac{\sigma_{1,i}\sigma_{2,i}\sigma_{1,j}\sigma_{2,j}}{\hat{\sigma}_{1,n}^2\hat{\sigma}_{2,n}^2}\mathfrak{f}\left(\frac{i-j}{q_n}\right)(Z_i^{(2l-1)} - \frac{\hat{\sigma}_{1,n}\hat{\sigma}_{2,n}}{\sigma_{1,i}\sigma_{2,i}}\tilde{\rho}_i^{(5)})(Z_j^{(2l-1)} - \frac{\hat{\sigma}_{1,n}\hat{\sigma}_{2,n}}{\sigma_{1,j}\sigma_{2,j}}\tilde{\rho}_j^{(5)})\\
&= \hat{D}_{2l-1,n}\left(1 + O_P(n^{-(\delta_{\sigma,1,n}\wedge\delta_{\sigma,2,n})})\right)
\end{aligned}$$

since $|\frac{\sigma_{1,i}\sigma_{2,i}\sigma_{1,j}\sigma_{2,j}}{\hat{\sigma}_{1,n}^2\hat{\sigma}_{2,n}^2} - 1| = O_P(n^{-(\delta_{\sigma,1,n}\wedge\delta_{\sigma,2,n})})$ and where $\hat{D}_{1,n}$ and $\hat{D}_{2,n}$ are the LRV estimates of Theorem 4.1.25 and 4.1.26, respectively. Applying these theorems implies the claim. □

Remark 6.1.6. *In the previous theorem we used general mean and variance estimates where it is clear that the variance estimates may depend on the mean estimates. In the following theorem we will present LRV estimates under the settings of Theorem 5.1.4 and Theorem 5.1.9, respectively. Therefore, we assume the same variance estimate as introduced in Sub-subsection 5.1.3, whereas we replace the means by some mean estimates.*

Theorem 6.1.7. *Let one of the assumptions of Theorem 4.1.25 or 4.1.26 and one of the assumptions on the Theorems 5.1.18 or 5.1.20 be fulfilled. Then, it holds for $l_1, l_2 = 1; 2$ that $\hat{D}_{8+l_1+2l_2,n} = \hat{D}_0 + \hat{R}_n^{(8+l_1+2l_2)}$ with*

$$\hat{D}_{10+l,n} = \frac{1}{n}\sum_{i=1}^{n}\sum_{j=1}^{n} \mathfrak{f}\left(\frac{i-j}{q_n}\right)(Z_i^{(10+l)} - \tilde{\rho}_i)(Z_j^{(10+l)} - \tilde{\rho}_j)$$

and

$$\hat{R}_n^{(10+l)} = O_P\left(n^{-1} \max_{1\le l_1,l_2\le 13} \sum_{i\in M_{l_1,n}} \sum_{j\in M_{l_2,n}} b_{l_1,i}b_{l_2,j}\left[(\#I_{l_1,i} \vee \#I_{l_2,j}) \wedge q_n\right] \#I_{l_1,i} \wedge \#I_{l_2,j}\right)$$

where

$$(b_{1,j},\dots,b_{12,j}) = \Big(b_{\mu,1j}, b_{\mu,2,j}, b_{\sigma,1,j_1} \vee b_{\sigma,2,j_2}, b_{\sigma,1,j_1}, b_{\sigma,2,j_2}, b_{\mu,1,j_1}b_{\mu,2,j_2},$$

$$b^2_{\mu,1,j}, b^2_{\mu,2,j}, b_{\mu,1j_1}b_{\sigma,1,j_2}, b_{\mu,2,j_1}b_{\sigma,2,j_2}, b^2_{\mu,1,j_1}b_{\sigma,1,j_2}, b^2_{\mu,2,j_1}b_{\sigma,2,j_2}\Big),$$

$$(M_{1,n},\dots,M_{12,n}) = \Big((0,m_{\mu,1,n}], (0,m_{\mu,2,n}], (0,m_{\sigma,1,n}] \times (0,m_{\sigma,2,n}], (0,m_{\sigma,1,n}], (0,m_{\sigma,2,n}],$$

$$(0,m_{\mu,1,n}] \times (0,m_{\mu,2,n}], (0,m_{\mu,1,n}], (0,m_{\mu,2,n}], (0,m_{\mu,1,n}] \times (0,m_{\sigma,1,n}],$$

$$(0,m_{\mu,2,n}] \times (0,m_{\sigma,2,n}], (0,m_{\mu,1,n}] \times (0,m_{\sigma,1,n}], (0,m_{\mu,2,n}] \times (0,m_{\sigma,2,n}]\Big),$$

$$(I_{1,j},\times,I_{12,j}) = (I_{\mu,1,j},\dots,I_{\mu,2,j_1} \cap I_{\sigma,2,j_2}).$$

Proof. Firstly, we obtain that

$$\left(Z_i^{(8+l_1+2l_2)} - \hat{\rho}_i - \frac{(X_i - \hat{\mu}_{1,i}^{(2l_1-1)})^2 - (\hat{\sigma}_{1,i,n}^{(l_2)})^2}{2(\hat{\sigma}_{1,i,n}^{(l_2)})^2}\hat{\rho}_i - \frac{(Y_i - \hat{\mu}_{2,i}^{(2l_1-1)})^2 - (\hat{\sigma}_{2,i,n}^{(l_2)})^2}{2(\hat{\sigma}_{2,i,n}^{(l_2)})^2}\hat{\rho}_i\right)$$
$$= (Z_i^{(0)} - \rho_i) - \frac{1}{2}\rho_i(\epsilon_{1,i}^2 - 1) - \frac{1}{2}\rho_i(\epsilon_{2,i}^2 - 1) + R_i = \tilde{W}_i + R_i$$

with

$$R_i = -\Bigg[(1 - \frac{\sigma_{1,i}\sigma_{2,i}}{\hat{\sigma}_{1,i,n}\hat{\sigma}_{2,i,n}})(Z_i^{(0)} - \rho_i) + (\frac{1}{2} - \frac{\sigma_{1,i}^2}{2\hat{\sigma}_{1,i,n}^2})\rho_i(\epsilon_{1,i}^2 - 1) - (\frac{1}{2} - \frac{\sigma_{2,i}^2}{2\hat{\sigma}_{2,i,n}^2})\rho_i(\epsilon_{2,i}^2 - 1)$$
$$+ \frac{\sigma_{1,i}^2}{2\hat{\sigma}_{1,i,n}^2}(\hat{\rho}_i - \rho_i)(\epsilon_{1,i}^2 - 1) + \frac{\sigma_{2,i}^2}{2\hat{\sigma}_{2,i,n}^2}(\hat{\rho}_i - \rho_i)(\epsilon_{2,i}^2 - 1)$$
$$+ \frac{\sigma_{1,i}^2}{2\hat{\sigma}_{1,i,n}^2}(\hat{\rho}_i - \rho_i) + \frac{\sigma_{2,i}^2}{2\hat{\sigma}_{2,i,n}^2}(\hat{\rho}_i - \rho_i) + \rho_i\left(\frac{\sigma_{1,i}^2}{2\hat{\sigma}_{1,n}^2} + \frac{\sigma_{2,i}^2}{2\hat{\sigma}_{2,i,n}^2} - 1\right)\Bigg]$$
$$+ \Bigg[\sum_{l=1}^{3} R_{liknZ^{(l_1)}} + \sum_{l=1}^{3} R_{liknZ^{(l_1)}}\left(\frac{\sigma_{1,i}\sigma_{2,i}}{\hat{\sigma}_{1,\cdot}\hat{\sigma}_{2,\cdot}} - 1\right)$$
$$+ \sum_{v=1}^{2}\left(\frac{\sigma_{v,i}(\mu_{v,i} - \hat{\mu}_{v,i})\hat{\rho}_i}{\hat{\sigma}_{v,i,n}^2}\epsilon_{v,i} + \frac{(\mu_{v,i} - \hat{\mu}_{v,i})^2\hat{\rho}_i}{2\hat{\sigma}_{v,i,n}^2}\right)\Bigg],$$

where we drop the indices l_1 and l_2. Hence, we get that

$$\hat{D}_{8+l_1+2l_2,n} = \hat{D}_0 + \frac{1}{n}\sum_{i=1}^{n}\sum_{j=1}^{n} \mathfrak{f}\left(\frac{i-j}{q_n}\right)\left[\tilde{W}_i R_j + R_i\tilde{W}_j + R_i R_j\right].$$

Furthermore, we obtain that

$$\frac{1}{n}\sum_{i=1}^{n}\sum_{j=1}^{n} \mathfrak{f}\left(\frac{i-j}{q_n}\right)\left[\tilde{W}_i R_j + R_i\tilde{W}_j + R_i R_j\right]$$
$$= O_P\left(\frac{1}{n}\sum_{i=1}^{n}\sum_{j=1}^{n} \mathfrak{f}\left(\frac{i-j}{q_n}\right)\left[|\tilde{W}_i R_j| + |R_i\tilde{W}_j| + |R_i R_j|\right]\right)$$

We set

$$\begin{aligned}
(b_{1,j}, \ldots, b_{12,(j_1,j_2)}) &= \Big(b_{\mu,1j}, b_{\mu,2,j}, b_{\sigma,1,j_1} \vee b_{\sigma,2,j_2}, b_{\sigma,1,j_1}, b_{\sigma,2,j_2}, b_{\mu,1,j_1} b_{\mu,2,j_2}, \\
&\qquad b^2_{\mu,1,j}, b^2_{\mu,2,j}, b_{\mu,1j_1} b_{\sigma,1,j_2}, b_{\mu,2,j_1} b_{\sigma,2,j_2}, b^2_{\mu,1,j_1} b_{\sigma,1,j_2}, b^2_{\mu,2,j_1} b_{\sigma,2,j_2} \Big), \\
(M_{1,n}, \ldots, M_{12,n}) &= ((0,m_{\mu,1,n}], \ldots, (0,m_{\mu,2,n}] \times (0,m_{\sigma,2,n}]), \\
(I_{1,j}, \ldots, I_{12,(j_1,j_2)}) &= (I_{\mu,1,j}, \ldots, I_{\mu,2,j_1} \cap I_{\sigma,2,j_2}).
\end{aligned}$$

Additionally, we define

$$b_{13,j} = n^{-\delta_1}, \quad M_{13,n} = \{1\}, \quad I_{13,j} = (0,n]$$

under case (A) and

$$b_{13,j} = n^{-\delta_j}, \quad M_{13,n} = (0,m], \quad I_{13,j} = C_j$$

under case (E). Then, we obtain by Markov's inequality that

$$\begin{aligned}
&\frac{1}{n} \sum_{i=1}^{n} \sum_{j=1}^{n} \mathfrak{f}\left(\frac{i-j}{q_n}\right) \left[|\tilde{W}_i R_j| + |R_i \tilde{W}_j| + |R_i R_j| \right] = o_P(1) \\
&\quad + O_P\left(n^{-1} \max_{1 \le l_1, l_2 \le 13} \sum_{i \in M_{l_1,n}} \sum_{j \in M_{l_2,n}} b_{l_1,i} b_{l_2,j} \left[(\#I_{l_1,i} \vee \#I_{l_2,j}) \wedge q_n \right] \#I_{l_1,i} \wedge \#I_{l_2,j} \right).
\end{aligned}$$

The different cases (A) to (H) have no influence since the correlation estimates are bounded by a term of order $O_P(1)$. In the cases of (B), (C) and (D) we add the rates $o_P(q_n n^{-1/2})$, $O_P(q_n n^{-1/2})$ and $O_P(q_n)$, respectively. The same rates are added as well in the cases of (F), (G) and (H). Hence, the claim follows by combining the above rates. □

6.2 Sequential Analysis under General Dependency Framework and General Variance and Mean Estimates

In this subsection, we consider the asymptotic behavior of the stopping times, where the means and variances are unknown. We define for $k = 1, \dots$

$$\hat{\rho}^{n}_{4+\psi_1+4\psi_2,k,1} = \frac{1}{n}\sum_{i=1}^{n} Z^{(4+\psi_1+4\psi_2)}_{i,k,n} \quad \text{and} \quad \hat{\rho}^{n+k}_{4+\psi_1+4\psi_2,k,n+1} = \frac{1}{k}\sum_{i=n+1}^{n+k} Z^{(4+\psi_1+4\psi_2)}_{i,k,n}, \tag{6.2.1}$$

where $\psi_1,\psi_2 = 1,\dots,4$ are design indices for different mean and variance estimate types and

$$Z^{(4+\psi_1+4\psi_2)}_{i,k,n} = \frac{(X_i - \hat{\mu}^{(\psi_1)}_{1,i,k})(Y_i - \hat{\mu}^{(\psi_1)}_{2,i,k})}{\hat{\sigma}^{(\psi_2)}_{1,i,k,n}\hat{\sigma}^{(\psi_2)}_{2,i,k,n}}.$$

Now, we distinguish between the mean and variance estimates as presented in the Subsection 4.2 and Subsection 5.2. To avoid repetition of similar results we use the following notation for $l = 1, 2$

$$\begin{aligned}
m_{l,n} &= m_{\mu,l,1,n} \vee m_{\sigma,l,1,n} \vee m_{\sigma,2,1,n}, \quad M_{n,l} = (0,m_{\mu,l,1,n}] \times (0,m_{\sigma,1,1,n}] \times (0,m_{\sigma,2,1,n}],\\
M_{n,0} &= (0,m_{\mu,1,1,n}] \times (0,m_{\mu,2,1,n}] \times (0,m_{\sigma,1,1,n}] \times (0,m_{\sigma,2,1,n}],\\
I^{(l)}_{\mathbf{i}} &= I_{\mu,l,j,i_1,n} \cap I_{\sigma,1,j,i_2} \cap I_{\sigma,2,j,i_3},, \quad \mathbf{i} \in M_{n,l},\\
I^{(0)}_{\mathbf{i}} &= I_{\mu,1,j,i_1,n} \cap I_{\mu,2,j,i_2,n} \cap I_{\sigma,1,j,i_3} \cap I_{\sigma,2,j,i_4}, \quad \mathbf{i} \in M_{n,0},\\
b^{(l)}_{\mathbf{i}} &= b_{\mu,l,i_1,1,n}(b_{\sigma,1,i_2,1} \vee b_{\sigma,2,i_3,1}), \quad \mathbf{i} \in M_{n,l},\\
b^{(0)}_{\mathbf{i}} &= b_{\mu,1,i_1,1,n} b_{\mu,2,i_2,1,n}(b_{\sigma,1,i_3,1} \vee b_{\sigma,2,i_4,1}), \quad \mathbf{i} \in M_{n,0},
\end{aligned}$$

where we set for each $\xi \in \{\mu,\sigma\}$ and $l = 1, 2$

$$m_{\xi,l,1,n} = \begin{cases} 1, & \text{if } \xi_l \text{ fulfills (PEE5) or (PEE6)},\\ m_{\xi_l,n} + m_{\xi_l,2,n}, & \text{if } \xi_l \text{ fulfills (PEE7)},\\ m_{\xi_l,n}, & \text{if } \xi_l \text{ fulfills (PEE8)}, \end{cases}$$

$$I_{\xi,l,j,1,n} = \begin{cases} (0,n+[nm]], & \text{if } \xi_l \text{ fulfills (PEE5) or (PEE6)},\\ \begin{cases} I_{\xi_l,j}, & j \le m_{\xi_l,n},\\ I_{\xi_l,2,j-m_{\xi_l,n}}, & j > m_{\xi_l,n}, \end{cases} & \text{if } \xi_l \text{ fulfills (PEE7)},\\ I_{x,j}, & \text{if } \xi_l \text{ fulfills (PEE8)}, \end{cases}$$

$$b_{\xi,l,j,1,n} = \begin{cases} 1, & \text{if } \xi_l \text{ fulfills (PEE5) or (PEE6)},\\ \begin{cases} b_{\xi_l,j}, & j \le m_{\xi_l,n},\\ b_{\xi_l,2,j-m_{\xi_l,n}}, & j > m_{\xi_l,n}, \end{cases} & \text{if } \xi_l \text{ fulfills (PEE7)},\\ b_{x,j}, & \text{if } \xi_l \text{ fulfills (PEE8)}. \end{cases}$$

6.2.1 Closed-end Procedure with Unknown Means and Variances

Theorem 6.2.1. *Let one of the assumptions on the Theorems 4.2.1, 4.2.3, 4.2.5, or 4.2.7 and one of the assumptions on the Theorems 5.2.1, 5.2.3, 5.2.5, or 5.2.7 be fulfilled where we assume that* $d_{l,i} \equiv 0$. *Let additionally*

$$m_{l,n}\left(\sum_{\mathbf{i}\in M_{n,l}} (b^{(l)}_{\mathbf{i}})^{r_{3-l}} \# I^{(l)}_{\mathbf{i}}\right)^{1/r_{3-l}} = o(n^{1/2}) \quad \text{and} \quad \sum_{\mathbf{i}\in M_{n,0}} b^{(0)}_{\mathbf{i}} \# I^{(0)}_{\mathbf{i}} = o(n^{1/2}) \tag{6.2.2}$$

as $n \to \infty$ *and for* $l = 1, 2$. *Then, Theorem 2.2.3 holds true if we replace* $\tau^{(c)}_{n,\iota,0,0,\gamma}$ *by* $\tau^{(c)}_{n,\iota,4+l_1+4l_2,0,\gamma}$.

Proof. Firstly, we obtain that

$$\tilde{B}_n^{4+l_1+4l_2,0,\gamma}(k) = \tilde{B}_n^{l_2,0,\gamma}(k) + \frac{n}{n+[n\cdot]}\left(\frac{1}{\sqrt{n}}\sum_{i=n+1}^{n+[n\cdot]} R_{i,[n\cdot]} - \frac{[n\cdot]}{n\sqrt{n}}\sum_{i=1}^{n} R_{i,[n\cdot]}\right)$$

with

$$R_{i,[n\cdot]} = \frac{\sigma_{1,i}\sigma_{2,i}}{\hat{\sigma}_{1,i,[n\cdot]}^{(l_2)}\hat{\sigma}_{2,i,[n\cdot]}^{(l_2)}}\sum_{l=1}^{3} R_{li([n\cdot]+n)nZ^{(l_1)}}.$$

Due to the assumptions of Theorem 4.2.1, 4.2.3, 4.2.5, or 4.2.7, we obtain that

$$\left\|\frac{n}{n+[n\cdot]}\left(\frac{1}{\sqrt{n}}\sum_{i=n+1}^{n+[n\cdot]}\sum_{l=1}^{3} R_{li([n\cdot]+n)nZ^{(l_1)}} - \frac{[n\cdot]}{n\sqrt{n}}\sum_{i=1}^{n}\sum_{l=1}^{3} R_{li([n\cdot]+n)nZ^{(l_1)}}\right)\right\| = o_P(1).$$

Since

$$\begin{aligned}&\sum_{l=1}^{3}\left\|\sum_{i=1}^{n+[n\cdot]}\left(1-\frac{\sigma_{1,i}\sigma_{2,i}}{\hat{\sigma}_{1,i,[n\cdot]}^{(l_2)}\hat{\sigma}_{2,i,[n\cdot]}^{(l_2)}}\right) R_{li([n\cdot]+n)nZ^{(l_1)}}\right\| \\ &= \sum_{l=1}^{2} O_P\left(m_{l,n}\left(\sum_{\mathrm{i}\in M_{n,l}} (b_{\mathrm{i}}^{(l)})^{r_{3-l}}\#I_{\mathrm{i}}^{(l)}\right)^{1/r_{3-l}}\right) + O_P\left(\sum_{\mathrm{i}\in M_{n,0}} b_{\mathrm{i}}^{(0)}\#I_{\mathrm{i}}^{(0)}\right),\end{aligned}$$

which is equal to $o_P(\sqrt{n})$ by (6.2.2), we get by Slutsky's Theorem that

$$\tilde{B}_n^{4+l_1+4l_2,0,\gamma}([n\cdot]) = \tilde{B}_n^{l_2,0,\gamma}([n\cdot]) + o_P(1).$$

Thus, the claim follows from Theorem 5.2.1, 5.2.3, 5.2.5, or 5.2.7. □

6.2.2 Open-end with Unknown Means and Variances

Theorem 6.2.2. *Let one of the assumptions of Theorems 4.2.9, 4.2.10, 4.2.11, or 4.2.12 and one of the assumptions on the Theorems 5.2.9, 5.2.10 , 5.2.11, or 5.2.12 be fulfilled where we assume that* $d_{\mu,l,i} = d_{\sigma,l,i} \equiv 0$. *Let additionally for* $l = 1,2$ *and some* $\lambda' > 0$ *hold that*

$$m_{l,n}\left(\sum_{j=1}^{\infty}\sum_{\mathrm{i}\in M_{n,l}} (b_{\mathrm{i}}^{(l)})^{r_{3-l}}\frac{\#(I_{\mathrm{i}}^{(l)}\cap[1,n+nm2^{i+1}])}{(2^{r_l})^j}\right)^{1/r_{3-l}} = O(n^{1/2}m^{-\lambda'}) \tag{6.2.3}$$

and

$$\sum_{\mathrm{i}\in M_{n,0}} b_{\mathrm{i}}^{(0)}\max_{k>[nm]}\frac{\#I_{\mathrm{i}}^{(0)}\cap[1,n+k]}{k-[nm]} = O(n^{1/2}m^{-\lambda'}). \tag{6.2.4}$$

as $n\to\infty$, *followed by* $m\to\infty$. *Then, Theorem 2.2.5 holds true if we replace* $\tau_{n,\iota,0,0,\gamma}^{(o)}$ *by* $\tau_{n,\iota,4+l_1+4l_2,0,\gamma}^{(o)}$.

Proof. Firstly, we obtain that

$$\begin{aligned}P(\tau_{n,\iota,4+l_1+4l_2,\gamma}^{(o)}<\infty) &= P\left(\sup_{z\in[0,1]} u_\iota((\cdot)^{-\gamma}W(\cdot))(z)\ge c_\alpha\right)+o(1)\\ &+P\left(\max_{1\le k\le mn} u_\iota(\tilde{B}_n^{4+l_1+4l_2,0,\gamma})(k)<c_\alpha, \max_{k>nm} u_\iota(\tilde{B}_n^{4+l_1+4l_2,0,\gamma})(k)\ge c_\alpha\right).\end{aligned}$$

Using the notation of Theorem 6.2.1 we get

$$\tilde{B}_n^{4+l_1+4l_2,0,\gamma}(k) = \tilde{B}_n^{l_2,0,\gamma}(k) + \frac{n}{n+[n\cdot]}\left(\frac{1}{\sqrt{n}}\sum_{i=n+1}^{n+[n\cdot]} R_{i,[n\cdot]} - \frac{[n\cdot]}{n\sqrt{n}}\sum_{i=1}^{n} R_{i,[n\cdot]}\right).$$

Using the arguments of the proof of Theorem 4.2.9, 4.2.10, 4.2.11, or 4.2.12 it remains to show that

$$\begin{aligned}
&\max_{k>mn}\left|\sum_{l=1}^{3}\sum_{i=1}^{n+[n\cdot]}\left(1-\frac{\sigma_{1,i}\sigma_{2,i}}{\hat{\sigma}^{(l_2)}_{1,i,[n\cdot]}\hat{\sigma}^{(l_2)}_{2,i,[n\cdot]}}\right)R_{li([n\cdot]+n)nZ^{(l_1)}}\right| \\
&= \sum_{l=1}^{2} O_P\left(m_{l,n}\left(\sum_{j=1}^{\infty}\sum_{\mathbf{i}\in M_{n,l}}(b_{\mathbf{i}}^{(l)})^{r_{3-l}}\frac{\#(I_{\mathbf{i}}^{(l)}\cap[1,n+nm2^{i+1}])}{(2^{r_l})^j}\right)^{1/r_{3-l}}\right) \\
&\quad + O_P\left(\sum_{\mathbf{i}\in M_{n,0}} b_{\mathbf{i}}^{(0)}\max_{k>[nm]}\frac{\#I_{\mathbf{i}}^{(0)}\cap[1,n+k]}{k-[nm]}\right)
\end{aligned}$$

which is equal to $o_P(\sqrt{n})$ with (6.2.3) and (6.2.4). Thus, we get with the arguments of Theorem 6.2.1 that

$$\begin{aligned}
&P\left(\max_{1\le k\le mn} u_\iota(\tilde{B}_n^{4+l_1+4l_2,0,\gamma})(k) < c_\alpha, \max_{k>nm} u_\iota(\tilde{B}_n^{4+l_1+4l_2,0,\gamma})(k) \ge c_\alpha\right) \\
&= P\left(\sup_{x>m} u_\iota\left(w_\gamma G\mathbb{1}_{[0,m]}(\cdot) - [w_\gamma(\cdot)\frac{\cdot}{\cdot+1}W(1) + \delta N m^{-\lambda}]\mathbb{1}_{\{\cdot\ge m\}}\right)(x)\right. \\
&\quad \left. - \sup_{0\le x\le m} u_\iota(w_\gamma G)(x) \ge \epsilon\right) + o(1)
\end{aligned}$$

as $n\to\infty$, followed by $m\to\infty$, $N\to\infty$, and $\epsilon\to 0$, where $G(\cdot)=\frac{1}{1+\cdot}(W(1+\cdot)-(1+\cdot)W(1))$, $\delta=\operatorname{sign}(W(1))$, and $\lambda\in(0,\lambda'\wedge\frac{1}{2}]$. Here we use the arguments of the proof of the Theorem 5.2.9, Theorem 5.2.10, Theorem 5.2.11, or Theorem 5.2.12 combined with the arguments of the proof of Theorem 2.2.5 the claim follows. □

6.3 Examples

In this subsection, we continue with the two examples (IID2) and (NED2). Thereby, we focus on different parameter settings which we will investigate by Monte–Carlo simulations in the next section. We assume the following model

$$\mu_{l,i} = \mu_{l0} + \sum_{j=1}^{R_{\mu,l,n}} \Delta_{\mu,l,j} \mathbb{1}_{\{i \le k^*_{\mu,l,j}\}} \qquad \text{and} \qquad \sigma^2_{l,i} = \sigma^2_{l0} + \sum_{j=1}^{R_{\sigma,l,n}} \Delta_{\sigma,l,j} \mathbb{1}_{\{i \le k^*_{\sigma,l,j}\}}, \tag{6.3.1}$$

where it holds for $u \in \{\mu,\sigma\}$ that $R_{u,l,n} \ge 0$, $\Delta_{u,l,j} \neq 0$ and $0 = k^*_{u,l,0} < k^*_{u,l,1} < \ldots < k^*_{u,l,R_{u,l,n}} < k^*_{u,l,R_{u,l,n}+1} = N_n$. Here, $N_n \in \{n, n(1+m),\infty\}$ depends on the considered procedure: a posteriori, closed-end, or open-end. In addition, we assume that $k^*_{u,l,i+1} - k^*_{u,l,i} \to \infty$ as $n \to \infty$ for all $i = 1,\ldots,R_{u,l,n} - 1$.

6.3.1 Constant Parameters

In this sub-subsection, we discuss some examples under the assumption that the parameters $\mu_{l,n}$ and $\sigma_{l,n}$ are constant, i.e., $R_{\mu,l,n} \equiv R_{\sigma,l,n} \equiv 0$, such that the main results of this section can be applied.

Using the weighting sample mean and weighting sample variance,

$$\hat{\mu}^{(1)}_{l,n} = n^{-1} \sum_{i=1}^{n} w_{i,n} Z_{l,i} \quad \text{and} \quad (\hat{\sigma}^{(1)}_{l,n})^2 = n^{-1} \sum_{i=1}^{n} w_{i,n} (Z_{l,i} - \hat{\mu}^{(1)}_{l,n})^2 \tag{6.3.2}$$

with deterministic, positive weights $\{w_{i,n}\}$ fulfilling $\sum_{i=1}^{n} w_{i,n} = n$ for all $n \in \mathbb{N}$ and $\sum_{i=1}^{n} w^2_{i,n} \sim n$.

Under (IID) the assumptions of Theorem 4.1.2 and Theorem 5.1.1 are fulfilled, see Sub-subsections 4.3.1 and 5.3.1. In particular, since $\hat{\mu}_l - \mu_l = O_P(n^{-1/2})$ and $\sigma_l - \hat{\sigma}_l = o_P(1)$, the rate assumptions (6.1.3) and (6.1.4) of Theorem 6.1.1 are satisfied. The necessary consistent LRV estimate is given by Theorem 6.1.5, where its assumptions are fulfilled if we additionally provide Assumption (LRV). Thus, we can apply Theorem 6.1.1 and use $\phi^{\gamma}_{l,9,9}$ to test whether the correlation is constant or not. With the same line of arguments we get the same result under Assumption (NED2), see p. 115.

In the next example we consider the sequences of weighted sample mean and of sample variances,

$$\left\{\hat{\mu}^{(2)}_{l,k}\right\} = \left\{k^{-1} \sum_{i=1}^{k} w_{i,k} Z_{l,i}\right\} \quad \text{and} \quad \left\{(\hat{\sigma}^{(2)}_{l,n})^2\right\} = \left\{k^{-1} \sum_{i=1}^{k} (Z_{l,i} - \hat{\mu}^{(2)}_{l,k})^2\right\} \tag{6.3.3}$$

for $k = 1,\ldots,n$, $n \in \mathbb{N}$, where the weights $\{w_{i,k}\}$ fulfill the previous conditions. To avoid that the test statistic $B^{14}_n(z)$ was not defined $(0,1/n]$, since the denominator of

$$\hat{\rho}_{14,[nz]} = \frac{\sum_{i=1}^{[nz]} (X_i - \hat{\mu}^{(2)}_{1,[nz]})(Y_i - \hat{\mu}^{(2)}_{2,[nz]})}{\sqrt{\sum_{i=1}^{k} (Z_{l,i} - \hat{\mu}^{(2)}_{l,[nz]})^2 \sum_{i=1}^{k} (Y_{1,i} - \hat{\mu}^{(2)}_{2,[nz]})^2}}$$

would be zero, we redefine w. l. o. g.

$$B^{14}_n(z) = \begin{cases} 0, & \text{if } z \in [0,2/n], \\ w_\gamma\left(\frac{[n\cdot]}{n}\right) \frac{[n\cdot]}{\sqrt{n}} \left(\hat{\rho}_{14,[\cdot n]} - \hat{\rho}_{14,n}\right), & \text{if } z \in (2/n,1]. \end{cases}$$

Then, under Assumption (IID2) or (NED2) the assumptions of Theorem 4.1.6 and Theorem 5.1.4 are fulfilled. In particular Assumption (PEE2) is fulfilled with $N = N_n = n^{1/2-\epsilon}$, $\epsilon \in (0,\frac{1}{2})$ for each parameter μ_l and σ_l such that $\delta_{\mu,l} = \delta_{\sigma,l} = \frac{1-2\epsilon}{4}$. Thus, (6.1.3) follows by Remark 6.1.2 and (6.1.4) is satisfied by setting

$$I^{(0,2)}_{\mathbf{j}} = \begin{cases} (0,n^{1/2-\epsilon}], & \text{if } \mathbf{j} = (1,1,1,1), \\ (n^{1/2-\epsilon},n], & \text{if } \mathbf{j} = (2,2,2,2), \\ \emptyset, & \text{else} \end{cases} \quad \text{and} \quad b^{(3-l)}_{1,\mathbf{j}} = \begin{cases} 1, & \text{if } \mathbf{j} = (1,1,1), \\ n^{-1/2+\epsilon}, & \text{if } \mathbf{j} = (2,2,2,2), \\ 1, & \text{else.} \end{cases}$$

We get the corresponding consistent LRV estimate $\hat{D}_{10}$ by Theorem 6.1.7 with $\tilde{\rho}_i \equiv \hat{\rho}_{14,n}$ and $q = o(n^{1/2})$ if Assumption (LRV2) is satisfied.

In analogy, we can show under Assumption (IID2) or (NED2) combined with Assumption (LRV) that $\tau^{(c)}_{n,\iota,9,9,\gamma}$ and $\tau^{(o)}_{n,\iota,9,9,\gamma}$ converge as given in Theorem 6.2.1 and Theorem 6.2.2, respectively. If we combine Assumption (IID2) or (NED2) with Assumption (LRV2) we obtain the convergence of $\tau^{(c)}_{n,\iota,14,10,\gamma}$ and $\tau^{(o)}_{n,\iota,14,10,\gamma}$, too.

6.3.2 Non-constant Parameters

In this sub-subsection, we present examples where the time series $\{X_n\}$ and $\{Y_n\}$ possess structural breaks in the means and the variances. Firstly, we consider the case where the change-points are known before we consider the case of unknown change-points.

Known Change-Point in the Parameter Firstly, we consider the estimates

$$\hat{\mu}^{(3')}_{l,i,n} = (k^*_{\mu,l,r} - k^*_{\mu,l,r-1})^{-1} \sum_{k=k^*_{\mu,l,r-1}+1}^{k^*_{\mu,l,r}} w_{i,k_{\mu,l,r}-k_{\mu,l,r-1}} Z_{l,k} \tag{6.3.4}$$

for $i \in (k^*_{\mu,l,r-1}, k^*_{\mu,l,r}]$ and

$$\left\{(\hat{\sigma}^{(4b)}_{l,i,k})^2\right\} = \left\{(k \wedge k^*_{\sigma,l,r} - k^*_{\sigma,l,r-1})^{-1} \sum_{v=k^*_{\sigma,l,r-1}+1}^{k\wedge k^*_{\sigma,l,r}} (Z_{l,v} - \hat{\mu}^{(3')}_{l,i,n})^2\right\} \tag{6.3.5}$$

for $i \in (k^*_{\sigma,l,r-1}, k^*_{\sigma,l,r}]$. We already know from Sub-subsection 4.3.2 that under (IID2) or (NED2) the assumptions of Theorem 4.1.13 are fulfilled if $R_{\mu,l,n} = o(n^{1/4})$. Furthermore, under these assumptions we obtain that

$$\frac{1}{\sqrt{n}} \sum_{i=1}^{[n\cdot]} ((\hat{\sigma}^{(4b)}_{l,i,k})^2 - \sigma^2_{l,i}) = \frac{1}{\sqrt{n}} \sum_{i=1}^{[n\cdot]} ((\hat{\sigma}^{(4')}_{l,i,k})^2 - \sigma^2_{l,i}) + o_P(1),$$

where $(\hat{\sigma}^{(4')}_{l,i,k})^2$ is defined as in Sub-subsection 5.3.2 and differs from $(\hat{\sigma}^{(4b)}_{l,i,k})^2$ by using the exact mean $\mu_{l,i}$. The above equation immediately follows by the arguments used in the proofs of the results of Sub-subsection 4.1.1. Thus, if $R_{\sigma,l,n} = o(n^{1/4})$, we get under Assumption (IID2) or Assumption (NED2) combined with Assumption (LRV2) that Theorem 6.1.1 is available, which follows by Sub-subsections 4.3.2 and 5.3.2.

Next, we consider the example where we replace $\hat{\mu}^{(3)}_{l,i,n}$ by

$$\hat{\mu}^{(4')}_{l,i,k,n} = (k \wedge k^*_{\mu,l,r} - k^*_{\mu,l,r-1})^{-1} \sum_{k=k^*_{\mu,l,r-1}+1}^{k^*_{\mu,l,r}\wedge k} w_{i,k\wedge k^*_{\mu,l,r}-k^*_{\mu,l,r-1}} Z_{l,k}$$
$$\forall i \in (k^*_{\mu,l,r-1}, k \wedge k^*_{\mu,l,r}], k = 2,\ldots,n, r = 1,\ldots,R_l + 1$$

which has also been considered in Sub-subsection 4.3.2. With the same arguments and assumptions as in Sub-subsection 4.3.2 and Sub-subsection 5.3.2 we get the assumed conditions of Theorem 6.1.1 under Assumption (IID2) or Assumption (NED2) combined with Assumption (LRV2).

Unknown Change-Point in the Parameter Now, we suppose that the change-points of the parameters are finite, unknown, and estimated. Furthermore, we suppose that Assumption (IID2) or Assumption (NED2) is fulfilled. Then, using the corresponding settings of Sub-subsection 4.3.2 and 5.3.2, we can replace the above exact parameters by their estimates as given by Theorem 2.1.22 to obtain that the assumptions of Theorem 6.1.1 are fulfilled.

7 Simulation Study

In this section, we present the finite sample behavior of the tests $\phi^{\gamma}_{\iota,\psi,l}$, of the stopping times $\tau^{(c)}_{n,\iota,\psi,l,\gamma}$, and of the change-point estimates $\hat{k}^{(\psi)}$ under known and unknown, constant or non-constant parameters, and under H_0 or some alternatives.

Simulation Setup The results, which will be presented in figures or tables, are based on 1000 repetitions, unless otherwise stated. Therefore, we consider the main model

$$\begin{pmatrix} X_i \\ Y_i \end{pmatrix} = \begin{pmatrix} \mu_{1,i} \\ \mu_{2,i} \end{pmatrix} + \begin{pmatrix} \sigma_{1,i} & 0 \\ 0 & \sigma_{2,i} \end{pmatrix} \begin{pmatrix} 1 & 0 \\ \rho_i & \sqrt{1-\rho_i^2} \end{pmatrix} \begin{pmatrix} \tilde{\epsilon}_{1,i} \\ \tilde{\epsilon}_{2,i} \end{pmatrix}$$

with independent, centered, and normalized sequences $\{\tilde{\epsilon}_{1,n}\}$ and $\{\tilde{\epsilon}_{2,n}\}$. Since the behavior of the test statistics, of the change-point estimates, and of the stopping-times are similar for ρ_1 and ρ_2 with $\rho_1 = -\rho_2$, we just consider the negative correlations. Furthermore, we are talking about a strong or high correlation if $|\rho| \in (0.5,1]$, a moderate or middle correlation if $|\rho| \in (0.3,0.5]$, and a weak or small correlation if $|\rho| \in (0,0.3]$. Moreover, in the following we distinguish between

IID) $\tilde{\epsilon}_{1,1},\ldots,\tilde{\epsilon}_{1,n},\tilde{\epsilon}_{2,1},\ldots,\tilde{\epsilon}_{2,n}$ are i.i.d. N(0,1);

AR) $\{\tilde{\epsilon}_{1,n}\}$ and $\{\tilde{\epsilon}_{2,n}\}$ are independent and both AR(1) processes, i.e.

$$\tilde{\epsilon}_{l,t} = \phi_l \epsilon_{l,t-1} + \xi_{l,t} \quad \text{with } \{\xi_{l,n}\} \text{ i.i.d. N(0,1)}, \ \phi_1 = 0.4, \phi_2 = 0.5;$$

GARCH) $\{\tilde{\epsilon}_{1,n}\}$ and $\{\tilde{\epsilon}_{2,n}\}$ are independent and both GARCH(1,1) processes,

$$\tilde{\epsilon}_{l,t} = \tilde{\sigma}_{l,t}\xi_{l,t} \quad \text{and} \quad \tilde{\sigma}^2_{l,t} = a_{l,0} + a_{l,1}\tilde{\epsilon}^2_{l,t-1} + b_{l,1}\tilde{\sigma}^2_{l,t-1}$$

$$\{\xi_{l,n}\} \text{ i.i.d. N(0,1)}, \ a_{1,0} = 0.8, \ a_{1,1} = 0.15, \ b_{1,1} = 0.05, \ a_{2,0} = 0.75, \ a_{2,1} = 0.15, \ b_{2,1} = 0.1;$$

and between

i) $\mu_{1,i} \equiv \mu_{2,i} \equiv 0$ and $\sigma_{1,i} \equiv \sigma_{2,i} \equiv 1$;

ii) $\mu_{l,i} = \sum_{j=1}^{R_\mu} \Delta_{\mu,l,j} \mathbb{1}_{\{[n\theta_{\mu,l,j-1}] \le i < [n\theta_{\mu,l,j}]\}}$ and $\sigma_{1,i} \equiv \sigma_{2,i} \equiv 1$;

iii) $\sigma_{l,i} = 1 + \sum_{j=1}^{R_\sigma} \Delta_{\sigma,l,j} \mathbb{1}_{\{[n\theta_{\sigma,l,j-1}] \le i < [n\theta_{\sigma,l,j}]\}}$ and $\mu_{1,i} \equiv \mu_{2,i} \equiv 0$;

iv) $\mu_{l,i} = \sum_{j=1}^{R_\mu} \Delta_{\mu,l,j} \mathbb{1}_{\{[n\theta_{\mu,l,j-1}] \le i < [n\theta_{\mu,l,j}]\}}$ and $\sigma_{l,i} = 1 + \sum_{j=1}^{R_\sigma} \Delta_{\sigma,l,j} \mathbb{1}_{\{[n\theta_{\sigma,l,j-1}] \le i < [n\theta_{\sigma,l,j}]\}}$.

In the following sections we will specify ii) to iv) and if we talk about the constant parameter case, we actually mean case i).

7.1 A Posteriori Testing

In this subsection we consider the behavior of the testing procedures $\phi^{\gamma}_{\iota,\psi,l}$ based on the test statistics

$$T_n^{\iota,\psi,l,\gamma} = f_\iota\left(\hat{D}_l^{-1/2} w_\gamma\left(\frac{[n\cdot]}{n}\right) \frac{[n\cdot]}{\sqrt{n}} \left(\hat{\rho}_{\psi,[\cdot n]} - \hat{\rho}_{\psi,n}\right)\right) \quad \text{with} \quad \hat{\rho}_{\psi,k} = \frac{1}{k}\sum_{i=1}^{k} Z^{(\psi)}_{i,k}$$

where

$$Z^{(\psi)}_{i,k} = \begin{cases} \frac{(X_i-\mu_{1,i})\cdot(Y_i-\mu_{2,i})}{\sigma_{1,i}\sigma_{2,i}}, & \text{if } \psi = 0, \\ \frac{(X_i-\hat{\mu}^{(\psi)}_{1,i,k})\cdot(Y_i-\hat{\mu}^{(\psi)}_{2,i,k})}{\sigma_{1,i}\sigma_{2,i}}, & \text{if } \psi = 1,\ldots,4, \\ \frac{(X_i-\mu_{1,i})\cdot(Y_i-\mu_{2,i})}{\hat{\sigma}^{(\psi)}_{1,i,k}\hat{\sigma}^{(\psi)}_{2,i,k}}, & \text{if } \psi = 5,\ldots,8 \end{cases}$$

and

$$Z_{i,k}^{(\psi)} = Z_{i,k}^{(4+\psi_1+4\psi_2)} = \frac{(X_i - \hat{\mu}_{1,i,k}^{(\psi_1)}) \cdot (Y_i - \hat{\mu}_{2,i,k}^{(\psi_1)})}{\hat{\sigma}_{1,i,k}^{(\psi_2)} \hat{\sigma}_{2,i,k}^{(\psi_2)}} \quad \text{for } \psi_1, \psi_2 = 1, \ldots, 4.$$

Firstly, we specify the parameter estimates, then the corresponding LRV estimates, the weighting function w_γ, and finally the detection function f_ι, before we will present the behavior of the test statistics.

Parameter Estimates In the following, we focus on the mean estimates $\hat{\mu}_{l,n}^{(1)}$, $\hat{\mu}_{l,k}^{(2)}$, $\hat{\mu}_{l,i,n}^{(3)}$, and $\hat{\mu}_{l,i,k}^{(4)}$ which are defined in (4.3.2) on page 85 (sample mean), in (4.3.3) on page 85 (cumulative average), in (4.3.5) on page 88 (piecewise sample means), and in (4.3.11) on page 89 (piecewise cumulative averages) with weights $w_{i,k} \equiv 1$, respectively. For the variance we focus on the estimates defined in (5.3.2) on page 115 (sample variance), in (5.3.3) on page 116 (cumulative average of the squares), in (5.3.5) on page 116 (piecewise sample variances), and in (5.3.6) on page 116 (piecewise cumulative averages of the squares) with weights $w_{i,k} \equiv 1$, respectively. Furthermore, we use the exact means for $Z_{i,k}^{(\psi)}$, $\psi = 5, \ldots, 8$, and their corresponding variance estimates. The piecewise estimates depend on change-point estimates for the corresponding change-points in the parameters. In this subsection, we use change-point estimates for the change-points in the parameter types of the form defined in (2.1.33) with $\beta_n = \sqrt{n}$. In particular, in the case of unknown means and variances, $\psi > 8$, the change-point estimates of structural breaks in the variances depend on the change-point estimates of the structural breaks in the means. In addition to these estimates we will use the sliding window estimate type with $h = n^{-3/5}$ for the means which is defined in (4.3.14) in the case of $\psi = *$.

LRV Estimates The LRV estimates depend on the choice of the parameter estimates, i.e., the index l depends on ψ. More precisely, we use

$$\hat{D}_l = n^{-1} \sum_{i=1}^{n} \sum_{j=1}^{n} k\left(\frac{i-j}{q_n}\right) (W_i^{(l)} - \overline{W^{(l)}})(W_j^{(l)} - \overline{W^{(l)}}) \tag{7.1.1}$$

with

$$W_i^{(\psi)} = \left\{ Z_i^{(\psi)}, \text{ if } \psi = 0,1,3,5,9,\ldots,12, \quad W_i^{(\psi)} = \left\{ Z_i^{(\psi-1)}, \text{ if } \psi = 2,4, \right.\right.$$

$$W_i^{(\psi)} = \begin{cases} Z_i^{(\psi-1)} - [(X_i - \mu_{1,i})/(\hat{\sigma}_{1,i}^{(\psi-5)})^2 + (Y_i - \mu_{2,i})/(\hat{\sigma}_{2,n}^{(\psi-5)})^2]/2, & \text{if } l = 6,8, \\ Z_i^{(\psi-4)} - \frac{[(X_i - \hat{\mu}_{1,i}^{(\psi_1)})/(\hat{\sigma}_{1,i}^{(\psi_2-1)})^2 + (Y_i - \hat{\mu}_{2,i}^{(\psi_1)})/(\hat{\sigma}_{2,n}^{(\psi_2-1)})^2]}{2}, & \text{if } \begin{array}{l}\psi = 4 + \psi_1 + 4\psi_2, \\ \psi_2 = 2,4,\end{array} \end{cases}$$

the Bartlett kernel k, and the bandwidth $q_n = \log(n)$. For $\psi = *$ we use $\hat{D}_1$. Thus, some LRV estimates are equal to each other, e.g. $\hat{D}_1 = \hat{D}_2$. Since we use only one LRV estimate for one ψ, we will omit the index l of the test statistic and detector in the following.

Weighting Function Figure 3 shows the weighting function

$$w_\gamma(z) = (z \cdot (1-z))^{-\gamma}, \quad \gamma \in [0, 0.5),$$

which will be used in the following. With this weighting function we can highlight the beginning and the end of the process to detect early and late changes better. $\gamma = 0.5$ is excluded since $\sup_{z \in (0,1)} w_\gamma(z)|B(z)|$ would be infinite almost surely.

Detection Function Later, we will consider the asymptotic size of the detectors. Therefore, we have to specify the function f_ι. In the following, we focus on the three functions

$$f_1(g(\cdot)) = \sup_{z \in [0,1]} |g(z)|, \quad f_2(g(\cdot)) = \left(\int_0^1 |g(z)|^2 dz\right)^{\frac{1}{2}}, \quad f_3(g(\cdot)) = \int_0^1 |g(z)| dz, \tag{7.1.2}$$

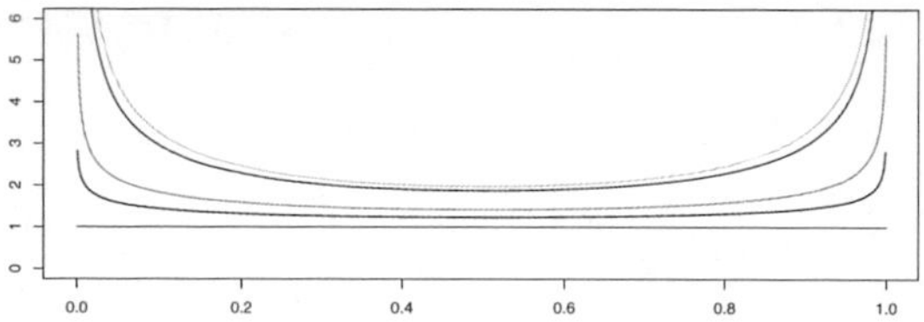

Figure 3: The graph of w_γ with $\gamma = 0$ (black), $\gamma = 0.15$ (red), $\gamma = 0.25$ (green), $\gamma = 0.45$ (blue), and $\gamma = 0.49$ (cyan).

where it holds that $f_1(g) \geq f_2(g) \geq f_3(g)$. Furthermore, the test statistic $\phi_{i_1,\cdot}(X,Y)$ is more sensitive for single outliers than $\phi_{i_2,\cdot}(X,Y)$ for $i_2 > i_1$, $i_1, i_2 \in \{1,2,3\}$.

In the Sections 2, 4, 5, and 6 we have seen that the test statistics converge towards $f_\iota(B_\gamma(\cdot))$ with $B_\gamma(\cdot) = w_\gamma(\cdot)B(\cdot)$, where $B(\cdot)$ is a standard Brownian bridge and $w_\gamma(z) = (z(1-z))^{-\gamma}$. Table 1 gives the approximate critical values such that $P(f_\iota(B_\gamma) > c_{\alpha,\gamma}) = \alpha$.

$1-\alpha$ \ γ	0	0.25	0.45	0.49	0	0.25	0.45	0.49	0	0.25	0.45	0.49
0.9	1.215	1.804	2.678	3.037	0.59	0.886	1.267	1.372	0.499	0.767	1.113	1.205
0.95	1.351	1.982	2.898	3.272	0.68	1.018	1.445	1.558	0.584	0.893	1.289	1.396
0.99	1.616	2.34	3.338	3.751	0.859	1.279	1.808	1.946	0.75	1.149	1.652	1.788

Table 1: Critical values $c_{\alpha,\gamma}$ based on 40000 replications of $f_1(B_\gamma)$, $f_2(B_\gamma)$, and $f_3(B_\gamma)$ (from left to right), where B_γ was approximated on a grid of 10000 equi-spaced points in $[0;1]$.

7.1.1 Influence of the Mean Estimates under H_0

In this subsection we consider the influence of the mean estimates on the test statistics. That there is an influence through the mean estimation becomes obvious in the following IID) example with $\rho_i \equiv 0.5$, $\sigma_{1,i} = \sigma_{2,i} \equiv 1$, and $n = 1000$ presented in Figure 4. Since the LRV is equal to $1 + \rho_0^2$ for each process, we drop this factor. On the one hand, the influence of structural breaks in the mean is small if the structural breaks are only in one time series, on the other hand, it is high if shifts in the means occur simultaneously in both time series, cf. the last row of Figure 4. Furthermore, it is obvious that the two processes using change-point estimates in the means behave like the process which depends on the exact means.

Marginal Distributions To begin with, we look at the constant parameter case i). Figure 5 and Figure 6 show the empirical marginal distribution of a Brownian bridge, $B_n^{0,0}$, $B_n^{1,0}$, $B_n^{2,0}$, $B_n^{3,0}$, $B_n^{4,0}$, and $B_n^{*,0}$ in the cases of IID) and of GARCH), respectively.
Firstly, we see that all the marginal distributions of the test statistics approximate those of the Brownian bridge similarly well. In particular, the test statistics, which are based on change-point estimates for the mean, are nearly the same as the corresponding ones of $B_n^{1,0}$ and $B_n^{2,0}$.
Secondly, we recognize that all the test statistics seem to converge. Moreover, we identify that a high correlation causes that there are more outliers at the beginning and the end of the processes $B_n^{\cdot,0}$ which implies the left-skewed curves. Furthermore, we note that the curves for a positive correlation are symmetric to the plotted one. In particular, the curves are a little right-skewed in cases of high correlation.

In the case of GARCH) the empirical marginal distributions are quite similar to the IID) case. In the case of AR) with $n = 100$ the mean estimates untruly register structural breaks in the mean. As a consequence the detectors $B_n^{1,0}$ and $B_n^{2,0}$ differ from the detectors $B_n^{3,0}$ and $B_n^{4,0}$. In particular,

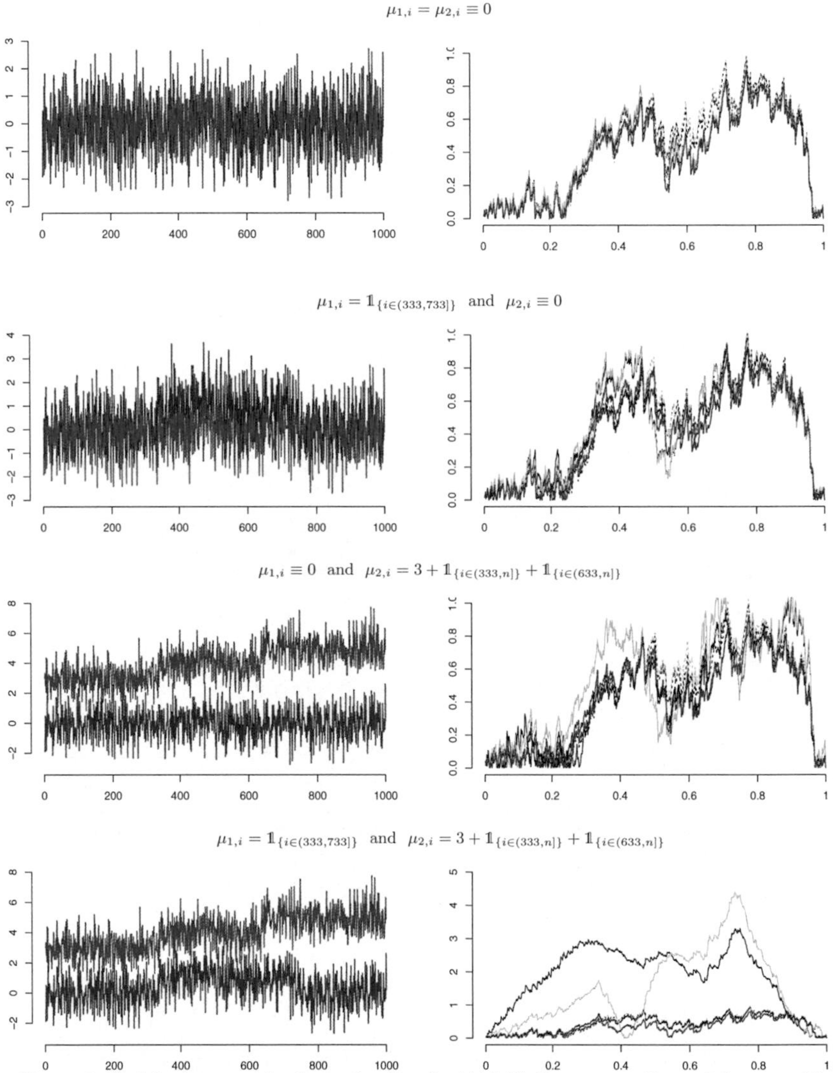

Figure 4: The left column contains the graphs of two fixed i.i.d. N(·,1)–processes with correlation $\rho_0 = 0.5$. The right column contains the processes $D^{-1/2}\frac{k}{\sqrt{n}}|\hat{\rho}_{\psi,k} - \hat{\rho}_{\psi,n}|$ with ψ equal to 0 (red), 1 (black), 2 (gray), 3 (black dotted), 4 (gray dotted), and $*$ (blue).

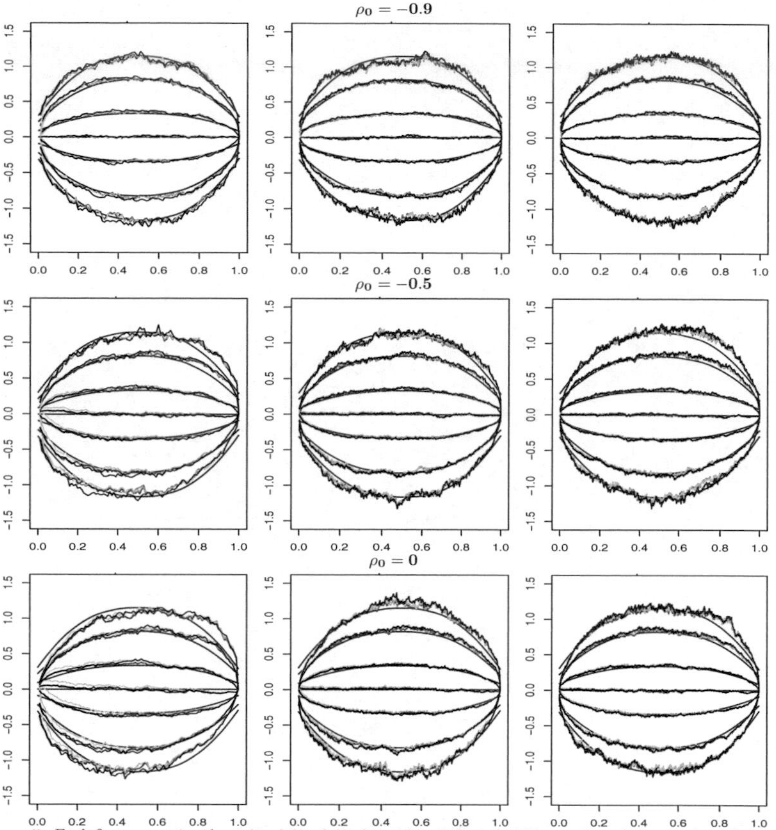

Figure 5: Each figure contains the 0.01, 0.05, 0.25 0.5, 0.75, 0.95 and 0.99-quantiles of the marginal distribution of the Brownian Bridge (red), $B_n^{0,0}$ (orange), $B_n^{1,0}$ (yellow), $B_n^{2,0}$ (green), $B_n^{3,0}$ (blue), $B_n^{4,0}$ (cyan), and $B_n^{*,0}$ (black) on [0,1] calculated by 1000 repetitions with i.i. N(0,1)-distributed innovations. n is taken as 100, 500, and 1000 in the left, middle, and right column. Each Brownian bridge is approximated by a Fourier series with 1000 supporting points and the quantile curve is smoothed by the 'R'-function 'LOESS'. For $B_n^{3,0}$ and $B_n^{4,0}$ we use $\beta_n = \sqrt{n}$. For $B_n^{*,0}$ we use $h = n^{-3/5}$.

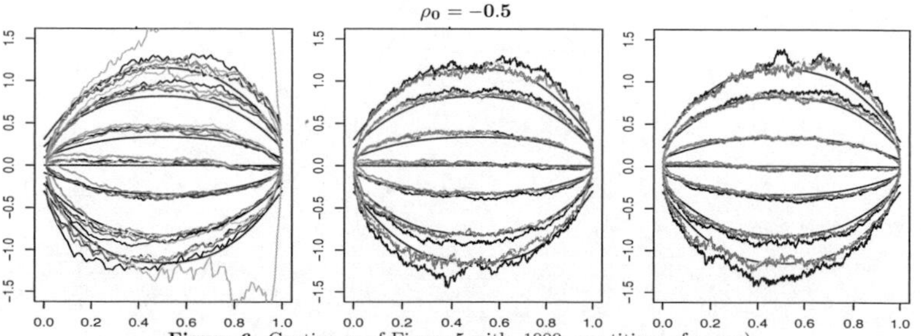

Figure 6: Caption as of Figure 5 with 1000 repetitions of case c).

$B_n^{4,0}$ has many outliers. This is pandered by high positive or high negative correlation. For higher sample sizes like $n = 500$ or $n = 1000$ the change-point detectors for the mean work almost exactly such that $B_n^{1,0}$ and $B_n^{2,0}$ are nearly equal to $B_n^{3,0}$ and $B_n^{4,0}$, respectively.

Now, we consider the cases of non-constant means ii) but known constant variances $\sigma_{l,i} = 1$ under H_0 with the mean settings

$$\mu_{1,i} = \begin{cases} 1, & \text{if } i \le n/4, \\ 0, & \text{else} \end{cases} \qquad \text{and} \qquad \mu_{2,i} \equiv 0, \tag{I$_\mu$}$$

$$\mu_{1,i} = \begin{cases} 1, & \text{if } i \le n/4, \\ 0, & \text{else} \end{cases} \qquad \text{and} \qquad \mu_{2,i} = \begin{cases} 1, & \text{if } i \le n/4, \\ 0, & \text{else,} \end{cases} \tag{II$_\mu$}$$

$$\mu_{1,i} = \begin{cases} 4, & \text{if } i \le n/2, \\ 0, & \text{else} \end{cases} \qquad \text{and} \qquad \mu_{2,i} = \begin{cases} 1, & \text{if } i \le n/4, \\ 0, & \text{else.} \end{cases} \tag{III$_\mu$}$$

We already know from Remark 4.1.5 that the tests based on $B_n^{1,0}$ and $B_n^{2,0}$ are not useful in this

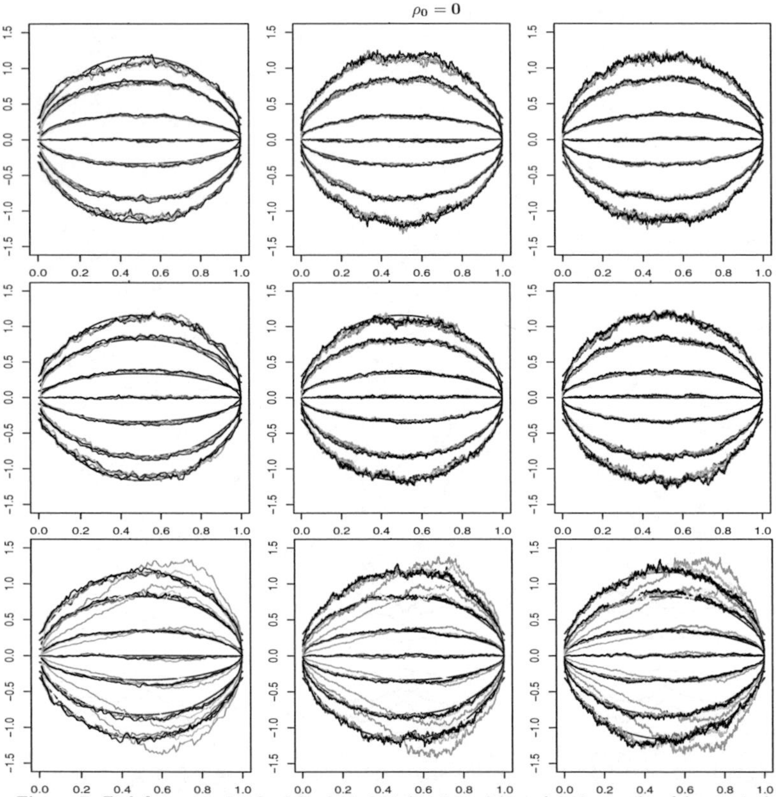

Figure 7: Each figure contains the 0.01, 0.05, 0.25 0.5, 0.75, 0.95 and 0.99-quantiles of the marginal distribution of the Brownian Bridge (red), $B_n^{0,0}$ (orange), $B_n^{1,0}$ (yellow), $B_n^{2,0}$ (green), $B_n^{3,0}$ (blue), $B_n^{4,0}$ (cyan), and $B_n^{*,0}$ (black) are calculated by 1000 repetitions with i.i. N(0,1)-distributed innovations. n is taken as 100, 500, and 1000 in the left, middle, and right column. A column contains the mean models (I$_\mu$), (II$_\mu$), and (III$_\mu$). Each Brownian bridge is approximated by a Fourier series with 1000 supporting points and the quantile curve is smoothed by the 'R'-function 'LOESS'. For $B_n^{3,0}$ and $B_n^{4,0}$ we use $\beta_n = \sqrt{n}$. For $B_n^{*,1,0}$ we use $h = n^{-3/5}$.

setting since the limit depends on the unknown structural breaks in the means. Nevertheless, we will also consider their empirical marginal distribution to verify the influence.
Since the correlation ρ has the same influence on the detectors as under the constant mean setting, we will just consider the case of $\rho_i \equiv 0$. Figure 7 shows that a small structural break in the mean induces nearly no problems in the asymptotic behavior in the case of $n = 1000$. Thus, all quantiles of the marginal distribution of the five processes are near the exact quantiles in the cases displayed in (I_μ) and (II_μ). In cases of high structural breaks in the mean such as in (III_μ) only the processes based on the sliding window estimation or based on the change-point estimation work well, cf. row three of Figure 7.

Empirical Size Since the empirical sizes of the tests in the cases of IID), AR), and GARCH) are quite similar, we will only focus on the case of AR). In this case combined with case i) the empirical size of each detector, with few exceptions, is near the postulated 5%, cf. Table 2. Furthermore, the detectors based on f_1 are slightly conservative for $n = 100$. We get similar results in the cases of IID) and of GARCH). In particular, the value of the constant correlation ρ is independent of the behavior of the test statistics.

		f_1				f_2				f_3			
n \ γ		0	.25	.45	.49	0	.25	.45	.49	0	.25	.45	.49
		$\rho = -0.9$											
100	$\phi^{\gamma}_{\iota,0}$	.031	.03	.037	.03	.047	.043	.041	.036	.053	.048	.036	.031
	$\phi^{\gamma}_{\iota,1}$	.028	.031	.035	.028	.052	.046	.037	.034	.049	.044	.031	.026
	$\phi^{\gamma}_{\iota,6}$	.031	.034	.028	.02	.054	.047	.039	.038	.052	**.05**	.037	.034
	$\phi^{\gamma}_{\iota,23}$	.029	.027	.031	.027	.046	.042	.033	.031	.049	.047	.033	.029
	$\phi^{\gamma}_{\iota,24}$	.034	.038	.034	.026	**.049**	.046	.043	.041	**.05**	.048	.038	.036
	$\phi^{*,\gamma}_{i}$	**.05**	.051	.054	.04	.081	.077	.067	.065	.078	.072	.062	.061
500	$\phi^{\gamma}_{\iota,0}$	.048	.053	.082	.077	.062	.061	.058	.057	.058	.061	.056	.053
	$\phi^{\gamma}_{\iota,1}$	**.051**	.054	.082	.073	.063	.059	.057	**.056**	.06	.06	.054	**.052**
	$\phi^{\gamma}_{\iota,6}$	.054	.054	.064	.057	.06	.063	.057	.056	.062	.062	.055	.053
	$\phi^{\gamma}_{\iota,23}$	.052	.054	.082	.073	.063	.059	.057	.056	.06	.06	.054	**.052**
	$\phi^{\gamma}_{\iota,24}$	.055	.054	.064	.057	.06	.063	.057	.056	.062	.062	.055	.053
	$\phi^{*,\gamma}_{i}$	.074	.088	.109	.09	.086	.085	.087	.087	.082	.083	.079	.076
		$\rho = -0.5$											
100	$\phi^{\gamma}_{\iota,0}$	.028	.028	.029	.024	.055	.051	.04	.039	.056	**.05**	.043	.04
	$\phi^{\gamma}_{\iota,1}$	.027	.023	.029	.022	.053	.051	.042	.039	.052	.048	.042	.041
	$\phi^{\gamma}_{\iota,6}$	.036	.032	.026	.015	.061	.054	.044	.042	.053	**.05**	.044	.04
	$\phi^{\gamma}_{\iota,23}$	.023	.02	.025	.018	.049	**.05**	.036	.034	.044	.041	.034	.034
	$\phi^{\gamma}_{\iota,24}$	.035	.031	.027	.019	**.05**	.046	.038	.035	.046	.044	.036	.032
	$\phi^{*,\gamma}_{i}$	.043	**.045**	.043	.037	.087	.074	.061	.061	.079	.065	.053	.049
500	$\phi^{\gamma}_{\iota,0}$	.055	**.05**	.068	.055	.059	.06	.055	.055	.058	.057	.054	.053
	$\phi^{\gamma}_{\iota,1}$	.054	**.05**	.06	.056	.055	.056	.055	.051	.055	.056	.053	.053
	$\phi^{\gamma}_{\iota,6}$	.057	.053	.053	.038	.054	.054	**.049**	.047	.052	.052	.048	.047
	$\phi^{\gamma}_{\iota,23}$	.054	**.05**	.06	.056	.054	.055	.054	.051	.054	.055	.053	.053
	$\phi^{\gamma}_{\iota,24}$	.057	.052	.053	.038	.053	.053	**.049**	.047	.052	.052	**.049**	.047
	$\phi^{*,\gamma}_{i}$	.079	.086	.099	.087	.084	.089	.086	.082	.083	.085	.078	.076
		$\rho = 0$											
100	$\phi^{\gamma}_{\iota,0}$	.026	.023	.016	.013	.053	.051	.042	.04	.053	**.049**	.043	.04
	$\phi^{\gamma}_{\iota,1}$	.028	.025	.017	.01	.058	**.05**	.04	.037	.052	.048	.04	.038
	$\phi^{\gamma}_{\iota,6}$	.031	.027	.015	.01	.056	.051	.035	.035	.047	.041	.035	.034
	$\phi^{\gamma}_{\iota,23}$	.027	.022	.012	.009	.052	.043	.037	.035	.048	.045	.035	.031
	$\phi^{\gamma}_{\iota,24}$	.031	.029	.018	.011	.051	.045	.034	.032	.044	.04	.031	.029
	$\phi^{*,\gamma}_{i}$	**.032**	.028	.024	.016	.094	.087	.072	.068	.088	.081	.068	.066
500	$\phi^{\gamma}_{\iota,0}$	.048	.047	.053	.048	.055	.057	.056	.054	.055	.057	.053	.051
	$\phi^{\gamma}_{\iota,1}$	.047	.045	.046	.043	.055	.053	.052	.049	.053	.052	.053	.049
	$\phi^{\gamma}_{\iota,6}$	**.049**	.043	.044	.04	.052	.053	.049	.047	.048	.049	.047	.043
	$\phi^{\gamma}_{\iota,23}$	.047	.045	.045	.042	.054	.053	.052	.049	.053	.052	.052	.048
	$\phi^{\gamma}_{\iota,24}$	**.049**	.044	.048	.042	.052	.055	**.05**	.048	.049	**.05**	.048	.044
	$\phi^{*,\gamma}_{i}$	.079	.083	.087	.07	.078	.08	.075	.072	.069	.07	.067	.064

Table 2: Empirical sizes of the detectors in the case of AR) under constant parameters.

Now, we focus on the empirical size under structural breaks in the mean, i.e. under the case ii). Firstly, no detector unjustified rejects H_0 very often in the cases of (I_μ), (II_μ), or (III_μ). Therefore, we just focus on case (III_μ), cf. Table 3. It turns out that the test $\phi^\gamma_{\iota,1}$, which is based on the sample mean, is too conservative except for $\phi^\gamma_{1,1}$ in the case of small correlations. The other tests, and even $\phi^\gamma_{\iota,3}$, work well.

		f_1				f_2				f_3			
n \ γ		0	.25	.45	.49	0	.25	.45	.49	0	.25	.45	.49
100	$\phi^\gamma_{\iota,0}$	.025	.024	.032	.03	.04	.037	.035	.034	.043	.037	.035	.034
	$\phi^\gamma_{\iota,23}$	.026	.024	.028	.025	.038	.036	.03	.03	.04	.036	.03	.029
	$\phi^\gamma_{\iota,24}$	.032	**.033**	.029	.017	**.047**	.041	.035	.034	**.049**	.046	.036	.036
	$\phi^{*,\gamma}_{i}$	.028	.027	.029	.021	.046	.042	.033	.032	.043	.042	.033	.03
1000	$\phi^\gamma_{\iota,0}$	.038	.038	.062	.056	**.051**	**.051**	**.049**	.046	**.05**	.051	.047	.046
	$\phi^\gamma_{\iota,23}$	.043	.041	.065	.069	.053	.054	.048	.047	.056	.057	.051	**.05**
	$\phi^\gamma_{\iota,24}$	.044	**.045**	.061	.059	.058	.057	.052	.053	.058	.057	.053	.052
	$\phi^{*,\gamma}_{i}$	.044	.044	.059	.062	.047	.048	.046	.047	.053	.055	**.05**	**.05**
100	$\phi^\gamma_{\iota,0}$	.024	.024	.023	.018	.04	.036	.031	.031	.04	.036	.03	.029
	$\phi^\gamma_{\iota,23}$	.025	.021	.018	.015	.041	.041	.034	.032	.043	.042	.034	.032
	$\phi^\gamma_{\iota,24}$	.028	**.032**	.028	.022	**.051**	.044	.039	.038	**.049**	.047	.039	.037
	$\phi^{*,\gamma}_{i}$	.02	.021	.016	.012	.039	.037	.033	.033	.042	.039	.036	.032
1000	$\phi^\gamma_{\iota,0}$	.044	.044	.055	.056	.052	.053	.047	.046	**.049**	**.049**	.046	.046
	$\phi^\gamma_{\iota,23}$	.041	.041	.057	.057	.054	.053	**.05**	.045	.055	.057	.053	.052
	$\phi^\gamma_{\iota,24}$	.053	.057	.062	.046	.059	.062	.059	.057	.062	.063	.058	.056
	$\phi^{*,\gamma}_{i}$	**.052**	.046	.06	.057	.056	.058	**.05**	.049	.054	**.051**	**.051**	**.051**
100	$\phi^\gamma_{\iota,0}$	.019	.013	.01	.005	.035	.03	.025	.023	.032	.03	.024	.023
	$\phi^\gamma_{\iota,23}$	.017	.007	.004	.002	.038	.034	.028	.025	.038	.033	.029	.024
	$\phi^\gamma_{\iota,24}$	.023	.018	.01	.007	**.048**	.045	.031	.027	**.047**	.043	.035	.027
	$\phi^{*,\gamma}_{i}$	**.027**	.023	.008	.002	.039	.034	.029	.027	.035	.03	.026	.023
1000	$\phi^\gamma_{\iota,0}$	.044	.045	.045	.038	.047	.044	.041	.041	.043	.046	.042	.041
	$\phi^\gamma_{\iota,23}$	.041	.044	.037	.034	.051	.056	.053	.049	.048	.048	.046	.046
	$\phi^\gamma_{\iota,24}$	**.05**	.054	.045	.04	.063	.063	.058	.056	.06	.06	.057	.054
	$\phi^{*,\gamma}_{i}$	.053	.051	.042	.033	.055	.054	.051	**.05**	.049	**.05**	.048	.048

Table 3: Empirical sizes of the detectors in the case of IID) combined mean setting (III_μ)

7.1.2 Influence of the Variance Estimates under H_0

In this subsection, we consider the influence of the variance estimates on the test statistics. An influence of the variance estimates becomes obvious through Figure 8 which presents an IID) example with $\rho_i \equiv 0.5$, $\mu_{1,i} = \mu_{l,i} \equiv 0$, $n = 1000$, and different variances. As the theoretical results postulate, if the variances are constant, the four processes $B_n^{\psi,0}$ with $\psi = 5,\ldots,8$ behave as $B_n^{0,0}$. In contrast, if the variances change, only $B_n^{8,0}$ is useful and behaves as $B_n^{0,0}$. To get a more precise

$$\sigma_{1,i}^2 = \sigma_{2,i}^2 \equiv 1$$

$$\sigma_{1,i}^2 = 1 + \mathbb{1}_{\{i \in (250,n]\}} - 1.5 \cdot \mathbb{1}_{\{i \in (650,n]\}} \quad \text{and} \quad \sigma_{2,i}^2 \equiv 1$$

$$\sigma_{1,i}^2 \equiv 1 \quad \text{and} \quad \sigma_{2,i}^2 = 1 - 0.5 \cdot \mathbb{1}_{\{i \in (450,n]\}} + 1.5 \cdot \mathbb{1}_{\{i \in (850,n]\}}$$

$$\sigma_{1,i}^2 = 1 + \mathbb{1}_{\{i \in (250,n]\}} - 1.5 \cdot \mathbb{1}_{\{i \in (650,n]\}} \quad \text{and} \quad \sigma_{2,i}^2 = 1 - 0.5 \cdot \mathbb{1}_{\{i \in (450,n]\}} + 1.5 \cdot \mathbb{1}_{\{i \in (850,n]\}}$$

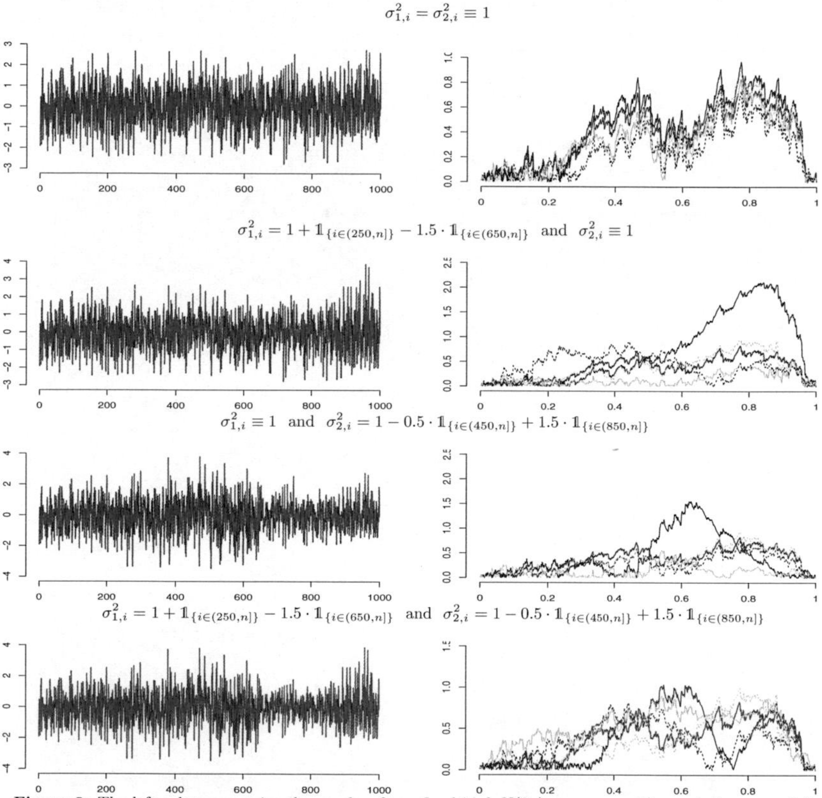

Figure 8: The left column contains the graphs of two fixed i.i.d. N(0,·)-processes with correlation $\rho_0 = 0.5$. The right column contains the processes $D_\psi^{-1/2} \frac{k}{\sqrt{n}} |\hat{\rho}_{\psi,k} - \hat{\rho}_{\psi,n}|$ with ψ equal to 0 (red), 5 (black), 6 (gray), 7 (black dotted), and 8 (gray dotted).

impression on how good the different testing procedures work, we firstly analyze the (asymptotic) marginal distributions of the processes.

Marginal Distribution Firstly, we look at case i). Figure 9 and Figure 10 show the empirical marginal distribution of a Brownian bridge, $B_n^{0,0}$, $B_n^{5,0}$, $B_n^{6,0}$, $B_n^{7,0}$, and $B_n^{8,0}$ in cases of IID) and of GARCH), respectively.
The marginal distributions of $B_n^{6,0}$ and of $B_n^{8,0}$ do not approximate the marginal distribution of the Brownian bridge well in the case of $\rho = -0.9$ at the beginning. In contrast, each approximation works well for $\rho = 0$. In the latter case, the main influence of the variance estimates is asymptotically not visible, cf. Remark 5.1.6.

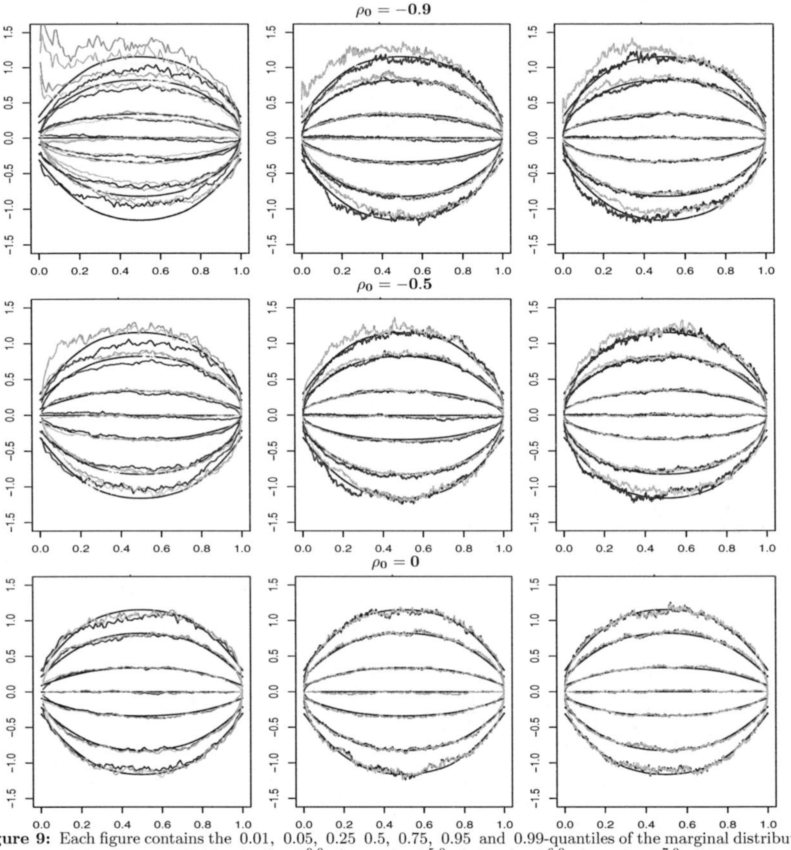

Figure 9: Each figure contains the 0.01, 0.05, 0.25 0.5, 0.75, 0.95 and 0.99-quantiles of the marginal distribution of the Brownian Bridge (red), $B_n^{0,0}$ (orange), $B_n^{5,0}$ (yellow), $B_n^{6,0}$ (green), $B_n^{7,0}$ (blue), and $B_n^{8,0}$ (cyan) on [0,1] calculated by 1000 repetitions with i.i. N(0,1)-distributed innovations. n is taken as 100, 500, and 1000 in the left, middle, and right column. Each Brownian bridge is approximated by a Fourier series with 1000 supporting points and the quantile curve is smoothed by the 'R'-function 'LOESS'.

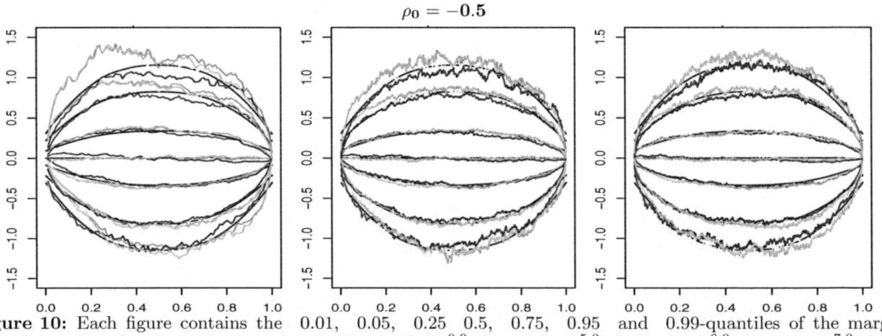

Figure 10: Each figure contains the 0.01, 0.05, 0.25 0.5, 0.75, 0.95 and 0.99-quantiles of the marginal distribution of the Brownian Bridge (red), $B_n^{0,0}$ (orange), $B_n^{5,0}$ (yellow), $B_n^{6,0}$ (green), $B_n^{7,0}$ (blue), and $B_n^{8,0}$ (cyan) on [0,1] calculated by 1000 repetitions of AR-processes defined in the case of c). n is taken as 100, 500, and 1000 in the left, middle, and right column. Each Brownian bridge is approximated by a Fourier series with 1000 supporting points and the quantile curve is smoothed by the 'R'-function 'LOESS'.

Secondly, the convergence of each test statistic seems to take place, even in the case of $\rho = -0.9$. In case GARCH) the empirical marginal distributions is quite similar to case IID). In case AR) we notice a little more fluctuation in the empirical marginal distribution.

Now, we consider case iii) with known constant means $\mu_{l,i} = 0$ but non-constant variances

$$\sigma_{1,i}^2 = \begin{cases} 2, & \text{if } i \le n/2, \\ 1, & \text{else} \end{cases} \quad \text{and} \quad \sigma_{2,i}^2 \equiv 1, \tag{I$_\sigma$}$$

$$\sigma_{1,i}^2 = \begin{cases} 1, & \text{if } i \le n/2, \\ 2, & \text{else} \end{cases} \quad \text{and} \quad \sigma_{2,i}^2 = \begin{cases} 1, & \text{if } i \le n/4, \\ 2, & \text{else,} \end{cases} \tag{II$_\sigma$}$$

or

$$\sigma_{1,i}^2 = \begin{cases} 1, & \text{if } i \le n/2, \\ 2, & \text{else} \end{cases} \quad \text{and} \quad \sigma_{2,i}^2 = \begin{cases} 0.5, & \text{if } i \le n/4, \\ 1, & \text{else.} \end{cases} \tag{III$_\sigma$}$$

Figure 11 shows the empirical marginal distribution of our processes under the structural breaks in the variance. The process $B_n^{6,0}$ is more concentrated in zero than the standard Brownian bridge in cases of (I_σ) and (III_σ). In case of (II_σ) the process $B_n^{6,0}$, (green), possesses a higher fluctuation than the Brownian bridge. The process $B_n^{5,0}$, (yellow), behaves quite similarly to $B_n^{6,0}$. Additionally, before the change in the variance appears, $B_n^{5,0}$ is a little more [less] fluctuated than a standard Brownian bridge in cases of (I_σ) and (III_σ) [in the case of (II_σ)]. The empirical marginal distribution

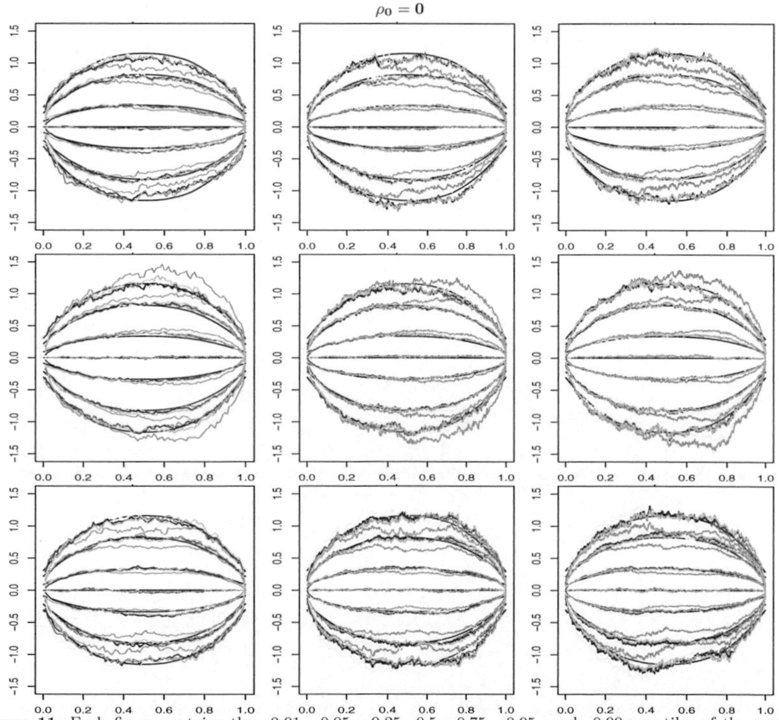

Figure 11: Each figure contains the 0.01, 0.05, 0.25 0.5, 0.75, 0.95 and 0.99-quantiles of the marginal distribution of the Brownian Bridge (red), $B_n^{0,0}$ (orange), $B_n^{5,0}$ (yellow), $B_n^{6,0}$ (green), $B_n^{7,0}$ (blue), and $B_n^{8,0}$ (cyan) on [0,1] calculated by 1000 repetitions with i.i. N(0,1)-distributed innovations. n is taken as 100, 500, and 1000 in the left, middle, and right column. Each Brownian bridge is approximated by a Fourier series with 1000 supporting points and the quantile curve is smoothed by the 'R'-function 'LOESS'.

curves of the processes $B_n^{7,0}$ and $B_n^{8,0}$, which are based on change-point estimates for the variances, run closely along those of the standard Brownian bridge.

Empirical Size Now, we focus on the empirical sample size of the tests which are based on the variance estimates. Firstly, we note that the cases a), b), and c) are quite similar. Thus, we will only focus on the AR-case b) in the following.
$\phi^\gamma_{\iota,0}$ and $\phi^\gamma_{\iota,1}$ are equal, which is no surprise since the variance estimates are canceled by the LRV estimates. Secondly, we note that $\phi^\gamma_{1,6}$ and $\phi^\gamma_{1,8}$ falsely reject H_0 quite often in the case of high correlations and small sample sizes. For $n = 1000$ and $\rho = -0.9$, these two tests only work well for $\gamma = 0$. This behavior has already been suggested by the empirical marginal distribution curves. $\phi^\gamma_{\iota,23}$ is conservative in the case of high correlations or $\rho = 0$ with small sample size. Overall, the tests based on f_2 or f_3 hold a better approximation for the given level of 5%.

n \ γ		f_1: 0	f_1: .25	f_1: .45	f_1: .49	f_2: 0	f_2: .25	f_2: .45	f_2: .49	f_3: 0	f_3: .25	f_3: .45	f_3: .49
		$\rho = -0.9$											
100	$\phi^\gamma_{\iota,0}$	.026	.027	**.035**	.031	**.052**	**.048**	.04	.036	.054	**.05**	.039	.036
	$\phi^\gamma_{\iota,1}$	.026	.027	**.035**	.031	**.052**	**.048**	.04	.036	.054	**.05**	.039	.036
	$\phi^\gamma_{i,6,5}$	.17	.249	.293	.278	.093	.097	.103	.107	.073	.068	.065	.062
	$\phi^\gamma_{\iota,23}$	.006	.007	.006	.006	.01	.006	.006	.004	.009	.008	.004	.003
	$\phi^\gamma_{i,8}$	.195	.275	.308	.294	.087	.088	.094	.098	.075	.072	.059	.058
1000	$\phi^\gamma_{\iota,0}$	.041	**.043**	.063	.068	.046	.045	.04	.039	.045	.045	.043	.039
	$\phi^\gamma_{\iota,1}$	.041	**.043**	.063	.068	.046	.045	.04	.039	.045	.045	.043	.039
	$\phi^\gamma_{i,6,5}$	.06	.103	.253	.264	.058	.057	.051	**.05**	.053	.053	.051	**.05**
	$\phi^\gamma_{\iota,23}$	.026	.026	.038	.042	.028	.026	.026	.025	.028	.028	.026	.024
	$\phi^\gamma_{i,8}$	.06	.105	.25	.263	.056	.057	.053	.053	.052	.052	.048	.048
		$\rho = -0.5$											
100	$\phi^\gamma_{\iota,0}$	.02	.02	.029	.023	.042	.038	.031	.025	.045	.04	.028	.027
	$\phi^\gamma_{\iota,1}$	.02	.02	.029	.023	.042	.038	.031	.025	.045	.04	.028	.027
	$\phi^\gamma_{i,6,5}$	.076	.081	.071	**.056**	.075	.069	.063	.061	.072	.063	.054	.052
	$\phi^\gamma_{\iota,23}$	.009	.008	.011	.008	.022	.019	.016	.013	.018	.017	.015	.014
	$\phi^\gamma_{i,8}$	.058	.057	.059	.039	**.05**	.049	.038	.037	**.049**	.046	.037	.035
1000	$\phi^\gamma_{\iota,0}$	.046	.046	.072	.065	.047	.046	.042	.042	.043	.045	.042	.041
	$\phi^\gamma_{\iota,1}$	.046	.046	.072	.065	.047	.046	.042	.042	.043	.045	.042	.041
	$\phi^\gamma_{i,6,5}$	**.052**	.054	.074	.067	.052	.051	.047	.047	.046	.046	.043	.042
	$\phi^\gamma_{\iota,23}$	.037	.033	.044	.044	.038	.037	.032	.032	.036	.038	.034	.033
	$\phi^\gamma_{i,8}$	.055	.054	.069	.061	**.05**	.052	.049	.049	**.049**	.045	.042	.04
		$\rho = 0$											
100	$\phi^\gamma_{\iota,0}$	.024	.028	.029	.019	.054	.052	.045	.04	.057	.053	.041	.039
	$\phi^\gamma_{\iota,1}$	.024	.028	.029	.019	.054	.052	.045	.04	.057	.053	.041	.039
	$\phi^\gamma_{i,6,5}$	.051	.057	.037	.022	.062	.057	.053	.048	.064	.058	.048	.048
	$\phi^\gamma_{\iota,23}$	.024	.021	.016	.008	**.051**	.047	.038	.037	**.05**	.044	.038	.036
	$\phi^\gamma_{i,8}$	**.05**	.042	.02	.013	.059	**.051**	.041	.038	.052	**.05**	.04	.038
1000	$\phi^\gamma_{\iota,0}$	.054	.055	.054	**.049**	.059	.059	.059	.058	.056	.058	.055	.054
	$\phi^\gamma_{\iota,1}$	.054	.055	.054	**.049**	.059	.059	.059	.058	.056	.058	.055	.054
	$\phi^\gamma_{i,6,5}$	.06	.059	.047	.04	.059	.06	.058	.058	.054	.058	.056	.054
	$\phi^\gamma_{\iota,23}$	.052	.053	**.049**	.039	**.054**	.056	.058	.056	.054	.056	.053	.053
	$\phi^\gamma_{i,8}$	.058	.057	.046	.032	.057	.058	.056	**.054**	.055	.056	.054	**.051**

Table 4: Empirical sizes of the detectors in the AR-case b) combined with case i)

Now, we focus on the behavior of tests under the case of iii) with the settings displayed in (I_σ), (II_σ), and (III_σ). From Section 5, we already know that only the test $\phi^\gamma_{i,8}$ is useful. Again, we remark that the behaviors are similar under cases IID), AR), and GARCH). Hence, we will only analyze the GARCH-case c) in the following, cf. Table 5. The tests $\phi^\gamma_{i,8}$, $i = 1, 2, 3$, are conservative for small correlations ($\rho = 0$), while $\phi^\gamma_{1,8}$ can not hold the given level $\alpha = 0.05$ for high correlations ($\rho = -0.9$). Overall, the test $\phi^\gamma_{i,8}$ works well.

		f_1				f_2				f_3			
n \ γ		0	.25	.45	.49	0	.25	.45	.49	0	.25	.45	.49
						$\rho=-0.9$							
100		.023	.026	.035	.031	.048	.043	.036	.034	.046	.041	.037	.034
	(I_σ)	.215	.312	.332	.31	.091	.097	.102	.104	.073	.074	.064	.063
	(II_σ)	.193	.285	.297	.276	.077	.082	.083	.082	.065	.059	**.05**	.047
	(III_σ)	.185	.275	.296	.282	.084	.088	.101	.098	.068	.067	.062	.06
1000		.039	.054	.076	.076	.054	.056	.056	.055	.057	.057	.054	.053
	(I_σ)	.087	.128	.289	.288	.069	.066	.063	.063	.058	.06	.057	.054
	(II_σ)	.073	.119	.278	.274	.06	.061	.059	.059	.057	.06	.058	.055
	(III_σ)	.079	.131	.279	.277	.065	.067	.066	.063	.063	.062	.061	.059
						$\rho=-0.5$							
100		.025	.025	.029	.026	.046	.04	.034	.032	.043	.042	.037	.036
	(I_σ)	.049	.064	.059	.047	.047	.045	.039	.035	.046	.046	.038	.037
	(II_σ)	.06	.058	.054	.043	.061	.06	.052	.051	.057	.053	.043	.039
	(III_σ)	.077	.09	.089	.078	.056	.053	.051	.049	.057	.051	.044	.042
1000		.041	.043	.065	.06	.049	**.05**	.051	**.05**	.055	.054	.052	.051
	(I_σ)	.048	.057	.079	.076	.061	.058	.056	.054	.054	.058	.053	.051
	(II_σ)	.047	.053	.061	.06	.047	.048	**.05**	.049	**.05**	.052	.052	**.05**
	(III_σ)	.041	.041	**.05**	.057	.046	.047	.046	.045	.047	**.05**	**.05**	.049
						$\rho=0$							
100		.014	.015	.018	.015	.037	.029	.029	.028	.041	.037	.03	.028
	(I_σ)	.031	.027	.013	.007	.039	.035	.031	.029	.038	.036	.028	.028
	(II_σ)	.043	.043	.023	.012	.052	.048	.038	.036	.053	.048	.038	.034
	(III_σ)	.06	.055	.026	.012	.056	.056	.049	.048	.056	.056	**.05**	.046
1000		.039	.049	.062	.052	**.05**	.049	**.05**	.047	.044	.046	.041	.04
	(I_σ)	.043	.054	.06	.039	.058	.056	.052	.049	.045	.045	.04	.041
	(II_σ)	.035	.039	.03	.021	.034	.035	.037	.036	.029	.031	.028	.027
	(III_σ)	.035	.035	.03	.018	.037	.037	.035	.036	.036	.036	.035	.033

Table 5: Empirical sizes in the GARCH-case c) in the settings (I_σ), (II_σ), and (III_σ) of the detectors $\phi^\gamma_{\iota,0}$ and $\phi^\gamma_{i,8}$. The empirical sizes of $\phi^\gamma_{\iota,0}$ is in the first row of each block and is independent of the settings.

7.1.3 Influence of the Combination of Mean and Variance Estimates under H_0

In this sub-subsection, we consider the test statistics which use the mean and variance estimates $B^{9,\gamma}_n$, $B^{14,\gamma}_n$, $B^{23,\gamma}_n$, and $B^{24,\gamma}_n$ and compare them with the test statistic $B^{0,\gamma}_n$ based on the known parameters. The marginal distributions behave as in the sections before. If the means are constant, the marginal distribution curves are alike the ones in the cases of Sub-subsection 7.1.2. If structural breaks in the means or variances occur, the same impact as described in Sub-subsection 7.1.1 and Sub-subsection 7.1.2 shows up.

Now, we consider the empirical size. Table 6 shows that the tests $\phi^\gamma_{2,\cdot}$ and $\phi^\gamma_{3,\cdot}$ work well with the exception of $\phi^\gamma_{1,\cdot}$ in the case of high correlations ($\rho=-0.9$) and a large γ. In case of high correlation only $\phi^\gamma_{1,9}$ can hold the given level of $\alpha=0.05$ for all $\gamma\in\{0,0.25,0.45,0.49\}$. If additionally the sample size is high ($n=1000$), then every $\phi^0_{1,\cdot}$ works well. Overall, it seems that the tests $\phi^\gamma_{2,\psi}$ and $\phi^\gamma_{3,\psi}$ hold the empirical size of 5% better than $\phi^\gamma_{1,\psi}$.

Now, we allow for structural breaks in the parameters. Table 7 presents the empirical sizes under H_0 of the non-useful tests $\phi^\gamma_{\iota,9}$ and $\phi^\gamma_{\iota,14}$ in the case of non-constant parameters with the settings (II_μ) and (II_σ). It is confirmed that the tests are non-useful since they often untruly reject the null hypothesis.

In the following, we focus on the empirical sizes of the tests $\phi^\gamma_{\iota,23}$ and $\phi^\gamma_{\iota,24}$, which respect changes in the means and variances by change-point estimates. Tables 8–11 present their empirical sizes under the mean setting (II_μ) combined with constant variances or variances fulfilling (I_σ)–(III_σ). Comparing their empirical sizes with the constant parameter setting, cf. Table 6, the structural breaks in the means and variances are sufficiently well estimated such that the sizes are merely a little higher. Additionally, we assert that in most cases the empirical sizes of $\phi^\gamma_{\iota,23}$ are a little lower than the ones of $\phi^\gamma_{\iota,24}$. However, we can summarize that both test types work well.

n \ γ		f_1				f_2				f_3			
		0	.25	.45	.49	0	.25	.45	.49	0	.25	.45	.49
						$\rho = -0.9$							
100	$\phi^{\gamma}_{\iota,0}$	.029	.029	.063	.06	.043	.041	.037	.037	.042	.04	.034	.033
	$\phi^{\gamma}_{\iota,9}$	.027	.032	.056	**.049**	**.047**	.044	.037	.037	.04	.041	.034	.03
	$\phi^{\gamma}_{\iota,14}$	.125	.241	.345	.337	.092	.099	.109	.109	.079	.077	.073	.072
	$\phi^{\gamma}_{\iota,23}$	.1	.15	.255	.271	.066	.07	.074	.076	.056	.058	.058	**.052**
	$\phi^{\gamma}_{\iota,24}$	.168	.271	.368	.35	.102	.108	.122	.124	.087	.087	.086	.082
1000	$\phi^{\gamma}_{\iota,0}$	.055	.057	.077	.079	**.052**	.053	**.052**	**.052**	.055	.056	.053	**.051**
	$\phi^{\gamma}_{\iota,9}$	**.054**	.057	.077	.081	.056	.057	.055	.057	.053	.058	.057	.055
	$\phi^{\gamma}_{\iota,14}$	.077	.081	.321	.342	.074	.078	.069	.067	.072	.07	.065	.062
	$\phi^{\gamma}_{\iota,23}$	.076	.086	.161	.203	.071	.076	.069	.067	.066	.066	.06	.058
	$\phi^{\gamma}_{\iota,24}$	.081	.085	.317	.331	.077	.078	.071	.067	.069	.072	.065	.062
						$\rho = -0.5$							
100	$\phi^{\gamma}_{\iota,0}$	.028	.035	.056	**.049**	.048	**.051**	.046	.043	.042	.041	.036	.035
	$\phi^{\gamma}_{\iota,9}$	.027	.038	.052	.046	.042	.043	.04	.036	.042	.04	.032	.032
	$\phi^{\gamma}_{\iota,14}$	.091	.116	.174	.169	.098	.098	.092	.087	.091	.094	.083	.079
	$\phi^{\gamma}_{\iota,23}$	.057	.069	.069	.065	.062	.057	**.051**	**.051**	.055	.055	.048	**.046**
	$\phi^{\gamma}_{\iota,24}$	.087	.119	.159	.149	.085	.088	.076	.073	.08	.077	.067	.062
1000	$\phi^{\gamma}_{\iota,0}$	.056	.06	.079	.079	.061	.063	.057	.056	.051	**.05**	**.05**	.044
	$\phi^{\gamma}_{\iota,9}$	**.05**	.06	.078	.074	.059	.056	.054	**.053**	**.05**	.052	.046	.043
	$\phi^{\gamma}_{\iota,14}$	.067	.063	.076	.081	.065	.065	.059	.058	.061	.063	.061	.06
	$\phi^{\gamma}_{\iota,23}$	.069	.061	.073	.071	.062	.063	.061	.062	.06	.06	.057	.057
	$\phi^{\gamma}_{\iota,24}$	.07	.064	.077	.076	.067	.067	.062	.06	.062	.063	.061	.059
						$\rho = 0$							
100	$\phi^{\gamma}_{\iota,0}$	.019	.023	.041	.034	.057	.053	.048	.044	.048	**.049**	.046	.045
	$\phi^{\gamma}_{\iota,9}$	.023	.025	.038	.032	.052	.055	**.049**	.044	.054	.051	.047	.042
	$\phi^{\gamma}_{\iota,14}$	.068	.075	.062	.042	.078	.081	.084	.082	.086	.086	.077	.073
	$\phi^{\gamma}_{\iota,23}$	.04	.041	.038	.023	.061	.063	.058	**.051**	.066	.066	.059	.056
	$\phi^{\gamma}_{\iota,24}$	**.054**	.062	.055	.04	.077	.077	.066	.063	.071	.072	.067	.062
1000	$\phi^{\gamma}_{\iota,0}$	**.05**	**.05**	.07	.063	.053	.052	.048	.045	.048	.048	.048	.045
	$\phi^{\gamma}_{\iota,9}$	.047	.047	.067	.065	.052	.053	.051	.048	**.05**	.052	.049	.047
	$\phi^{\gamma}_{i,14}$	.048	**.05**	.054	.047	.051	**.05**	.047	.046	.052	.051	**.05**	.048
	$\phi^{\gamma}_{\iota,23}$	.049	.044	.047	.037	**.05**	.048	.047	.043	.052	.053	.049	.048
	$\phi^{\gamma}_{\iota,24}$	**.05**	.048	.048	.039	.048	.048	.043	.043	.051	.052	.049	.046

Table 6: Empirical sizes of the detectors in the AR-case b) with constant parameters.

n \ γ		f_1				f_2				f_3			
		0	.25	.45	.49	0	.25	.45	.49	0	.25	.45	.49
						$\rho = -0.5$							
100	$\phi^{\gamma}_{\iota,0}$	.018	.025	**.048**	.037	.056	**.052**	.047	.045	.053	**.05**	.042	.038
	$\phi^{\gamma}_{\iota,9}$	.135	.127	.075	.054	.301	.282	.257	.245	.279	.268	.245	.235
	$\phi^{\gamma}_{\iota,14}$	.275	.279	.231	.181	.221	.219	.2	.197	.188	.18	.16	.154
1000	$\phi^{\gamma}_{\iota,0}$	.06	**.058**	.075	.072	.055	.062	**.059**	.059	.055	.056	.055	**.054**
	$\phi^{\gamma}_{\iota,9}$	.997	.997	.994	.989	.997	.997	.996	.996	.997	.997	.997	.996
	$\phi^{\gamma}_{\iota,14}$	.778	.784	.699	.616	.689	.702	.697	.698	.666	.669	.663	.658

Table 7: Empirical sizes of the detectors in the AR-case b) with mean and variance setting (II_μ) and (II_σ)

n \ γ		f_1				f_2				f_3			
		0	.25	.45	.49	0	.25	.45	.49	0	.25	.45	.49
						$\rho = -0.75$							
100	$\phi^{\gamma}_{\iota,0}$	.026	.033	.067	**.066**	**.05**	.047	.046	.046	**.051**	.048	.046	.044
	$\phi^{\gamma}_{\iota,23}$	.143	.172	.207	.189	.119	.111	.101	.102	.098	.094	.089	.084
	$\phi^{\gamma}_{\iota,24}$	.211	.255	.32	.299	.153	.143	.136	.135	.123	.119	.109	.108
1000	$\phi^{\gamma}_{\iota,0}$	**.051**	.054	.092	.098	**.051**	**.051**	.048	.045	**.047**	.046	.046	.044
	$\phi^{\gamma}_{\iota,23}$	.09	.097	.157	.187	.085	.081	.077	.074	.072	.075	.072	.069
	$\phi^{\gamma}_{\iota,24}$	.097	.103	.31	.324	.081	.082	.078	.077	.07	.073	.068	.067
						$\rho = -0.5$							
100	$\phi^{\gamma}_{\iota,0}$	.034	.036	.061	.06	.059	.058	.057	.055	.069	.065	.057	**.053**
	$\phi^{\gamma}_{\iota,23}$	.091	.086	.064	**.048**	.083	.078	.066	.064	.072	.069	.059	.055
	$\phi^{\gamma}_{\iota,24}$	.118	.124	.126	.107	.102	.1	.096	.094	.086	.081	.072	.069
1000	$\phi^{\gamma}_{\iota,0}$	.053	**.05**	.074	.088	.055	.055	.055	**.054**	.055	.058	.057	**.053**
	$\phi^{\gamma}_{\iota,23}$	.068	.068	.068	.067	.065	.063	.062	.06	.064	.065	.061	.059
	$\phi^{\gamma}_{\iota,24}$	.067	.069	.082	.077	.068	.069	.062	.062	.061	.061	.059	.058
						$\rho = 0$							
100	$\phi^{\gamma}_{\iota,0}$	.037	.03	.039	.027	.053	.052	**.051**	.047	**.052**	.054	.045	.042
	$\phi^{\gamma}_{\iota,23}$	**.057**	.055	.035	.023	.07	.069	.063	.06	.069	.066	.058	.057
	$\phi^{\gamma}_{\iota,24}$	.082	.085	**.057**	.04	.088	.09	.086	.083	.08	.08	.079	.077
1000	$\phi^{\gamma}_{\iota,0}$	.056	.053	.067	.067	**.057**	.059	.058	.058	.061	.061	.059	**.057**
	$\phi^{\gamma}_{\iota,23}$	.057	.055	.043	.036	.062	.061	.065	.064	.061	.064	.061	.059
	$\phi^{\gamma}_{\iota,24}$	.06	.056	**.047**	.037	.063	.066	.067	.065	.062	.063	.063	.06

Table 8: Empirical sizes of the detectors in the AR-case b) with mean setting (II_μ)

n \ γ		f_1				f_2				f_3			
		0	.25	.45	.49	0	.25	.45	.49	0	.25	.45	.49
						$\rho = -0.75$							
100	$\phi^{\gamma}_{\iota,0}$	.021	.033	.06	**.053**	**.05**	.052	.054	**.052**	.053	.054	**.052**	.046
	$\phi^{\gamma}_{\iota,23}$	.11	.137	.196	.194	.083	.079	.075	.075	.068	.067	.065	.062
	$\phi^{\gamma}_{\iota,24}$	.182	.239	.339	.324	.107	.11	.112	.113	.09	.088	.08	.077
1000	$\phi^{\gamma}_{\iota,0}$	.047	**.048**	.09	.09	**.05**	.052	.049	.048	.049	**.05**	.049	.047
	$\phi^{\gamma}_{\iota,23}$	.083	.084	.152	.181	.064	.065	.061	.061	.059	.059	.055	.053
	$\phi^{\gamma}_{\iota,24}$	.084	.085	.301	.33	.061	.06	.059	.058	.056	.054	.055	.053
						$\rho = -0.5$							
100	$\phi^{\gamma}_{\iota,0}$	.026	.038	.057	**.046**	.053	.052	.048	.047	.052	**.051**	.045	.042
	$\phi^{\gamma}_{\iota,23}$	.063	.061	.07	.058	.063	.063	.052	**.05**	.063	.059	.053	**.049**
	$\phi^{\gamma}_{\iota,24}$	.087	.098	.132	.123	.071	.071	.065	.062	.069	.067	.063	.056
1000	$\phi^{\gamma}_{\iota,0}$	.039	.047	.081	.084	.043	.046	.044	.044	.043	.045	.045	.043
	$\phi^{\gamma}_{\iota,23}$	**.052**	.053	.071	.065	.06	.057	**.052**	**.052**	.055	.055	**.05**	.049
	$\phi^{\gamma}_{\iota,24}$	**.055**	.057	.075	.077	.059	.058	.053	.053	.055	.056	.051	**.05**
						$\rho = 0$							
100	$\phi^{\gamma}_{\iota,0}$	.018	.025	.029	.026	**.051**	**.049**	.04	.039	**.05**	.051	.04	.037
	$\phi^{\gamma}_{\iota,23}$	.037	.043	.03	.019	.06	.06	.052	**.049**	.057	.056	.047	.042
	$\phi^{\gamma}_{\iota,24}$	**.05**	.055	.043	.03	.076	.072	.065	.061	.072	.068	.059	.054
1000	$\phi^{\gamma}_{\iota,0}$	.035	.036	.054	.063	.047	.048	.042	.043	.047	.046	.046	.042
	$\phi^{\gamma}_{\iota,23}$	.045	.045	**.049**	.037	.047	.052	.051	.048	.049	.052	**.05**	.047
	$\phi^{\gamma}_{\iota,24}$	.046	**.049**	.048	.035	.051	.051	.052	**.05**	.051	.055	**.05**	**.05**

Table 9: Empirical sizes of the detectors in the AR-case b) with mean and variance setting (II_μ) and (I_σ)

n \ γ		f_1				f_2				f_3			
		0	.25	.45	.49	0	.25	.45	.49	0	.25	.45	.49
						$\rho = -0.75$							
100	$\phi^{\gamma}_{\iota,0}$	.027	.04	.056	**.047**	.052	**.05**	.047	.045	**.05**	.047	.045	.045
	$\phi^{\gamma}_{\iota,23}$	.152	.168	.199	.183	.115	.107	.096	.098	.092	.088	.078	.076
	$\phi^{\gamma}_{\iota,24}$	.231	.281	.329	.311	.129	.134	.133	.132	.107	.106	.095	.091
1000	$\phi^{\gamma}_{\iota,0}$	.058	**.055**	.083	.086	**.051**	.052	**.051**	**.049**	.047	**.051**	**.049**	.047
	$\phi^{\gamma}_{\iota,23}$	.094	.085	.136	.165	.074	.077	.077	.073	.073	.074	.07	.067
	$\phi^{\gamma}_{\iota,24}$	.092	.095	.298	.319	.074	.077	.076	.074	.068	.069	.067	.063
						$\rho = -0.5$							
100	$\phi^{\gamma}_{\iota,0}$	.018	.025	**.048**	.037	.056	**.052**	.047	.045	.053	**.05**	.042	.038
	$\phi^{\gamma}_{\iota,23}$	.085	.089	.072	.059	.075	.069	.068	.062	.06	.059	.057	.052
	$\phi^{\gamma}_{\iota,24}$	.121	.122	.139	.116	.081	.082	.076	.076	.076	.077	.069	.067
1000	$\phi^{\gamma}_{\iota,0}$	.06	**.058**	.075	.072	**.055**	.062	.059	.059	.055	.056	.055	**.054**
	$\phi^{\gamma}_{\iota,23}$	.077	.075	.081	.069	.072	.069	.063	.061	.066	.065	.061	.061
	$\phi^{\gamma}_{\iota,24}$	.08	.085	.079	.08	.069	.068	.065	.064	.066	.066	.064	.06
						$\rho = 0$							
100	$\phi^{\gamma}_{\iota,0}$	.024	.022	.026	.025	.038	.039	.036	.033	.039	.039	.034	.031
	$\phi^{\gamma}_{\iota,23}$	.058	**.051**	.024	.013	.06	.057	.054	**.052**	.056	**.052**	.047	.045
	$\phi^{\gamma}_{\iota,24}$	.088	.079	.052	.034	.077	.079	.069	.067	.069	.07	.064	.061
1000	$\phi^{\gamma}_{\iota,0}$	.067	.063	.072	.068	**.069**	.076	.075	.072	.069	.071	.067	**.063**
	$\phi^{\gamma}_{\iota,23}$	.063	.067	**.051**	.039	.075	.08	.073	.072	.068	.071	.068	.066
	$\phi^{\gamma}_{\iota,24}$	.077	.071	.055	.046	.079	.078	.073	.072	.066	.07	.068	.065

Table 10: Empirical sizes of the detectors in the AR-case b) with mean and variance setting (II_μ) and (II_σ)

n \ γ		f_1				f_2				f_3			
		0	.25	.45	.49	0	.25	.45	.49	0	.25	.45	.49
						$\rho = -0.75$							
100	$\phi^{\gamma}_{\iota,0}$	.02	.025	**.051**	.045	**.051**	.045	.046	.044	.054	.055	.043	.038
	$\phi^{\gamma}_{\iota,23}$	.165	.178	.218	.19	.11	.108	.099	.098	.088	.085	.078	.073
	$\phi^{\gamma}_{\iota,24}$	.232	.28	.322	.295	.124	.123	.111	.109	.098	.092	.079	.076
1000	$\phi^{\gamma}_{\iota,0}$	**.049**	.054	.084	.096	**.05**	.052	.045	.045	.046	**.049**	.048	.046
	$\phi^{\gamma}_{\iota,23}$	.082	.091	.153	.184	.073	.072	.064	.064	.061	.061	.059	.057
	$\phi^{\gamma}_{\iota,24}$	.081	.087	.302	.323	.076	.071	.06	.059	.065	.066	.061	.059
						$\rho = -0.5$							
100	$\phi^{\gamma}_{\iota,0}$	.024	.023	.037	.033	.055	.056	.053	**.049**	.056	.054	**.051**	**.05**
	$\phi^{\gamma}_{\iota,23}$	.091	.092	.077	**.057**	.086	.081	.07	.065	.07	.064	.054	.054
	$\phi^{\gamma}_{\iota,24}$	.125	.132	.137	.118	.101	.094	.083	.081	.078	.076	.072	.068
1000	$\phi^{\gamma}_{\iota,0}$	.034	.041	.077	.08	**.051**	**.051**	.048	.047	.051	.054	.048	.044
	$\phi^{\gamma}_{\iota,23}$	**.053**	.055	.062	.067	.056	.053	.057	.055	.054	.053	.051	.051
	$\phi^{\gamma}_{\iota,24}$	.054	.062	.074	.07	.057	.053	.055	.055	**.05**	.054	.052	.051
						$\rho = 0$							
100	$\phi^{\gamma}_{\iota,0}$	.028	.028	.035	.026	.059	.059	.056	**.054**	.057	.057	**.051**	**.049**
	$\phi^{\gamma}_{\iota,23}$	.055	**.054**	.036	.015	.067	.064	.061	.059	.064	.062	.054	.052
	$\phi^{\gamma}_{\iota,24}$	.086	.082	.054	.033	.084	.083	.075	.074	.074	.074	.065	.062
1000	$\phi^{\gamma}_{\iota,0}$	.061	.056	.077	.074	.068	.071	.067	.066	.067	.062	.062	**.059**
	$\phi^{\gamma}_{\iota,23}$	.059	.057	.057	.046	.064	.066	.062	**.061**	.061	.063	.061	**.059**
	$\phi^{\gamma}_{\iota,24}$	.059	.058	.058	**.048**	.064	.066	.066	.063	.063	.065	.061	**.059**

Table 11: Empirical sizes of the detectors in the AR-case b) with mean and variance setting (II_μ) and (III_σ).

7.1.4 Asymptotic Power

In this subsection, we present the asymptotic power of the test statistics. In section 2, we have seen that the limit of the detectors is a mapping of a Brownian bridge plus a deterministic function under $\mathbf{H_{LA}}$. Under the local alternative of an AMOC model with a change-point at $[n\theta]$, $\theta \in (0,1)$, and a change size of $n^{-1/2}\Delta \neq 0$ we obtain that the deterministic functions are of the following form

$$h_\theta(x) = \Sigma_0^{-1/2}\Delta\left((x-\theta)^+ - x(1-\theta)\right) \quad \text{with} \quad \Sigma_0^{-1/2} > 0, \tag{7.1.3}$$

which are plotted in Figure 12 for $\Sigma_0 = 1$.

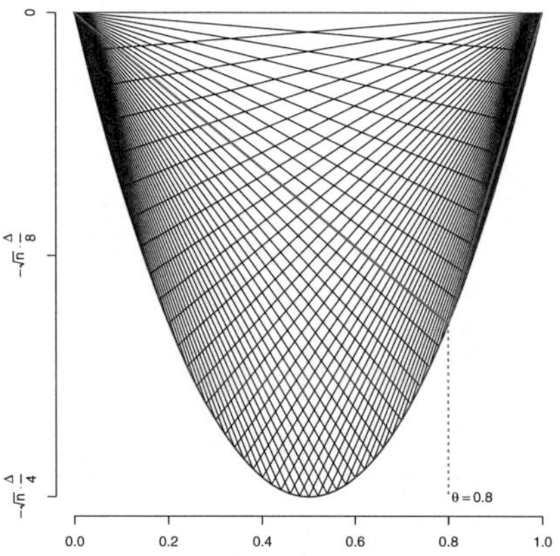

Figure 12: The family of functions h_θ as defined in (7.1.3) with $\Sigma_0 = 1$ and $\Delta > 0$.

In a local, epidemic change-point setting with change-points at θ_1, θ_2, and change size $n^{-1/2}\Delta$ we obtain a deterministic function

$$h_{(\theta_1,\theta_2)}(x) = \Sigma_0^{-1/2}\Delta\left((x \wedge \theta_2 - \theta_1)^+ - x(\theta_2-\theta_1)\right) \quad \text{with} \quad \Sigma_0^{-1/2} > 0. \tag{7.1.4}$$

This family is plotted in Figure 13 for $\Sigma_0 = 1$.

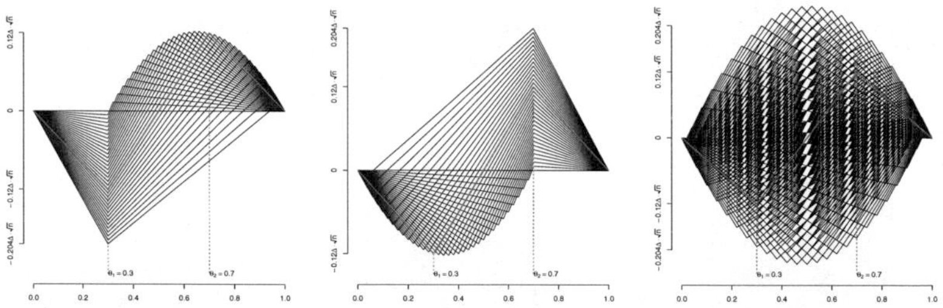

Figure 13: The family of functions $h_{(\theta_1,\theta_2)}$ as defined in (7.1.4), where the left figure shows a part of the family for a fixed $\theta_1 = 0.3$ and the middle one for a fixed $\theta_2 = 0.7$.

Finally, Figure 14 presents three examples for gradual change-point settings. All the previous alternatives have in common that the deterministic functions h_θ, $h_{(\theta_1,\theta_2)}$, and h_g are close to zero in the neighborhood of zero and of one, which is due to the detector's construction. While the maximum of

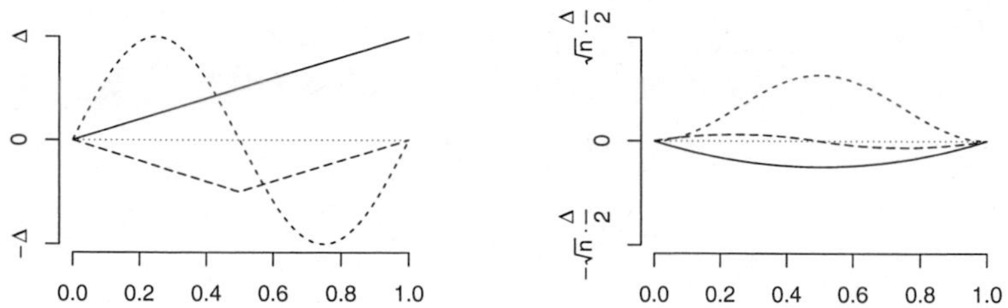

Figure 14: Three examples for gradual-change functions with zero line (dotted): $g_1(x) = \Delta_\rho x$ (solid), $g_2(x) = \Delta_\rho \sin(2\pi x)$ (dashed), and $g_3(x) = \Delta_\rho(x\mathbb{1}_{\{x\le 0.5\}} + (1-x)\mathbb{1}_{\{x\ge 0.5\}})$ (longdash) on the left-hand side and the corresponding graphs of h_g on the right-hand side.

the absolute value of the deterministic function is at a change-point in an AMOC (epidemic) model, there are different possibilities in a gradual change-point setting.

To imagine the curves of the empirical marginal distributions under alternatives, we just have to add the curves of the deterministic function h pointwisely. This implies that under certain alternatives certain procedures detect changes better or worse. For example, the detectors based on $B_n^{6,0}$ or $B_n^{8,0}$ can detect early changes for $\Delta_\rho < 0$ and $\rho_0 = -0.9$ better than late changes, while $B_n^{7,0}$ can detect late changes better than early ones for $\Delta_\rho < 0$ and $\rho_0 = -0.9$. Now, we focus on the empirical power and consider the following change-settings

$$\rho_i = \rho_0 + \Delta_\rho \mathbb{1}_{\{i > [n\theta_\rho]\}}, \tag{7.1.5}$$

with

$$\boxed{\Delta_\rho \in \{0.1, 0.25, 0.5, 0.75\} \text{ and } \theta_\rho \in \{0.1, 0.25, 0.5\}.}$$

Furthermore, we will only discuss the IID) case, since the power similarly behaves under the other dependency cases. Moreover, we already note here that the power of each test increases for the change size and the sample size, respectively. Additionally, the difference between some procedures is small such that we will only present the best and the worst performance of the tests using the different parameter estimates. More precisely, for each θ_ρ and $\iota = 1, 2, 3$ we plot for each Δ_ρ the maximum and the minimum of the empirical sizes of a group of tests. This group consists of tests which use different weighting factors γ and different mean, or variance estimates.

Tests based on Unknown Means In the case of constant parameters with $\theta = 0.5$, the power of the best and worst test slightly differs with the exception of the tests based on f_1, cf. Figure 15. In particular, each of the six tests based on f_2 and f_3 possesses almost the same power as the best test based on f_1. Furthermore, for $n = 1000$ each test has power one if the change size Δ_ρ is at least 0.5. Moreover, the tests using a small γ provide a higher power than the ones using a large γ.

In the case of early changes such as for $\theta = 0.1$, cf. Figure 17, the best tests are based on f_1 and a high γ such as $\gamma \in \{0.45, 0.49\}$. Additionally, the difference is bigger between the powers of the tests. Moreover, the test based on f_1 performs worst if it uses $\hat{\mu}_{l,k}^{(2)}$ as mean estimators. Not quite so clear, we can glean the worst mean estimate for the power if the tests depend on f_2 or f_3, but the tests based on the mean estimates $\hat{\mu}_{l,k}^{(2)}$ or the sliding window appear frequently under the worst power.

Finally, we focus on the power under structural breaks in the mean. The test using the exact means possesses the best power, cf. Figure 18. In general the structural breaks in the mean do not influence the power of the best and worst test.

$n\backslash\Delta_\rho$	0.1	.25	.5	.75	0.1	.25	.5	.75	0.1	.25	.5	.75
1000	ϕ_1^*	$\phi_{1,0}^0$	$\phi_{1,0}^0$	$\phi_{1,0}^0$	$\phi_{1,0}^0$	$\phi_{1,0}^0$	$\phi_{1,0}^0$	$\phi_{1,0}^0$	$\phi_{1,1}^0$	$\phi_{1,0}^0$	$\phi_{1,0}^0$	$\phi_{1,0}^0$
	$\phi_{1,2}^{0.49}$	$\phi_{1,2}^{0.49}$	$\phi_{1,0}^0$	$\phi_{1,0}^0$	$\phi_{1,2}^{0.49}$	$\phi_{1,2}^{0.49}$	$\phi_{1,0}^0$	$\phi_{1,0}^0$	$\phi_{1,2}^{0.49}$	$\phi_1^{*,0.49}$	$\phi_{1,0}^0$	$\phi_{1,0}^0$
1000	$\phi_2^{*,0}$	$\phi_{2,1}^0$	$\phi_{2,0}^0$	$\phi_{2,0}^0$	$\phi_{2,0}^0$	$\phi_{2,0}^0$	$\phi_{2,0}^0$	$\phi_{2,0}^0$	$\phi_{2,0}^0$	$\phi_{2,1}^0$	$\phi_{2,0}^0$	$\phi_{2,0}^0$
	$\phi_{2,2}^{0.49}$	$\phi_{2,2}^{0.49}$	$\phi_{2,0}^0$	$\phi_{2,0}^0$	$\phi_{2,2}^{0.49}$	$\phi_2^{*,0.49}$	$\phi_{2,0}^0$	$\phi_{2,0}^0$	$\phi_2^{*,0.49}$	$\phi_2^{*,0.49}$	$\phi_{2,0}^0$	$\phi_{2,0}^0$
1000	$\phi_3^{*,0.25}$	$\phi_{3,0}^0$	$\phi_{3,0}^0$	$\phi_{3,0}^0$	$\phi_3^{*,0.25}$	$\phi_{3,0}^0$	$\phi_{3,0}^0$	$\phi_{3,0}^0$	$\phi_{3,0}^{0.25}$	$\phi_3^{*,0}$	$\phi_{3,0}^0$	$\phi_{3,0}^0$
	$\phi_{3,2}^{0.49}$	$\phi_{3,2}^{0.49}$	$\phi_{3,0}^0$	$\phi_{3,0}^0$	$\phi_{3,2}^{0.49}$	$\phi_3^{*,0.49}$	$\phi_{3,0}^0$	$\phi_{3,0}^0$	$\phi_3^{*,0.49}$	$\phi_3^{*,0.49}$	$\phi_{3,0}^0$	$\phi_{3,0}^0$
200	$\phi_1^{*,0.45}$	ϕ_1^*	$\phi_{1,0}^0$	$\phi_{1,0}^0$	$\phi_{1,1}^{0.45}$	$\phi_{1,1}^0$	$\phi_{1,0}^0$	$\phi_{1,1}^0$	ϕ_1^*	$\phi_{1,0}^0$	$\phi_{1,0}^0$	$\phi_{1,0}^0$
	$\phi_{1,2}^{0.49}$	$\phi_{1,2}^{0.49}$	$\phi_{1,2}^{0.49}$	$\phi_{1,2}^{0.49}$	$\phi_{1,4}^{0.49}$	$\phi_{1,2}^{0.49}$	$\phi_{1,2}^{0.49}$	$\phi_1^{*,0.49}$	$\phi_{1,2}^{0.49}$	$\phi_{1,2}^{0.49}$	$\phi_{1,1}^{0.49}$	$\phi_1^{*,0.49}$
200	$\phi_2^{*,0}$	$\phi_2^{*,0}$	$\phi_{2,0}^0$	$\phi_{2,0}^0$	$\phi_2^{*,0}$	$\phi_2^{*,0}$	$\phi_{2,0}^0$	$\phi_{2,0}^0$	$\phi_2^{*,0}$	$\phi_2^{*,0}$	$\phi_{2,0}^0$	$\phi_{2,0}^0$
	$\phi_{2,2}^{0.49}$	$\phi_{2,2}^{0.49}$	$\phi_{2,2}^{0.49}$	$\phi_2^{*,0.49}$	$\phi_{2,2}^{0.49}$	$\phi_{2,2}^{0.49}$	$\phi_{2,2}^{0.49}$	$\phi_2^{*,0.49}$	$\phi_{2,0}^{0.49}$	$\phi_2^{*,0.49}$	$\phi_{2,1}^{0.49}$	$\phi_2^{*,0.49}$
200	$\phi_3^{*,0}$	$\phi_3^{*,0}$	$\phi_{3,0}^0$	$\phi_{3,0}^0$	$\phi_3^{*,0}$	$\phi_3^{*,0}$	$\phi_{3,0}^0$	$\phi_{3,0}^0$	$\phi_3^{*,0}$	$\phi_{3,0}^0$	$\phi_{3,2}^0$	$\phi_{3,0}^0$
	$\phi_{3,2}^{0.49}$	$\phi_{3,2}^{0.49}$	$\phi_{3,2}^{0.49}$	$\phi_3^{*,0.49}$	$\phi_{3,2}^{0.49}$	$\phi_{3,2}^{0.49}$	$\phi_{3,2}^{0.49}$	$\phi_3^{*,0.49}$	$\phi_{3,0}^{0.49}$	$\phi_3^{*,0.49}$	$\phi_3^{*,0.49}$	$\phi_3^{*,0.49}$

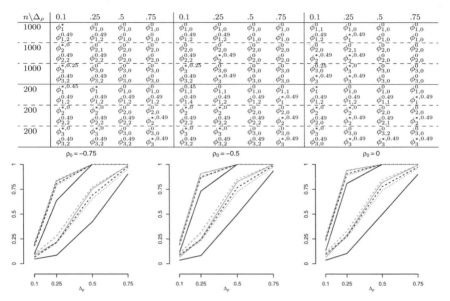

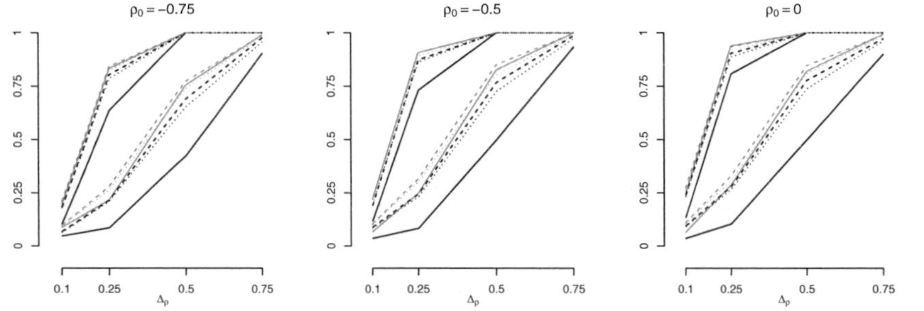

Figure 15: Best (green) and worst power (red) of $\phi_{\iota,0}^\gamma$, $\phi_{\iota,1}^\gamma$, $\phi_{\iota,2}^\gamma$, $\phi_{\iota,3}^\gamma$, $\phi_{\iota,4}^\gamma$, and $\phi_\iota^{*,\gamma}$ and for $\gamma \in \{0,0.25,0.45,0.49\}$ in the setting i) with $\theta = \mathbf{0.5}$. Here, we consider the best and worst power for each $\iota = 1$ (solid), $\iota = 2$ (dashed), and $\iota = 3$ (dotted), as well as for $n = 200$ and $n = 1000$, where the upper three green and the upper three red lines are the ones generated with $n = 1000$. Moreover, ρ_0 is equal to -0.9, -0.5, and 0 (from left to right). The overlying table contains the corresponding best and worst tests belonging to three figures (from left to right) for each n, ι, and $\Delta_\rho = 0.1, 0.25, 0.5, 0.75$. In each of these variations the upper row contains the best test and the lower one presents the worst of the 6x4 tests.

Tests based on Variance Estimates and Known Means In this paragraph, we consider the empirical powers of $\phi_{\iota,5}$, $\phi_{\iota,6}$, $\phi_{\iota,7}$, and $\phi_{\iota,8}$ under the case of constant variances as well as $\phi_{\iota,7}$ and $\phi_{\iota,8}$ in the case of structural breaks in the variances. In both cases we compare the powers with those of the test $\phi_{\iota,0}$, which uses the exact variances.

In case of constant variances, the power of the best test is higher than the best power in the corresponding case of known variances and unknown means, while the worst power stays nearly unchanged. Additionally, the test $\phi_{\iota,6}$ is often the one with the highest power. In the case of $n = 1000$, each test has power one if the change size is equal to 0.5 or 0.75.

In case of structural breaks in the variances, setting (II_σ), each test has power one if the change size is equal to 0.5 or 0.75 and $n = 1000$. Furthermore, if the change size is smaller, the test $\phi_{\iota,8}^\gamma$ possesses the best power for a suitable γ and ι. Again, the influence of the structural break in the parameter is not observable.

Tests Based on Mean and Variance Estimates In this paragraph, we present the powers of the tests $\phi_{\iota,23}^\gamma$ and $\phi_{\iota,24,10}^\gamma$ compared with $\phi_{\iota,0}^\gamma$. The powers of the $\phi_{\iota,9,9}^\gamma$, $\phi_{\iota,9,14}^\gamma$, $\phi_{\iota,23}^\gamma$, and $\phi_{\iota,24,10}^\gamma$ behave similarly as the tests described in the previous paragraph within the setting of constant parameters. Thus, we will focus on the powers of the tests under structural breaks in the mean and the variance, cf. Figure 21 and Figure 22. As in the paragraph before, the influence of the structural breaks are not really observable, since the means and variance estimates use change-point estimators. Hence, the properties of the tests recur.

$n\backslash\Delta_\rho$	0.1	.25	.5	.75	0.1	.25	.5	.75	0.1	.25	.5	.75
1000	$\phi_1^{*,0.45}$	$\phi_{1,0}^{0.25}$	$\phi_{1,0}^{0}$	$\phi_{1,0}^{0}$	$\phi_1^{*,0.45}$	$\phi_1^{*,0.25}$	$\phi_{1,0}^{0}$	$\phi_{1,0}^{0}$	$\phi_1^{*,0.25}$	$\phi_{1,1}^{0.25}$	$\phi_{1,0}^{0}$	$\phi_{1,0}^{0}$
	$\phi_{1,2}^{0}$	$\phi_{1,2}^{0.49}$	$\phi_1^{*,0.49}$	$\phi_{1,0}^{0}$	$\phi_{1,2}^{0}$	$\phi_{1,2}^{0.49}$	$\phi_1^{*,0.49}$	$\phi_{1,0}^{0}$	$\phi_{1,2}^{0.49}$	$\phi_{1,0}^{0.49}$	$\phi_{1,0}^{0}$	$\phi_{1,0}^{0}$
1000	$\phi_2^{*,0.45}$	$\phi_{2,0}^{0.25}$	$\phi_{2,0}^{0}$	$\phi_{2,0}^{0}$	$\phi_2^{*,0.25}$	$\phi_{2,1}^{0.45}$	$\phi_{2,0}^{0}$	$\phi_{2,0}^{0}$	$\phi_2^{*,0.25}$	$\phi_{2,2}^{0.45}$	$\phi_{2,0}^{0}$	$\phi_{2,0}^{0}$
	$\phi_{2,2}^{0.49}$	$\phi_{2,2}^{0}$	$\phi_2^{*,0}$	$\phi_{2,0}^{0}$	$\phi_{2,2}^{0}$	$\phi_{2,2}^{0}$	$\phi_2^{*,0}$	$\phi_{2,0}^{0}$	$\phi_{2,0}^{0}$	$\phi_2^{*,0}$	$\phi_{2,0}^{0}$	$\phi_{2,0}^{0}$
1000	$\phi_{3,1}^{0}$	$\phi_{3,0}^{0.25}$	$\phi_{3,0}^{0}$	$\phi_{3,0}^{0}$	$\phi_3^{*,0.25}$	$\phi_{3,0}^{0.45}$	$\phi_{3,0}^{0}$	$\phi_{3,0}^{0}$	$\phi_{3,1}^{0.25}$	$\phi_{3,0}^{0.25}$	$\phi_{3,0}^{0}$	$\phi_{3,0}^{0}$
	$\phi_{3,2}^{0.49}$	$\phi_3^{*,0}$	$\phi_{3,0}^{0}$	$\phi_{3,0}^{0}$	$\phi_{3,2}^{0.49}$	$\phi_{3,2}^{0}$	$\phi_3^{*,0.49}$	$\phi_{3,0}^{0}$	$\phi_{3,0}^{0.49}$	$\phi_{3,1}^{0}$	$\phi_{3,0}^{0}$	$\phi_{3,0}^{0}$
200	$\phi_1^{*,0.45}$	$\phi_1^{*,0.45}$	$\phi_{1,1}^{0.25}$	$\phi_{1,0}^{0.25}$	$\phi_{1,0}^{0.45}$	$\phi_1^{*,0.25}$	$\phi_{1,0}^{0.25}$	$\phi_{1,0}^{0.25}$	$\phi_1^{*,0.45}$	$\phi_1^{*,0.25}$	$\phi_1^{*,0.25}$	$\phi_{1,2}^{0.25}$
	$\phi_{1,2}^{0}$	$\phi_{1,2}^{0.49}$	$\phi_{1,2}^{0.49}$	$\phi_{1,2}^{0.49}$	$\phi_{1,2}^{0.49}$	$\phi_{1,2}^{0.49}$	$\phi_{1,2}^{0.49}$	$\phi_{1,2}^{0.49}$	$\phi_{1,2}^{0.49}$	$\phi_{1,2}^{0.49}$	$\phi_{1,0}^{0.49}$	$\phi_1^{*,0.49}$
200	$\phi_2^{*,0.25}$	$\phi_2^{*,0.25}$	$\phi_{2,0}^{0.25}$	$\phi_{2,0}^{0.25}$	$\phi_2^{*,0.25}$	$\phi_{2,0}^{0.25}$	$\phi_{2,0}^{0.25}$	$\phi_{2,0}^{0.45}$	$\phi_2^{*,0.25}$	$\phi_{2,2}^{0}$	$\phi_{2,2}^{0.25}$	$\phi_{2,2}^{0.25}$
	$\phi_{2,2}^{0.49}$	$\phi_{2,2}^{0.49}$	$\phi_{2,2}^{0.49}$	$\phi_{2,2}^{0}$	$\phi_{2,2}^{0.49}$	$\phi_{2,2}^{0.49}$	$\phi_{2,2}^{0.49}$	$\phi_2^{*,0}$	$\phi_{2,2}^{0.49}$	$\phi_2^{*,0.49}$	$\phi_2^{*,0.49}$	$\phi_2^{*,0}$
200	$\phi_3^{*,0.25}$	$\phi_{3,0}^{0.25}$	$\phi_{3,0}^{0}$	$\phi_{3,1}^{0.25}$	$\phi_3^{*,0}$	$\phi_{3,1}^{0.25}$	$\phi_{3,0}^{0.25}$	$\phi_{3,1}^{0.25}$	$\phi_3^{*,0}$	$\phi_{3,2}^{0}$	$\phi_{3,2}^{0.25}$	$\phi_{3,2}^{0.25}$
	$\phi_{3,2}^{0.49}$	$\phi_{3,2}^{0.49}$	$\phi_{3,2}^{0.49}$	$\phi_{3,2}^{0.49}$	$\phi_{3,2}^{0.49}$	$\phi_{3,2}^{0.49}$	$\phi_{3,2}^{0.49}$	$\phi_3^{*,0.49}$	$\phi_{3,2}^{0.49}$	$\phi_3^{*,0.49}$	$\phi_3^{*,0.49}$	$\phi_{3,0}^{0.49}$

$\rho_0 = -0.75$ $\rho_0 = -0.5$ $\rho_0 = 0$

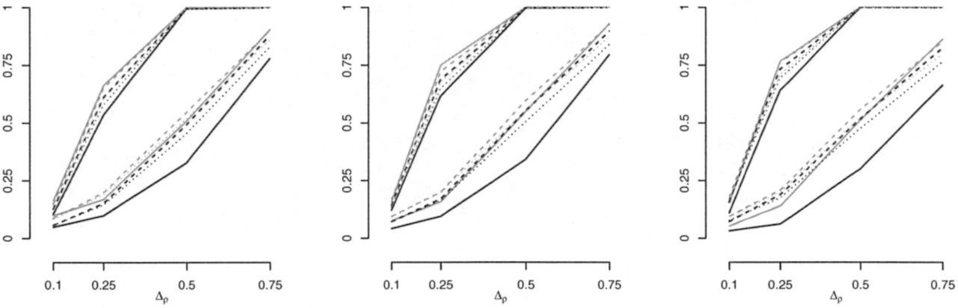

Figure 16: Caption of Figure 15 with $\theta = \mathbf{0.25}$ under constant parameters.

$n\backslash\Delta_\rho$	0.1	.25	.5	.75	0.1	.25	.5	.75	0.1	.25	.5	.75
1000	$\phi_1^{*,0.49}$	$\phi_1^{*,0.45}$	$\phi_{1,0}^{0.45}$	$\phi_1^{*,0.25}$	$\phi_1^{*,0.49}$	$\phi_1^{*,0.45}$	$\phi_1^{*,0.45}$	$\phi_{1,0}^{0.25}$	$\phi_1^{*,0.45}$	$\phi_1^{*,0.45}$	$\phi_{1,2}^{0.45}$	$\phi_{1,0}^{0.25}$
	$\phi_{1,2}^{0}$	$\phi_{1,2}^{0}$	$\phi_{1,2}^{0}$	ϕ_1^{*}	$\phi_{1,0}^{0}$	$\phi_{1,2}^{0}$	$\phi_{1,2}^{0}$	ϕ_1^{*}	$\phi_{1,0}^{0}$	$\phi_{1,0}^{0}$	$\phi_{1,1}^{0}$	$\phi_{1,0}^{0}$
1000	$\phi_{2,1}^{0.45}$	$\phi_2^{*,0.49}$	$\phi_{2,1}^{0.49}$	$\phi_{2,0}^{0.49}$	$\phi_2^{*,0.49}$	$\phi_2^{*,0.49}$	$\phi_{2,0}^{0.49}$	$\phi_{2,0}^{0.45}$	$\phi_{2,0}^{0.25}$	$\phi_{2,2}^{0.49}$	$\phi_{2,2}^{0.49}$	$\phi_2^{*,0.49}$
	$\phi_{2,1}^{0}$	$\phi_{2,2}^{0}$	$\phi_2^{*,0}$	$\phi_{2,2}^{0}$	$\phi_{2,2}^{0}$	$\phi_{2,2}^{0}$	$\phi_2^{*,0}$	$\phi_{2,2}^{0}$	$\phi_{2,0}^{0}$	$\phi_{2,0}^{0}$	$\phi_{2,1}^{0}$	$\phi_2^{*,0}$
1000	$\phi_{3,1}^{0.25}$	$\phi_3^{*,0.45}$	$\phi_{3,0}^{0.49}$	$\phi_{3,1}^{0.45}$	$\phi_3^{*,0.45}$	$\phi_3^{*,0.45}$	$\phi_{3,1}^{0.49}$	$\phi_3^{*,0.49}$	$\phi_{3,2}^{0.25}$	$\phi_{3,2}^{0.45}$	$\phi_{3,2}^{0.49}$	$\phi_{3,2}^{0.45}$
	$\phi_{3,2}^{0}$	$\phi_{3,2}^{0}$	$\phi_{3,2}^{0}$	$\phi_{3,0}^{0}$	$\phi_{3,0}^{0}$	$\phi_{3,2}^{0}$	$\phi_3^{*,0}$	$\phi_{3,1}^{0}$	$\phi_3^{*,0}$	$\phi_{3,0}^{0}$	$\phi_{3,0}^{0}$	$\phi_3^{*,0}$
200	$\phi_1^{*,0.45}$	$\phi_1^{*,0.45}$	$\phi_{1,1}^{0.45}$	$\phi_{1,0}^{0.45}$	$\phi_1^{*,0.45}$	$\phi_1^{*,0.45}$	$\phi_{1,0}^{0.45}$	$\phi_{1,0}^{0.45}$	$\phi_1^{*,0.49}$	$\phi_1^{*,0.45}$	$\phi_1^{*,0.45}$	$\phi_{1,2}^{0.45}$
	$\phi_{1,0}^{0}$	$\phi_{1,1}^{0}$	$\phi_{1,2}^{0}$	$\phi_{1,2}^{0}$	$\phi_{1,1}^{0}$	$\phi_{1,0}^{0}$	$\phi_{1,0}^{0}$	$\phi_{1,2}^{0}$	$\phi_{1,2}^{0.49}$	$\phi_{1,2}^{0.49}$	$\phi_{1,1}^{0}$	$\phi_{1,0}^{0}$
200	$\phi_2^{*,0.45}$	$\phi_2^{*,0.45}$	$\phi_2^{*,0.49}$	$\phi_{2,0}^{0.49}$	$\phi_2^{*,0.45}$	$\phi_2^{*,0.45}$	$\phi_{2,1}^{0.49}$	$\phi_{2,0}^{0.49}$	$\phi_2^{*,0.25}$	$\phi_2^{*,0.25}$	$\phi_2^{*,0.45}$	$\phi_{2,2}^{0.49}$
	$\phi_{2,2}^{0.49}$	$\phi_{2,2}^{0}$	$\phi_{2,2}^{0}$	$\phi_{2,2}^{0}$	$\phi_{2,2}^{0.49}$	$\phi_{2,2}^{0}$	$\phi_{2,2}^{0}$	$\phi_2^{*,0}$	$\phi_{2,0}^{0.49}$	$\phi_{2,0}^{0}$	$\phi_{2,1}^{0}$	$\phi_{2,1}^{0}$
200	$\phi_3^{*,0}$	$\phi_3^{*,0.25}$	$\phi_3^{*,0.45}$	$\phi_{3,1}^{0.45}$	$\phi_3^{*,0.25}$	$\phi_3^{*,0.25}$	$\phi_{3,1}^{0.45}$	$\phi_{3,0}^{0.45}$	$\phi_3^{*,0.25}$	$\phi_3^{*,0.25}$	$\phi_3^{*,0.25}$	$\phi_{3,2}^{0.45}$
	$\phi_{3,2}^{0.49}$	$\phi_{3,2}^{0.49}$	$\phi_{3,2}^{0}$	$\phi_{3,2}^{0}$	$\phi_{3,2}^{0.45}$	$\phi_{3,2}^{0.49}$	$\phi_{3,2}^{0.49}$	$\phi_3^{*,0}$	$\phi_{3,0}^{0.49}$	$\phi_{3,0}^{0}$	$\phi_{3,0}^{0}$	$\phi_{3,1}^{0}$

$\rho_0 = -0.75$ $\rho_0 = -0.5$ $\rho_0 = 0$

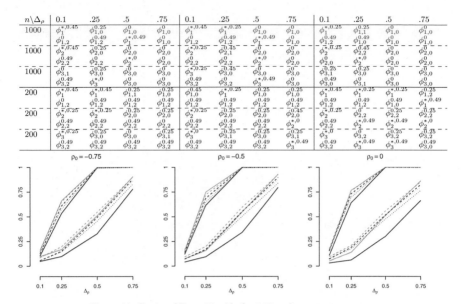

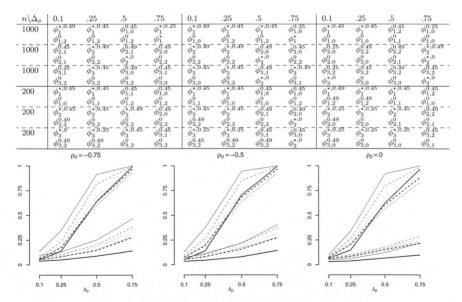

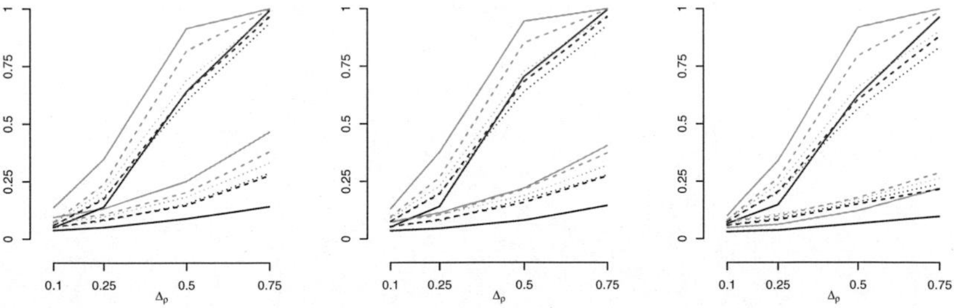

Figure 17: Caption of Figure 15 with $\theta = \mathbf{0.1}$ under constant parameters.

$n\backslash\Delta_\rho$	0.1	.25	.5	.75	0.1	.25	.5	.75	0.1	.25	.5	.75
1000	$\phi_{1,0}^{0.45}$	$\phi_{1,0}^{0.25}$	$\phi_{1,0}^{0}$	$\phi_{1,0}^{0}$	$\phi_{1}^{*,0.45}$	$\phi_{1,0}^{0.25}$	$\phi_{1,0}^{0}$	$\phi_{1,0}^{0}$	$\phi_{1,0}^{0.25}$	$\phi_{1,0}^{0.25}$	$\phi_{1,0}^{0}$	$\phi_{1,0}^{0}$
	$\phi_{1,4}^{0}$	$\phi_{1,4}^{0.49}$	$\phi_{1,4}^{0.49}$	$\phi_{1,0}^{0}$	$\phi_{1,3}^{0}$	$\phi_{1,4}^{0.49}$	$\phi_{1}^{*,0.45}$	$\phi_{1,0}^{0}$	$\phi_{1,4}^{0.49}$	$\phi_{1,4}^{0.49}$	$\phi_{1,0}^{0}$	$\phi_{1,0}^{0}$
	$\phi_{2,0}^{0.45}$	$\phi_{2,0}^{0.25}$	$\phi_{2,0}^{0}$	$\phi_{2,0}^{0}$	$\phi_{2,0}^{0.45}$	$\phi_{2,0}^{0.45}$	$\phi_{2,0}^{0}$	$\phi_{2,0}^{0}$	$\phi_{2,4}^{0.45}$	$\phi_{2,3}^{0.45}$	$\phi_{2,0}^{0}$	$\phi_{2,0}^{0}$
	$\phi_{2,4}^{0}$	$\phi_{2,4}^{0}$	$\phi_{2}^{*,0}$	$\phi_{2,0}^{0}$	$\phi_{2,4}^{0}$	$\phi_{2}^{*,0}$	$\phi_{2}^{*,0}$	$\phi_{2,0}^{0}$	$\phi_{2}^{*,0}$	$\phi_{2}^{*,0}$	$\phi_{2,0}^{0}$	$\phi_{2,0}^{0}$
	$\phi_{3,0}^{0.25}$	$\phi_{3,0}^{0.25}$	$\phi_{3,0}^{0}$	$\phi_{3,0}^{0}$	$\phi_{3,0}^{0.25}$	$\phi_{3,0}^{0.45}$	$\phi_{3,0}^{0}$	$\phi_{3,0}^{0}$	$\phi_{3,4}^{0.25}$	$\phi_{3,4}^{0.25}$	$\phi_{3,0}^{0}$	$\phi_{3,0}^{0}$
	$\phi_{3,4}^{0.49}$	$\phi_{3,4}^{0}$	$\phi_{3,4}^{0}$	$\phi_{3,0}^{0}$	$\phi_{3,4}^{0.49}$	$\phi_{3}^{*,0}$	$\phi_{3}^{*,0}$	$\phi_{3,0}^{0}$	$\phi_{3}^{*,0}$	$\phi_{3}^{*,0}$	$\phi_{3}^{*,0}$	$\phi_{3,0}^{0}$
200	$\phi_{1}^{*,0.45}$	$\phi_{1}^{*,0.45}$	$\phi_{1,0}^{0.25}$	$\phi_{1,0}^{0.25}$	$\phi_{1,0}^{0.45}$	$\phi_{1,0}^{0.25}$	$\phi_{1,0}^{0.25}$	$\phi_{1,0}^{0.25}$	$\phi_{1,4}^{0.25}$	$\phi_{1,0}^{0.25}$	$\phi_{1,4}^{0.25}$	$\phi_{1,4}^{0.25}$
	$\phi_{1,4}^{0}$	$\phi_{1,4}^{0}$	$\phi_{1,4}^{0.49}$	$\phi_{1,4}^{0.49}$	$\phi_{1,3}^{0}$	$\phi_{1,4}^{0.49}$	$\phi_{1,4}^{0.49}$	$\phi_{1}^{*,0.49}$	$\phi_{1,4}^{0.49}$	$\phi_{1,0}^{0.49}$	$\phi_{1,0}^{0.49}$	$\phi_{1}^{*,0.49}$
	$\phi_{2}^{*,0.25}$	$\phi_{2,0}^{0.25}$	$\phi_{2,0}^{0.25}$	$\phi_{2,0}^{0.25}$	$\phi_{2,0}^{0}$	$\phi_{2,0}^{0.25}$	$\phi_{2,0}^{0.25}$	$\phi_{2,0}^{0.45}$	$\phi_{2,0}^{0.25}$	$\phi_{2,0}^{0}$	$\phi_{2,4}^{0.25}$	$\phi_{2,4}^{0.25}$
	$\phi_{2,4}^{0.49}$	$\phi_{2,4}^{0}$	$\phi_{2,4}^{0}$	$\phi_{2,3}^{0}$	$\phi_{2,4}^{0.49}$	$\phi_{2,4}^{0.49}$	$\phi_{2,4}^{0.49}$	$\phi_{2}^{*,0}$	$\phi_{2,3}^{0.49}$	$\phi_{2}^{*,0}$	$\phi_{2}^{*,0}$	$\phi_{2}^{*,0}$
	$\phi_{3}^{*,0.25}$	$\phi_{3,0}^{0.25}$	$\phi_{3,0}^{0}$	$\phi_{3,0}^{0.25}$	$\phi_{3,0}^{0.25}$	$\phi_{3,0}^{0.25}$	$\phi_{3,0}^{0.25}$	$\phi_{3,0}^{0}$	$\phi_{3,0}^{0.25}$	$\phi_{3,4}^{0}$	$\phi_{3,4}^{0.25}$	$\phi_{3,4}^{0.25}$
	$\phi_{3,3}^{0.49}$	$\phi_{3,4}^{0.49}$	$\phi_{3,4}^{0.49}$	$\phi_{3,4}^{0.49}$	$\phi_{3,3}^{0.49}$	$\phi_{3,4}^{0.49}$	$\phi_{3,4}^{0.49}$	$\phi_{3}^{*,0.49}$	$\phi_{3,3}^{0.49}$	$\phi_{3}^{*,0.49}$	$\phi_{3}^{*,0.49}$	$\phi_{3}^{*,0.49}$

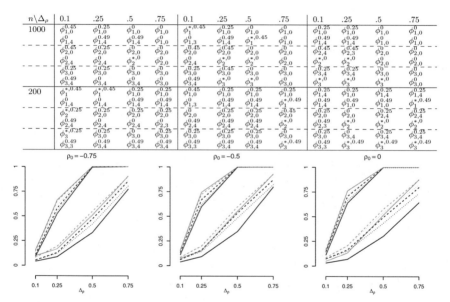

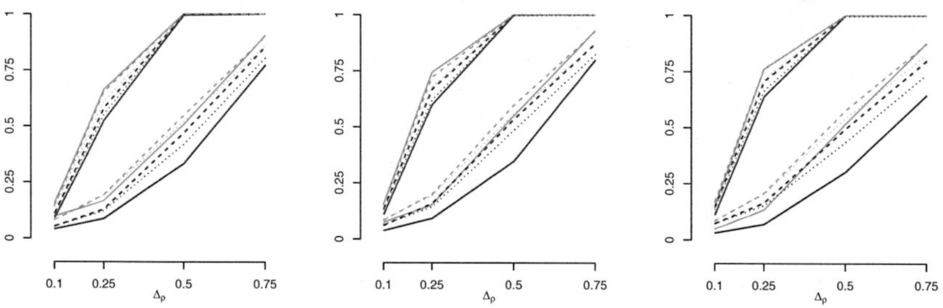

Figure 18: Best (green) and worst power (red) of $\phi_{\iota,0}^{\gamma}$, $\phi_{\iota,3}^{\gamma}$, $\phi_{\iota,4}^{\gamma}$, and $\phi_{\iota}^{*,\gamma}$ and for $\gamma \in \{0,0.25,0.45,0.49\}$ under the setting ii) with (II_μ) and with $\theta_\rho = \mathbf{0.25}$. Here, we consider the best and worst power for each $\iota = 1$ (solid), $\iota = 2$ (dashed), and $\iota = 3$ (dotted), as well as for $n = 200$ and $n = 1000$, where the upper three green and upper three red lines are the ones generated with $n = 1000$. Moreover, ρ_0 is equal to -0.9, -0.5, and 0 (from left to right). The overlying table contains the corresponding best and worst tests belonging to three figures (from left to right) for each n, ι, and $\Delta_\rho = 0.1, 0.25, 0.5, 0.75$. In each of these variations the upper row illustrates the best test and the lower one presents the worst of the 4x4 tests.

$n\backslash\Delta_\rho$	0.1	.25	.5	.75	0.1	.25	.5	.75	0.1	.25	.5	.75
1000	$\phi^{0.25}_{1,6}$	$\phi^{0.25}_{1,6}$	$\phi^{0}_{1,6}$	$\phi^{0}_{1,0}$	$\phi^{0.25}_{1,6}$	$\phi^{0.25}_{1,6}$	$\phi^{0}_{1,0}$	$\phi^{0}_{1,0}$	$\phi^{0.25}_{1,6}$	$\phi^{0.25}_{1,6}$	$\phi^{0}_{1,0}$	$\phi^{0}_{1,0}$
	$\phi^{0}_{1,7}$	$\phi^{0.49}_{1,7}$	$\phi^{0}_{1,0}$	$\phi^{0}_{1,0}$	$\phi^{0.49}_{1,8}$	$\phi^{0.49}_{1,7}$	$\phi^{0}_{1,0}$	$\phi^{0}_{1,0}$	$\phi^{0.49}_{1,8}$	$\phi^{0.49}_{1,7}$	$\phi^{0}_{1,0}$	$\phi^{0}_{1,0}$
	$\phi^{0.25}_{2,6}$	$\phi^{0.25}_{2,6}$	$\phi^{0}_{2,6}$	$\phi^{0}_{2,0}$	$\phi^{0.25}_{2,6}$	$\phi^{0.25}_{2,6}$	$\phi^{0}_{2,0}$	$\phi^{0}_{2,0}$	$\phi^{0.45}_{2,6}$	$\phi^{0.25}_{2,6}$	$\phi^{0}_{2,0}$	$\phi^{0}_{2,0}$
	$\phi^{0.49}_{2,0}$	$\phi^{0}_{2,0}$	$\phi^{0}_{2,0}$	$\phi^{0}_{2,0}$	$\phi^{0}_{2,0}$	$\phi^{0}_{2,0}$	$\phi^{0}_{2,0}$	$\phi^{0}_{2,0}$	$\phi^{0}_{2,6}$	$\phi^{0}_{2,7}$	$\phi^{0}_{2,0}$	$\phi^{0}_{2,0}$
	$\phi^{0.25}_{3,8}$	$\phi^{0.45}_{3,6}$	$\phi^{0}_{3,6}$	$\phi^{0}_{3,0}$	$\phi^{0.25}_{3,6}$	$\phi^{0.25}_{3,6}$	$\phi^{0}_{3,0}$	$\phi^{0}_{3,0}$	$\phi^{0.25}_{3,0}$	$\phi^{0.25}_{3,8}$	$\phi^{0}_{3,0}$	$\phi^{0}_{3,0}$
	$\phi^{0.49}_{3,0}$	$\phi^{0}_{3,7}$	$\phi^{0}_{3,0}$	$\phi^{0}_{3,0}$	$\phi^{0.49}_{3,0}$	$\phi^{0.49}_{3,7}$	$\phi^{0}_{3,0}$	$\phi^{0}_{3,0}$	$\phi^{0.49}_{3,0}$	$\phi^{0.49}_{3,0}$	$\phi^{0}_{3,0}$	$\phi^{0}_{3,0}$
200	$\phi^{0.49}_{1,6}$	$\phi^{0}_{1,8}$	$\phi^{0.25}_{1,8}$	$\phi^{0.25}_{1,8}$	$\phi^{0.45}_{1,0}$	$\phi^{0.45}_{1,0}$	$\phi^{0.25}_{1,8}$	$\phi^{0.25}_{1,8}$	$\phi^{0.45}_{1,0}$	$\phi^{0.25}_{1,6}$	$\phi^{0.25}_{1,8}$	$\phi^{0.25}_{1,8}$
	$\phi^{0}_{1,7}$	$\phi^{0.49}_{1,6}$	$\phi^{0.49}_{1,7}$	$\phi^{0.49}_{1,0}$	$\phi^{0.49}_{1,8}$	$\phi^{0.49}_{1,6}$	$\phi^{0.49}_{1,6}$	$\phi^{0.49}_{1,0}$	$\phi^{0.49}_{1,8}$	$\phi^{0.49}_{1,7}$	$\phi^{0.49}_{1,7}$	$\phi^{0.49}_{1,0}$
	$\phi^{0}_{2,6}$	$\phi^{0}_{2,6}$	$\phi^{0.25}_{2,6}$	$\phi^{0.45}_{2,6}$	$\phi^{0.45}_{2,0}$	$\phi^{0}_{2,8}$	$\phi^{0.25}_{2,8}$	$\phi^{0.25}_{2,8}$	$\phi^{0}_{2,6}$	$\phi^{0.25}_{2,6}$	$\phi^{0.25}_{2,6}$	$\phi^{0.45}_{2,6}$
	$\phi^{0.49}_{2,7}$	$\phi^{0.49}_{2,7}$	$\phi^{0.49}_{2,7}$	$\phi^{0}_{2,7}$	$\phi^{0}_{2,7}$	$\phi^{0.49}_{2,7}$	$\phi^{0.49}_{2,7}$	$\phi^{0.49}_{2,0}$	$\phi^{0.45}_{2,7}$	$\phi^{0.49}_{2,7}$	$\phi^{0}_{2,7}$	$\phi^{0}_{2,0}$
	$\phi^{0}_{3,6}$	$\phi^{0.25}_{3,6}$	$\phi^{0.25}_{3,6}$	$\phi^{0.25}_{3,6}$	$\phi^{0}_{3,8}$	$\phi^{0}_{3,6}$	$\phi^{0.25}_{3,6}$	$\phi^{0.25}_{3,8}$	$\phi^{0}_{3,6}$	$\phi^{0.25}_{3,6}$	$\phi^{0.25}_{3,8}$	$\phi^{0.25}_{3,8}$
	$\phi^{0.49}_{3,7}$	$\phi^{0.49}_{3,7}$	$\phi^{0.49}_{3,7}$	$\phi^{0.49}_{3,7}$	$\phi^{0}_{3,7}$	$\phi^{0.49}_{3,7}$	$\phi^{0.49}_{3,7}$	$\phi^{0.49}_{3,7}$	$\phi^{0.49}_{3,7}$	$\phi^{0.49}_{3,7}$	$\phi^{0.49}_{3,7}$	$\phi^{0.49}_{3,7}$

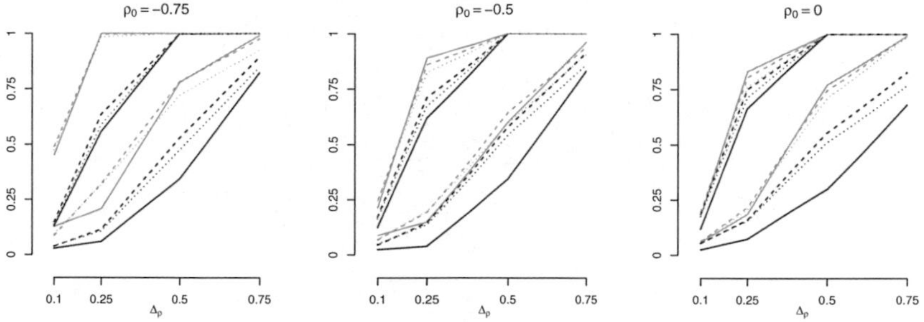

Figure 19: Best (green) and worst power (red) of $\phi^{\gamma}_{\iota,0}$, $\phi^{\gamma}_{\iota,5,5}$, $\phi^{\gamma}_{\iota,6}$, $\phi^{\gamma}_{\iota,7}$, and $\phi^{\gamma}_{\iota,8}$ and for $\gamma \in \{0,0.25,0.45,0.49\}$ under the setting i) with $\theta = \mathbf{0.25}$. Here, we consider the best and worst power for each $\iota = 1$ (solid), $\iota = 2$ (dashed), and $\iota = 3$ (dotted), as well as for $n = 200$ and $n = 1000$, where the upper three green and upper three red lines are the ones generated with $n = 1000$. Moreover, ρ_0 is equal to -0.9, -0.5, and 0 (from left to right). The overlying table contains the corresponding best and worst tests belonging to three figures (from left to right) for each n, ι, and $\Delta_\rho = 0.1, 0.25, 0.5, 0.75$. In each of these variations the upper row illustrates the best test and the lower one presents the worst of the 5x4 tests.

$n\backslash\Delta_\rho$	0.1	.25	.5	.75	0.1	.25	.5	.75	0.1	.25	.5	.75
1000	$\phi^{0.25}_{1,8}$	$\phi^{0.25}_{1,8}$	$\phi^{0}_{1,0}$	$\phi^{0}_{1,0}$	$\phi^{0.25}_{1,8}$	$\phi^{0.25}_{1,8}$	$\phi^{0}_{1,0}$	$\phi^{0}_{1,0}$	$\phi^{0.25}_{1,8}$	$\phi^{0.25}_{1,8}$	$\phi^{0}_{1,0}$	$\phi^{0}_{1,0}$
	$\phi^{0}_{1,0}$	$\phi^{0.49}_{1,0}$	$\phi^{0}_{1,0}$	$\phi^{0}_{1,0}$	$\phi^{0.49}_{1,8}$	$\phi^{0.49}_{1,0}$	$\phi^{0}_{1,0}$	$\phi^{0}_{1,0}$	$\phi^{0.49}_{1,8}$	$\phi^{0.49}_{1,0}$	$\phi^{0}_{1,0}$	$\phi^{0}_{1,0}$
	$\phi^{0.25}_{2,8}$	$\phi^{0.25}_{2,8}$	$\phi^{0}_{2,0}$	$\phi^{0}_{2,0}$	$\phi^{0.25}_{2,8}$	$\phi^{0.45}_{2,8}$	$\phi^{0}_{2,0}$	$\phi^{0}_{2,0}$	$\phi^{0.25}_{2,8}$	$\phi^{0.45}_{2,8}$	$\phi^{0}_{2,0}$	$\phi^{0}_{2,0}$
	$\phi^{0.49}_{2,0}$	$\phi^{0}_{2,0}$	$\phi^{0}_{2,0}$	$\phi^{0}_{2,0}$	$\phi^{0.49}_{2,0}$	$\phi^{0}_{2,0}$	$\phi^{0}_{2,0}$	$\phi^{0}_{2,0}$	$\phi^{0}_{2,0}$	$\phi^{0}_{2,0}$	$\phi^{0}_{2,0}$	$\phi^{0}_{2,0}$
	$\phi^{0.25}_{3,8}$	$\phi^{0.45}_{3,8}$	$\phi^{0}_{3,2}$	$\phi^{0}_{3,0}$	$\phi^{0.25}_{3,8}$	$\phi^{0.25}_{3,8}$	$\phi^{0}_{3,0}$	$\phi^{0}_{3,0}$	$\phi^{0.25}_{3,0}$	$\phi^{0.25}_{3,8}$	$\phi^{0}_{3,1}$	$\phi^{0}_{3,0}$
	$\phi^{0.49}_{3,0}$	$\phi^{0.49}_{3,0}$	$\phi^{0}_{3,0}$	$\phi^{0}_{3,0}$	$\phi^{0.49}_{3,0}$	$\phi^{0}_{3,0}$	$\phi^{0}_{3,0}$	$\phi^{0}_{3,0}$	$\phi^{0.49}_{3,8}$	$\phi^{0.49}_{3,0}$	$\phi^{0}_{3,0}$	$\phi^{0}_{3,0}$
200	$\phi^{0.49}_{1,1}$	$\phi^{0.25}_{1,8}$	$\phi^{0.25}_{1,8}$	$\phi^{0.25}_{1,8}$	$\phi^{0.45}_{1,0}$	$\phi^{0.25}_{1,8}$	$\phi^{0.25}_{1,8}$	$\phi^{0.25}_{1,8}$	$\phi^{0}_{1,8}$	$\phi^{0.25}_{1,8}$	$\phi^{0.25}_{1,8}$	$\phi^{0}_{1,2}$
	$\phi^{0}_{1,0}$	$\phi^{0.49}_{1,8}$	$\phi^{0.49}_{1,0}$	$\phi^{0.49}_{1,0}$	$\phi^{0.49}_{1,8}$	$\phi^{0.49}_{1,8}$	$\phi^{0.49}_{1,0}$	$\phi^{0.49}_{1,0}$	$\phi^{0.49}_{1,8}$	$\phi^{0.49}_{1,0}$	$\phi^{0.49}_{1,0}$	$\phi^{0.49}_{1,0}$
	$\phi^{0.25}_{2,8}$	$\phi^{0.25}_{2,8}$	$\phi^{0.45}_{2,8}$	$\phi^{0.45}_{2,8}$	$\phi^{0.25}_{2,0}$	$\phi^{0.25}_{2,8}$	$\phi^{0.25}_{2,8}$	$\phi^{0.25}_{2,8}$	$\phi^{0.25}_{2,0}$	$\phi^{0.25}_{2,8}$	$\phi^{0.45}_{2,8}$	$\phi^{0}_{2,8}$
	$\phi^{0.25}_{2,0}$	$\phi^{0.45}_{2,0}$	$\phi^{0}_{2,0}$	$\phi^{0}_{2,0}$	$\phi^{0.49}_{2,8}$	$\phi^{0}_{2,0}$	$\phi^{0.49}_{2,0}$	$\phi^{0}_{2,0}$	$\phi^{0.49}_{2,8}$	$\phi^{0.49}_{2,0}$	$\phi^{0.49}_{2,0}$	$\phi^{0}_{2,0}$
	$\phi^{0}_{3,8}$	$\phi^{0.25}_{3,8}$	$\phi^{0.25}_{3,8}$	$\phi^{0.25}_{3,8}$	$\phi^{0}_{3,8}$	$\phi^{0.25}_{3,8}$	$\phi^{0.25}_{3,8}$	$\phi^{0.25}_{3,8}$	$\phi^{0.25}_{3,0}$	$\phi^{0.25}_{3,8}$	$\phi^{0.25}_{3,8}$	$\phi^{0}_{3,2}$
	$\phi^{0.49}_{3,0}$	$\phi^{0.49}_{3,0}$	$\phi^{0.49}_{3,0}$	$\phi^{0.49}_{3,0}$	$\phi^{0.49}_{3,0}$	$\phi^{0.49}_{3,0}$	$\phi^{0.49}_{3,0}$	$\phi^{0.49}_{3,0}$	$\phi^{0.49}_{3,8}$	$\phi^{0.49}_{3,0}$	$\phi^{0.49}_{3,0}$	$\phi^{0.49}_{3,0}$

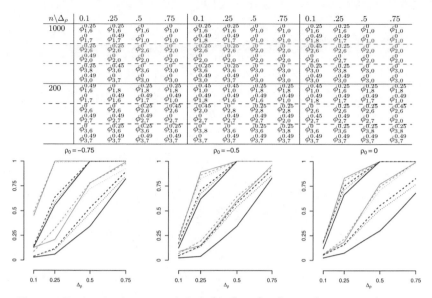

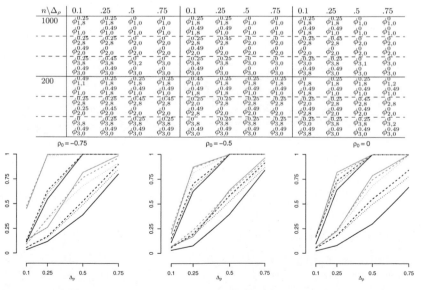

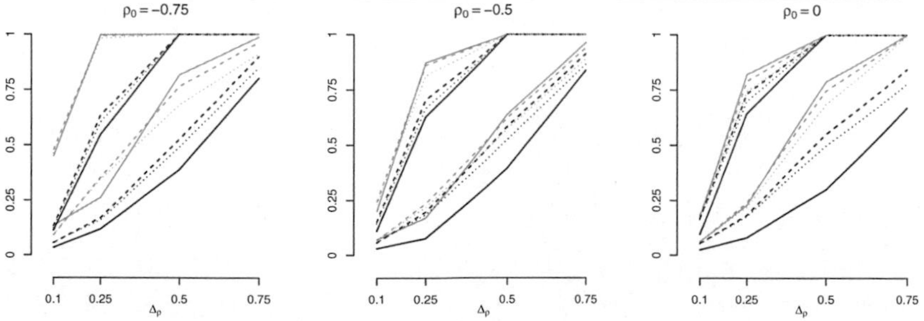

Figure 20: Caption of Figure 19 with $\theta = \mathbf{0.25}$ and under setting iv) with (II_σ).

$n\backslash\Delta_\rho$	0.1	.25	.5	.75	0.1	.25	.5	.75	0.1	.25	.5	.75
1000	$\phi_{1,b}^{0.45}$	$\phi_{1,b}^{0.25}$	$\phi_{1,b}^{0.49}$	$\phi_{1,b}^{0.49}$	$\phi_{1,a}^{0.25}$	$\phi_{1,b}^{0.25}$	$\phi_{1,b}^{0.49}$	$\phi_{1,b}^{0.49}$	$\phi_{1,b}^{0.25}$	$\phi_{1,b}^{0.25}$	$\phi_{1,b}^{0.49}$	$\phi_{1,b}^{0.49}$
	$\phi_{1,0}^{0}$	$\phi_{1,0}^{0.49}$	$\phi_{1,0}^{0.49}$	$\phi_{1,b}^{0.49}$	$\phi_{1,a}^{0.49}$	$\phi_{1,0}^{0.49}$	$\phi_{1,b}^{0.49}$	$\phi_{1,b}^{0.49}$	$\phi_{1,a}^{0.49}$	$\phi_{1,0}^{0.49}$	$\phi_{1,b}^{0.49}$	$\phi_{1,b}^{0.49}$
	$\phi_{2,b}^{0.25}$	$\phi_{2,b}^{0.49}$	$\phi_{2,b}^{0.49}$	$\phi_{2,b}^{0.49}$	$\phi_{2,b}^{0.25}$	$\phi_{2,b}^{0.45}$	$\phi_{2,b}^{0.49}$	$\phi_{2,b}^{0.49}$	$\phi_{2,b}^{0.45}$	$\phi_{2,b}^{0.45}$	$\phi_{2,b}^{0.49}$	$\phi_{2,b}^{0.49}$
	$\phi_{2,0}^{0}$	$\phi_{2,0}^{0}$	$\phi_{2,0}^{0.25}$	$\phi_{2,b}^{0.49}$	$\phi_{2,0}^{0}$	$\phi_{2,0}^{0}$	$\phi_{2,b}^{0.49}$	$\phi_{2,b}^{0.49}$	$\phi_{2,a}^{0}$	$\phi_{2,0}^{0}$	$\phi_{2,b}^{0.49}$	$\phi_{2,b}^{0.49}$
	$\phi_{3,b}^{0.25}$	$\phi_{3,b}^{0.49}$	$\phi_{3,b}^{0.49}$	$\phi_{3,b}^{0.49}$	$\phi_{3,b}^{0.25}$	$\phi_{3,b}^{0.25}$	$\phi_{3,b}^{0.49}$	$\phi_{3,b}^{0.49}$	$\phi_{3,0}^{0.25}$	$\phi_{3,b}^{0.25}$	$\phi_{3,b}^{0.49}$	$\phi_{3,b}^{0.49}$
	$\phi_{3,0}^{0}$	$\phi_{3,0}^{0.49}$	$\phi_{3,0}^{0}$	$\phi_{3,b}^{0.49}$	$\phi_{3,0}^{0}$	$\phi_{3,0}^{0}$	$\phi_{3,0}^{0}$	$\phi_{3,b}^{0.49}$	$\phi_{3,b}^{0.49}$	$\phi_{3,0}^{0.49}$	$\phi_{3,0}^{0}$	$\phi_{3,b}^{0.49}$
200	$\phi_{1,b}^{0.49}$	$\phi_{1,b}^{0.25}$	$\phi_{1,b}^{0.25}$	$\phi_{1,b}^{0.25}$	$\phi_{1,0}^{0.45}$	$\phi_{1,0}^{0.25}$	$\phi_{1,b}^{0.25}$	$\phi_{1,b}^{0.25}$	$\phi_{1,b}^{0.25}$	$\phi_{1,b}^{0.25}$	$\phi_{1,b}^{0.25}$	$\phi_{1,b}^{0.25}$
	$\phi_{1,a}^{0}$	$\phi_{1,a}^{0.49}$	$\phi_{1,a}^{0.49}$	$\phi_{1,0}^{0.49}$	$\phi_{1,a}^{0.49}$	$\phi_{1,a}^{0.49}$	$\phi_{1,a}^{0.49}$	$\phi_{1,0}^{0.49}$	$\phi_{1,a}^{0.49}$	$\phi_{1,0}^{0.49}$	$\phi_{1,0}^{0.49}$	$\phi_{1,0}^{0.49}$
	$\phi_{2,0}^{0.45}$	$\phi_{2,b}^{0.25}$	$\phi_{2,b}^{0.25}$	$\phi_{2,b}^{0.49}$	$\phi_{2,0}^{0.45}$	$\phi_{2,0}^{0.25}$	$\phi_{2,b}^{0.25}$	$\phi_{2,b}^{0.45}$	$\phi_{2,0}^{0.25}$	$\phi_{2,b}^{0.25}$	$\phi_{2,b}^{0.45}$	$\phi_{2,b}^{0.25}$
	$\phi_{2,a}^{0.49}$	$\phi_{2,0}^{0.49}$	$\phi_{2,0}^{0.49}$	$\phi_{2,0}^{0}$	$\phi_{2,b}^{0.49}$	$\phi_{2,a}^{0.49}$	$\phi_{2,a}^{0}$	$\phi_{2,a}^{0}$	$\phi_{2,a}^{0.49}$	$\phi_{2,0}^{0.49}$	$\phi_{2,0}^{0}$	$\phi_{2,0}^{0}$
	$\phi_{3,0}^{0.45}$	$\phi_{3,b}^{0.25}$	$\phi_{3,b}^{0.25}$	$\phi_{3,b}^{0.25}$	$\phi_{3,0}^{0.25}$	$\phi_{3,0}^{0.25}$	$\phi_{3,b}^{0.25}$	$\phi_{3,b}^{0.25}$	$\phi_{3,b}^{0.25}$	$\phi_{3,b}^{0.25}$	$\phi_{3,b}^{0.25}$	$\phi_{3,2}^{0.49}$
	$\phi_{3,a}^{0.49}$	$\phi_{3,0}^{0.49}$	$\phi_{3,0}^{0.49}$	$\phi_{3,0}^{0.49}$	$\phi_{3,a}^{0.49}$	$\phi_{3,a}^{0.49}$	$\phi_{3,a}^{0.49}$	$\phi_{3,0}^{0.49}$	$\phi_{3,a}^{0.49}$	$\phi_{3,0}^{0.49}$	$\phi_{3,0}^{0.49}$	$\phi_{3,0}^{0.49}$

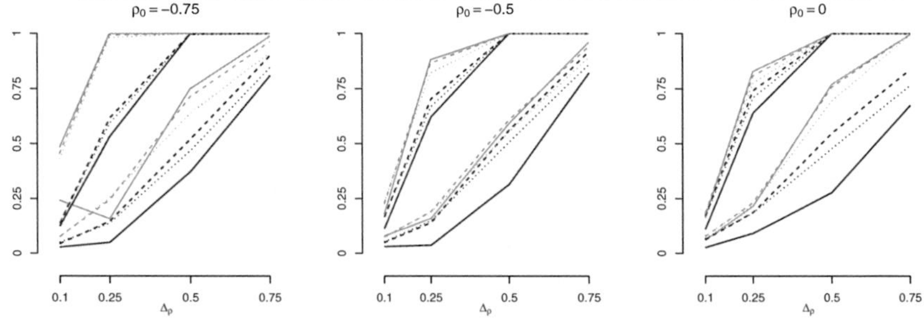

Figure 21: Best (green) and worst power (red) of $\phi^{\gamma}_{\iota,0}$, $\phi^{\gamma}_{\iota,a} = \phi^{\gamma}_{\iota,23}$, and $\phi^{\gamma}_{\iota,b} = \phi^{\gamma}_{\iota,24,10}$, and for $\gamma \in \{0{,}0.25{,}0.45{,}0.49\}$ under the setting ii) with (III_μ) and with $\theta = \mathbf{0.25}$. Here, we consider the best and worst power for each $\iota = 1$ (solid), $\iota = 2$ (dashed), and $\iota = 3$ (dotted), as well as for $n = 200$ and $n = 1000$, where the upper three green and upper three red lines are the ones generated with $n = 1000$. Moreover, ρ_0 is equal to -0.9, -0.5, and 0 (from left to right). The overlying table contains the corresponding best and worst tests belonging to three figures (from left to right) for each n, ι, and $\Delta_\rho = 0.1, 0.25, 0.5, 0.75$. In each of these variations the upper row illustrates the best test and the lower one presents the worst of the 3x4 tests.

$n\backslash\Delta_\rho$	0.1	.25	.5	.75	0.1	.25	.5	.75	0.1	.25	.5	.75
1000	$\phi_{1,b}^{0.45}$	$\phi_{1,b}^{0.25}$	$\phi_{1,b}^{0.49}$	$\phi_{1,b}^{0.49}$	$\phi_{1,b}^{0.25}$	$\phi_{1,b}^{0.25}$	$\phi_{1,b}^{0.49}$	$\phi_{1,b}^{0.49}$	$\phi_{1,b}^{0.25}$	$\phi_{1,a}^{0.25}$	$\phi_{1,b}^{0.49}$	$\phi_{1,b}^{0.49}$
	$\phi_{1,0}^{0}$	$\phi_{1,0}^{0.49}$	$\phi_{1,b}^{0.49}$	$\phi_{1,b}^{0.49}$	$\phi_{1,a}^{0.49}$	$\phi_{1,0}^{0.49}$	$\phi_{1,b}^{0.49}$	$\phi_{1,b}^{0.49}$	$\phi_{1,a}^{0.49}$	$\phi_{1,0}^{0.49}$	$\phi_{1,b}^{0.49}$	$\phi_{1,b}^{0.49}$
	$\phi_{2,b}^{0.25}$	$\phi_{2,b}^{0.49}$	$\phi_{2,b}^{0.49}$	$\phi_{2,b}^{0.49}$	$\phi_{2,b}^{0.45}$	$\phi_{2,b}^{0.45}$	$\phi_{2,b}^{0.49}$	$\phi_{2,b}^{0.49}$	$\phi_{2,b}^{0.45}$	$\phi_{2,b}^{0.45}$	$\phi_{2,b}^{0.49}$	$\phi_{2,b}^{0.49}$
	$\phi_{2,0}^{0.49}$	$\phi_{2,0}^{0.49}$	$\phi_{2,0}^{0.49}$	$\phi_{2,b}^{0.49}$	$\phi_{2,0}^{0}$	$\phi_{2,0}^{0}$	$\phi_{2,b}^{0.49}$	$\phi_{2,b}^{0.49}$	$\phi_{2,0}^{0}$	$\phi_{2,0}^{0}$	$\phi_{2,b}^{0.49}$	$\phi_{2,b}^{0.49}$
	$\phi_{3,b}^{0.25}$	$\phi_{3,b}^{0.49}$	$\phi_{3,b}^{0.49}$	$\phi_{3,b}^{0.49}$	$\phi_{3,b}^{0.25}$	$\phi_{3,b}^{0.25}$	$\phi_{3,b}^{0.49}$	$\phi_{3,b}^{0.49}$	$\phi_{3,b}^{0.25}$	$\phi_{3,b}^{0.25}$	$\phi_{3,b}^{0.49}$	$\phi_{3,b}^{0.49}$
	$\phi_{3,0}^{0.49}$	$\phi_{3,0}^{0.49}$	$\phi_{3,0}^{0.49}$	$\phi_{3,b}^{0.49}$	$\phi_{3,0}^{0.49}$	$\phi_{3,0}^{0.49}$	$\phi_{3,b}^{0.49}$	$\phi_{3,b}^{0.49}$	$\phi_{3,0}^{0.49}$	$\phi_{3,0}^{0.49}$	$\phi_{3,0}^{0.49}$	$\phi_{3,b}^{0.49}$
200	$\phi_{1,b}^{0.49}$	$\phi_{1,b}^{0.25}$	$\phi_{1,b}^{0.25}$	$\phi_{1,b}^{0.25}$	$\phi_{1,0}^{0.49}$	$\phi_{1,0}^{0.45}$	$\phi_{1,b}^{0.25}$	$\phi_{1,b}^{0.25}$	$\phi_{1,b}^{0.25}$	$\phi_{1,b}^{0.25}$	$\phi_{1,b}^{0.25}$	$\phi_{1,a}^{0.25}$
	$\phi_{1,a}^{0}$	$\phi_{1,a}^{0.49}$	$\phi_{1,0}^{0.49}$	$\phi_{1,0}^{0.49}$	$\phi_{1,a}^{0.49}$	$\phi_{1,a}^{0.49}$	$\phi_{1,a}^{0.49}$	$\phi_{1,0}^{0.49}$	$\phi_{1,a}^{0.49}$	$\phi_{1,0}^{0.49}$	$\phi_{1,0}^{0.49}$	$\phi_{1,0}^{0.49}$
	$\phi_{2,0}^{0}$	$\phi_{2,b}^{0.25}$	$\phi_{2,b}^{0.45}$	$\phi_{2,b}^{0.49}$	$\phi_{2,0}^{0}$	$\phi_{2,0}^{0.45}$	$\phi_{2,b}^{0.45}$	$\phi_{2,b}^{0.45}$	$\phi_{2,b}^{0.25}$	$\phi_{2,b}^{0.25}$	$\phi_{2,b}^{0.45}$	$\phi_{2,a}^{0.49}$
	$\phi_{2,a}^{0.49}$	$\phi_{2,0}^{0.49}$	$\phi_{2,0}^{0}$	$\phi_{2,0}^{0.49}$	$\phi_{2,a}^{0.49}$	$\phi_{2,a}^{0.49}$	$\phi_{2,a}^{0}$	$\phi_{2,0}^{0}$	$\phi_{2,a}^{0.49}$	$\phi_{2,0}^{0.49}$	$\phi_{2,0}^{0.49}$	$\phi_{2,0}^{0}$
	$\phi_{3,0}^{0.25}$	$\phi_{3,b}^{0.25}$	$\phi_{3,b}^{0.25}$	$\phi_{3,b}^{0.25}$	$\phi_{3,0}^{0.25}$	$\phi_{3,0}^{0.25}$	$\phi_{3,0}^{0.25}$	$\phi_{3,b}^{0.25}$	$\phi_{3,b}^{0.25}$	$\phi_{3,b}^{0.25}$	$\phi_{3,b}^{0.25}$	$\phi_{3,b}^{0.49}$
	$\phi_{3,a}^{0.49}$	$\phi_{3,0}^{0.49}$	$\phi_{3,0}^{0.49}$	$\phi_{3,0}^{0.49}$	$\phi_{3,a}^{0.49}$	$\phi_{3,a}^{0.49}$	$\phi_{3,a}^{0.49}$	$\phi_{3,0}^{0.49}$	$\phi_{3,a}^{0.49}$	$\phi_{3,0}^{0.49}$	$\phi_{3,0}^{0.49}$	$\phi_{3,0}^{0.49}$

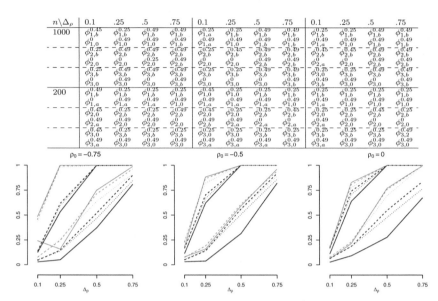

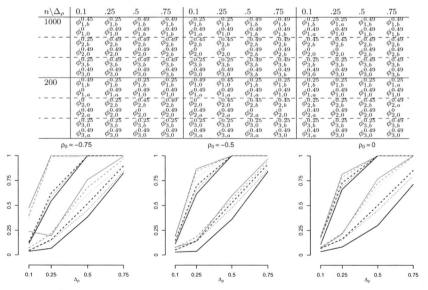

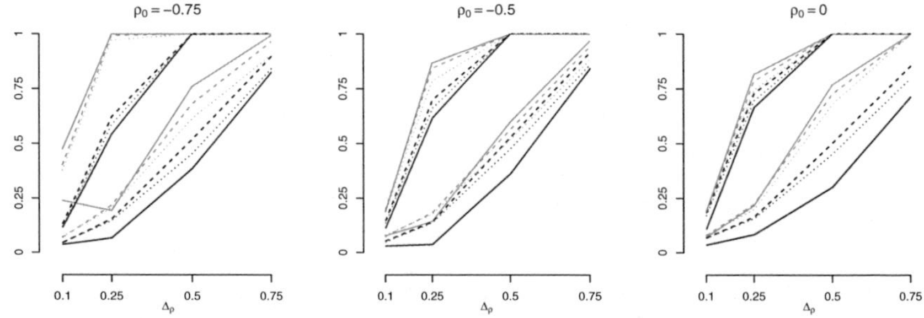

Figure 22: Caption of Figure 21 with $\theta = \mathbf{0.25}$ and under setting v) with (III_μ) and (II_σ).

7.2 Sequential Analysis

In this subsection, we present the behavior of the closed-end sequential testing procedures

$$\tau_{n,\psi,\gamma} = \inf\left\{1 \le k \le nm \ : \ u_1\left(\hat{D}_\psi^{-1/2} w_\gamma\left(\frac{k}{n}\right)\frac{n}{n+k}\frac{k}{\sqrt{n}}\left(\hat{\rho}_{\psi,k,n+1}^{n+k} - \hat{\rho}_{\psi,k,1}^{n}\right)\right) > c_{\alpha,\psi}\right\}$$

with $u_1(\cdot) = |\cdot|$,

$$\hat{\rho}_{\psi,k,j}^{j+h} = (h+1)^{-1}\sum_{i=j}^{j+h} Z_{i,k,h}^{(\psi)} \quad \text{for } 1 \le j \le j+h \le (m+1)n,\ k = 1,\dots,mn,$$

where

$$Z_{i,k}^{(\psi)} = \begin{cases} \frac{(X_i-\mu_{1,i})\cdot(Y_i-\mu_{2,i})}{\sigma_{1,i}\sigma_{2,i}}, & \text{if } \psi = 0, \\ \frac{(X_i-\hat{\mu}_{1,i,n+k}^{(\psi)})\cdot(Y_i-\hat{\mu}_{2,i,n+k}^{(\psi)})}{\sigma_{1,i}\sigma_{2,i}}, & \text{if } \psi = 1,\dots,4, \\ \frac{(X_i-\mu_{1,i})\cdot(Y_i-\mu_{2,i})}{\hat{\sigma}_{1,i,n+k}^{(\psi)}\hat{\sigma}_{2,i,n+k}^{(\psi)}}, & \text{if } \psi = 5,\dots,8, \end{cases}$$

and

$$Z_{i,k}^{(\psi)} = Z_{i,k}^{(4+\psi_1+4\psi_2)} = \frac{(X_i - \hat{\mu}_{1,i,n+k}^{(\psi_1)})\cdot(Y_i - \hat{\mu}_{2,i,n+k}^{(\psi_1)})}{\hat{\sigma}_{1,i,n+k}^{(\psi_2)}\hat{\sigma}_{2,i,n+k}^{(\psi_2)}} \quad \text{for } \psi_1,\psi_2 = 1,\dots,4.$$

Firstly, we specify the parameter estimates, the corresponding LRV estimates, and the weighting function w_γ, before we will present the behavior of the test statistics.

Parameter Estimates In the following we focus on these mean estimates:

$$\hat{\mu}_{1,i,n+k}^{(1)} \equiv n^{-1}\sum_{j=1}^{n} X_j, \quad \hat{\mu}_{1,i,n+k}^{(2)} = \begin{cases} n^{-1}\sum_{j=1}^{n} X_j, & \text{if } i \in [1,n], \\ k^{-1}\sum_{j=n+1}^{n+k} X_j, & \text{if } i \in [n+1],\ k = 1,\dots,nm, \end{cases}$$

$$\hat{\mu}_{1,i,k}^{(3)} = \begin{cases} (n \wedge \hat{k}_{\mu_1,k})^{-1}\sum_{j=1}^{n\wedge\hat{k}_{\mu_1,k}} X_j, & \text{if } 1 \le i \le \hat{k}_{\mu_1,k} \wedge n, \\ (n - \hat{k}_{\mu_1,k})^{-1}\sum_{j=1+\hat{k}_{\mu_1,k}}^{n} X_j, & \text{if } 1 + \hat{k}_{\mu_1,k} \le i \le n, \\ ((n+k) \wedge \hat{k}_{\mu_1,k})^{-1}\sum_{j=1}^{(n+k)\wedge\hat{k}_{\mu_1,k}} X_j, & \text{if } n+1 \le i \le (n+k) \wedge \hat{k}_{\mu_1,k}, \\ ((n+k) - \hat{k}_{\mu_1,k})^{-1}\sum_{j=1+\hat{k}_{\mu_1,k}}^{(n+k)} X_j, & \text{if } n+1 < \hat{k}_{\mu_1,k} \le i \le (n+k), \end{cases}$$

$\hat{\mu}_{1,i,k}^{(4)}$ is of the form $\hat{\mu}_{1,i,k}^{(3)}$,

whereas $\hat{\mu}_{1,i,k}^{(4)}$ uses another change-point estimate $\hat{k}_{\mu_1,k}$ as $\hat{\mu}_{1,i,k}^{(3)}$. In $\hat{\mu}_{1,i,k}^{(3)}$ the change-point estimate is defined as

$$\hat{k}_{\mu_1,k} = \begin{cases} n+k, & \text{if } n+k \le \tau_{\mu_1}, \\ \arg\max_{1\le i<\tau_{\mu_1}} w_{0.45}\left(\frac{i}{\tau_{\mu_1}}\right)\frac{i}{\sqrt{\tau_{\mu_1}}}\left|\overline{X}_1^i - \overline{X}_1^{\tau_{\mu_1}}\right|, & \text{if } n+k > \tau_{\mu_1} \end{cases}$$

with $w_{0.45}(z) = [z(1-z)]^{-0.45}$. In $\hat{\mu}_{1,i,k}^{(4)}$ the change-point estimate is defined as

$$\hat{k}_{\mu_1,k} = \begin{cases} n+k, & \text{if } n+k \le \tau_{\mu_1}, \\ \arg\max_{1\le i<n+k} w_{0.45}\left(\frac{i}{n+k}\right)\frac{i}{\sqrt{n+k}}\left|\overline{X}_1^i - \overline{X}_1^{n+k}\right|, & \text{if } n+k > \tau_{\mu_1}, \end{cases}$$

where for both estimates τ_{μ_1} is the stopping time which sequentially controls whether or not there is a change in the means, i.e.,

$$\tau_{\mu_1} = \inf\left\{1 \le i \le nm \ : \ \max\left(\hat{D}_\mu^{-1/2}\frac{n}{n+i}\frac{i}{\sqrt{n}}\left(\overline{X}_{n+1}^{n+i} - \overline{X}_1^{n}\right)\right) > \left(\frac{m-1}{m}\right)^{0.5} c_{0.82}\right\}.$$

Here, $c_{0.82}$ is the 0.82–quantile of $\|W\|_{[0,1]}$ where W is a standard Brownian motion. This quantile is chosen quite arbitrarily, but it is also a compromise between a high number of false stops

if no structural breaks occur in the mean and fast detection after changes appear. Heuristically, $\hat{\mu}^{(4)}_{1,i,k}$ depends on sequentially estimated change-point estimates if the stopping time stops once. The advantage over $\hat{\mu}^{(3)}_{1,i,k}$ is that the sequential estimation of the change-point could save the mean estimation in the case of early false detection of a change in the mean.
The mean estimates $\hat{\mu}^{(\psi)}_{2,i,k}$ are defined analogously. The variance estimates $(\hat{\sigma}^{(\psi)}_{1,i,k})^2$ and $(\hat{\sigma}^{(\psi)}_{2,i,k})^2$ are similarly defined for $\psi = 1{,}2$ as the mean estimates, whereas X_i and Y_i are just replaced by $(X_i - \mu_{1,i})^2$ and $(Y_i - \mu_{2,i})^2$ in the case of known means and by $(X_i - \hat{\mu}^{(\psi_1)}_{1,i,k})^2$ and $(Y_i - \hat{\mu}^{(\psi_1)}_{2,i,k})^2$ in the case of unknown means. Furthermore, we set

$$(\hat{\sigma}^{(3)}_{1,i,k})^2 = \begin{cases} n^{-1}\sum_{j=1}^{n}(X_j - \mu_{1,j})^2, & \text{if } 1 \le i \le n, \\ (k \wedge (\hat{k}_{\sigma_1,k} - n))^{-1}\sum_{j=n+1}^{(n+k)\wedge \hat{k}_{\sigma_1,k}}(X_j - \mu_{1,j})^2, & \text{if } n+1 \le i \le (n+k) \wedge \hat{k}_{\sigma_1,k}, \\ ((n+k) - \hat{k}_{\sigma_1,k})^{-1}\sum_{j=1+\hat{k}_{\sigma_1,k}}^{(n+k)}(X_j - \sigma_{1,j})^2, & \text{if } n+1 < \hat{k}_{\sigma_1,k} \le i \le (n+k) \end{cases}$$

and $(\hat{\sigma}^{(4)}_{1,i,k})^2$ analogously. They only differ in the change-point estimate $\hat{k}_{\sigma_1,k}$ in the same way as in the mean estimates. $(\hat{\sigma}^{(\psi)}_{2,i,k})^2$ is correspondingly defined for $\psi = 3{,}4$.

LRV Estimates The LRV estimates are the same as those used in the a posteriori procedure, which means in particular that the estimate only depends on the training period $1,\dots,n$.

Weighting Function In the following we focus on the weighting function

$$w_\gamma : (0,\infty) \to \mathbb{R}, \quad w_\gamma(z) = \left(\frac{1+z}{z}\right)^\gamma, \quad \gamma \in \left[0,\frac{1}{2}\right)$$

for all stopping times. By the proof of Theorem 2.2.5 we already know that under the assumptions of Theorem 2.2.3 and under H_0 it holds that

$$P(\tau_{n,\psi,\gamma} < \infty) \to P\left(\sup_{z\in[0,1]} \left|(z)^{-\gamma}W(z)\right| > c_\alpha \left(\frac{m+1}{m}\right)^{1/2-\gamma}\right),$$

where W is a standard Brownian motion. Table 12 contains the corresponding critical values, where we neglect the constant factor depending on m initially.

$1-\alpha$	0.9				0.95				0.975			
γ	0	.25	.45	.49	0	.25	.45	.49	0	.25	.45	.49
	1.946	2.095	2.505	2.732	2.221	2.367	2.758	2.990	2.485	2.621	2.977	3.208

Table 12: Critical values $c_{\alpha,\gamma}$ based on 40.000 replications of $u_1(W_\gamma)$, where W_γ is approximated on a grid of 10.000 equi-spaced points in $[0,1]$.

7.2.1 Influence of the Mean Estimates under $H_0^{(2)}$

In this sub-subsection, we focus on the finite sample behavior of the stopping times $\tau_{n,\psi,\gamma}$, $\psi = 0,\dots,4$, under $H_0^{(2)}$. Table 13 presents the relative frequency of false alarms under $H_0^{(2)}$, constant zero means, and normalized variances. Furthermore, the number of false alarms of $\tau_{n,1,\gamma}$ consistently lies slightly above the given level of $\alpha = 5\%$. Moreover, for higher γ false stops appear more frequently in most cases. Even the stopping times $\tau_{n,3,\gamma}$ and $\tau_{n,4,\gamma}$ work well, since they use change-point estimates for the mean if the corresponding stopping times falsely trigger an alarm that there is a change in the mean.

		$n = 200$						$n = 2000$					
		$m = 0.5$			$m = 1.5$			$m = 0.5$			$m = 1.5$		
ρ_0	γ	0	.25	.45	0	.25	.45	0	.25	.45	0	.25	.45
-0.9	$\tau_{n,0,\gamma}$	.132	.153	.151	.087	.089	.083	.07	.071	.094	.069	.065	.08
	$\tau_{n,1,\gamma}$	.134	.164	.15	.094	.097	.086	.069	.072	.09	.069	.066	.079
	$\tau_{n,2,\gamma}$	.06	.061	.059	.078	.076	**.044**	.044	.045	**.05**	.059	.063	**.043**
	$\tau_{n,3,\gamma}$	.057	.07	.088	.082	.09	.083	.043	.045	.069	.059	.06	.075
	$\tau_{n,4,\gamma}$	**.055**	.069	.086	.084	.09	.083	.043	.044	.068	.06	.061	.075
-0.5	$\tau_{n,0,\gamma}$	.116	.126	.13	.081	.075	.092	.074	.073	.09	.061	**.052**	.075
	$\tau_{n,1,\gamma}$	.118	.131	.139	.08	.08	.097	.072	.078	.093	.061	.054	.078
	$\tau_{n,2,\gamma}$	.054	.053	.053	.078	.066	**.045**	.042	.037	**.047**	.062	.053	.038
	$\tau_{n,3,\gamma}$	**.051**	.055	.079	.082	.085	.098	.042	.036	.066	.061	.055	.076
	$\tau_{n,4,\gamma}$	.052	.055	.08	.08	.087	.098	.041	.037	.065	.06	.055	.076
0	$\tau_{n,0,\gamma}$	.1	.114	.113	.069	.077	.089	.088	.087	.103	.079	.072	.089
	$\tau_{n,1,\gamma}$	.106	.117	.123	.073	.081	.091	.09	.09	.105	.082	.072	.09
	$\tau_{n,2,\gamma}$	.038	.038	.041	.068	**.064**	.034	.06	**.057**	.058	.075	.065	**.048**
	$\tau_{n,3,\gamma}$	.039	**.047**	.057	.077	.077	.095	.062	.058	.068	.075	.068	.083
	$\tau_{n,4,\gamma}$	.041	.046	.058	.076	.078	.095	.064	.059	.068	.076	.066	.083

Table 13: Relative frequency of a false alarm under $H_0^{(2)}$ and constant parameters.

Tables 14, 15, and 16 present the empirical size under $H_0^{(2)}$ and the following mean settings:

$$\mu_{1,i} = \begin{cases} 0, & \text{if } i \le n + mn/2, \\ 1, & \text{else,} \end{cases} \quad \text{and} \quad \mu_{2,i} \equiv 0, \tag{$\mathrm{I}_\mu^{(c)}$}$$

$$\mu_{1,i} = \begin{cases} 0, & \text{if } i \le n + mn/4, \\ 1, & \text{else,} \end{cases} \quad \text{and} \quad \mu_{2,i} = \begin{cases} 1, & \text{if } i \le n + mn/4, \\ 0, & \text{else,} \end{cases} \tag{$\mathrm{II}_\mu^{(c)}$}$$

or

$$\mu_{1,i} = \begin{cases} 2, & \text{if } i \le n + 3mn/4, \\ 0, & \text{else,} \end{cases} \quad \text{and} \quad \mu_{2,i} \equiv 0. \tag{$\mathrm{III}_\mu^{(c)}$}$$

As expected $\tau_{n,1,\gamma}$ and $\tau_{n,2,\gamma}$ are not suitable. In particular, this is at least obvious in the mean setting ($\mathrm{II}_\mu^{(c)}$), where the means change in both time series at the same time. In this setting the relative frequencies of the false alarms of the procedures $\tau_{n,3\gamma}$ and $\tau_{n,4,\gamma}$ amount over 10%. Moreover, the early change in the mean, setting ($\mathrm{I}_\mu^{(c)}$), is handled slightly better than the late changes of mean setting ($\mathrm{III}_\mu^{(c)}$).

		$n = 200$						$n = 2000$					
		$m = 0.5$			$m = 1.5$			$m = 0.5$			$m = 1.5$		
ρ_0	γ	0	.25	.45	0	.25	.45	0	.25	.45	0	.25	.45
-0.9	$\tau_{n,0,\gamma}$	.132	.153	.151	.087	.089	.083	.07	.071	.094	.069	.065	.08
	$\tau_{n,1,\gamma}$	.192	.199	.176	.091	.095	.089	.127	.116	.124	.084	.078	.082
	$\tau_{n,2,\gamma}$	.071	.07	.062	.083	.074	**.045**	.066	.06	.055	.075	.072	**.043**
	$\tau_{n,3,\gamma}$	.059	.067	.087	.084	.087	.08	**.05**	.054	.071	.065	.062	.074
	$\tau_{n,4,\gamma}$	**.055**	.065	.085	.082	.088	.08	.046	.049	.069	.063	.063	.074
-0.5	$\tau_{n,0,\gamma}$	.116	.126	.13	.081	.075	.092	.074	.073	.09	.061	.052	.075
	$\tau_{n,1,\gamma}$	.182	.184	.162	.088	.083	.097	.136	.134	.13	.085	.064	.08
	$\tau_{n,2,\gamma}$	.076	.076	.062	.083	.07	**.046**	.074	.059	.055	.069	.056	.038
	$\tau_{n,3,\gamma}$	.058	.059	.083	.082	.089	.094	**.047**	.04	.067	.057	**.052**	.076
	$\tau_{n,4,\gamma}$	.048	**.051**	.08	.079	.085	.094	.044	.041	.066	.058	.054	.077
0	$\tau_{n,0,\gamma}$	.1	.114	.113	.069	.077	.089	.088	.087	.103	.079	.072	.089
	$\tau_{n,1,\gamma}$	.168	.175	.16	.097	.092	.097	.176	.172	.147	.109	.09	.094
	$\tau_{n,2,\gamma}$	.063	.06	.057	.073	.069	**.037**	.095	.086	.074	.091	.076	**.05**
	$\tau_{n,3,\gamma}$	.043	**.051**	.064	.072	.079	.092	.065	.058	.067	.079	.069	.081
	$\tau_{n,4,\gamma}$	.042	.048	.06	.07	.078	.093	.058	**.055**	.067	.078	.067	.083

Table 14: Relative frequency of a false alarm under $H_0^{(2)}$ and mean setting ($\mathrm{I}_\mu^{(c)}$).

		$n = 200$						$n = 2000$					
		$m = 0.5$			$m = 1.5$			$m = 0.5$			$m = 1.5$		
ρ_0	γ	0	.25	.45	0	.25	.45	0	.25	.45	0	.25	.45
-0.9	$\tau_{n,0,\gamma}$	.132	.153	.151	.087	.089	**.083**	.07	.071	.094	.069	.065	.08
	$\tau_{n,1,\gamma}$	.979	.976	.964	.898	.856	.725	1	1	1	1	1	1
	$\tau_{n,2,\gamma}$	.322	.32	.267	.308	.274	.175	.948	.953	.92	.949	.936	.836
	$\tau_{n,3,\gamma}$	**.111**	.148	.167	.102	.114	.098	.1	.133	.16	.103	.092	.099
	$\tau_{n,4,\gamma}$	.117	.155	.174	.094	.105	.09	**.066**	.106	.137	.066	**.061**	.076
-0.5	$\tau_{n,0,\gamma}$	**.116**	.126	.13	.081	**.075**	.092	.074	**.073**	.09	.061	**.052**	.075
	$\tau_{n,1,\gamma}$	.995	.996	.992	.957	.949	.884	1	1	1	1	1	1
	$\tau_{n,2,\gamma}$	.387	.384	.324	.387	.345	.234	.989	.988	.982	.99	.985	.961
	$\tau_{n,3,\gamma}$	.113	.143	.161	.094	.1	.113	.089	.138	.166	.069	.057	.079
	$\tau_{n,4,\gamma}$	.123	.166	.176	.086	.093	.108	.082	.13	.161	.062	.055	.075
0	$\tau_{n,0,\gamma}$	.1	.114	.113	**.069**	.077	.089	.088	**.087**	.103	.079	.072	.089
	$\tau_{n,1,\gamma}$	.998	.998	.998	.984	.976	.945	1	1	1	1	1	1
	$\tau_{n,2,\gamma}$	.441	.441	.367	.463	.407	.238	.996	.996	.995	.996	.996	.988
	$\tau_{n,3,\gamma}$	**.097**	.136	.154	.085	.095	.108	.106	.17	.195	.077	.064	.089
	$\tau_{n,4,\gamma}$	.114	.171	.186	.081	.091	.107	.11	.163	.188	.075	**.063**	.084

Table 15: Relative frequency of a false alarm under $H_0^{(2)}$ and mean setting ($\mathrm{II}_\mu^{(c)}$).

		$n = 200$						$n = 2000$					
		$m = 0.5$			$m = 1.5$			$m = 0.5$			$m = 1.5$		
ρ_0	γ	0	.25	.45	0	.25	.45	0	.25	.45	0	.25	.45
-0.9	$\tau_{n,0,\gamma}$	.132	.153	.151	.087	.089	.083	.07	.071	.094	.069	.065	.08
	$\tau_{n,1,\gamma}$	.215	.213	.186	.1	.099	.09	.155	.138	.132	.078	.068	.081
	$\tau_{n,2,\gamma}$	.108	.1	.073	.081	.077	**.047**	.093	.077	.068	.068	.061	**.044**
	$\tau_{n,3,\gamma}$	.086	.09	.099	.083	.088	.083	.06	.059	.073	.063	.06	.076
	$\tau_{n,4,\gamma}$	**.064**	.074	.088	.083	.087	.083	.045	**.048**	.069	.061	.061	.075
-0.5	$\tau_{n,0,\gamma}$	.116	.126	.13	.081	.075	.092	.074	.073	.09	.061	**.052**	.075
	$\tau_{n,1,\gamma}$	.226	.222	.192	.086	.086	.099	.199	.18	.15	.076	.065	.079
	$\tau_{n,2,\gamma}$	.11	.099	.082	.079	.07	**.047**	.117	.105	.08	.073	.063	.036
	$\tau_{n,3,\gamma}$	.086	.083	.092	.083	.087	.098	.06	**.052**	.075	.061	.054	.076
	$\tau_{n,4,\gamma}$	**.065**	.07	.084	.081	.086	.097	**.048**	.04	.067	.06	.054	.076
0	$\tau_{n,0,\gamma}$	.1	.114	.113	.069	.077	.089	.088	.087	.103	.079	.072	.089
	$\tau_{n,1,\gamma}$	.225	.209	.186	.08	.086	.094	.242	.222	.19	.097	.081	.095
	$\tau_{n,2,\gamma}$	.114	.101	.074	.072	.069	**.038**	.151	.134	.092	.085	.069	**.051**
	$\tau_{n,3,\gamma}$	.075	.075	.077	.076	.078	.093	.08	.07	.072	.077	.068	.084
	$\tau_{n,4,\gamma}$	**.052**	.055	.065	.078	.078	.094	.066	**.061**	.068	.076	.066	.083

Table 16: Relative frequency of a false alarm under $H_0^{(2)}$ and mean setting ($\mathrm{III}_\mu^{(c)}$).

7.2.2 Influence of the Variance Estimates under $H_0^{(2)}$ and Known Means

In this sub-subsection, we focus on the finite sample behavior of the stopping times $\tau_{n,\psi,\gamma}$ with $\psi = 0, 5, \ldots, ,8$ under $H_0^{(2)}$. Table 17 presents the empirical sizes under $H_0^{(2)}$ and constant variances. In the case of high correlations and $n = 200$ each stopping time lies over 10%. If additionally γ is high, only $\tau_{n,5,\gamma}$ seems to approximate the given level $\alpha = 5\%$. The stopping times $\tau_{n,8,\gamma}$, which sequentially use change-point estimates for the mean, do not work well. Moreover, the relative frequencies lie over 10% in the case of $\rho_0 \in \{-0.5, -0.9\}$. In most cases the relative frequencies of the false alarms of $\tau_{2000,8,\gamma}$ are smaller than $\tau_{200,8,\gamma}$, but the differences are slight.

		$n = 200$						$n = 2000$					
		$m = 0.5$			$m = 1.5$			$m = 0.5$			$m = 1.5$		
ρ_0	γ	0	.25	.45	0	.25	.45	0	.25	.45	0	.25	.45
-0.9	$\tau_{n,0,\gamma}$	.132	.153	.151	**.087**	.089	.083	.07	.071	.094	.069	.065	.08
	$\tau_{n,5,\gamma}$	.134	.15	.165	.095	.104	.11	.076	.073	.086	.058	**.057**	.078
	$\tau_{n,6,\gamma}$	**.111**	.251	.279	.188	.232	.234	**.056**	.108	.299	.079	.216	.246
	$\tau_{n,7,\gamma}$	.148	.28	.301	.211	.256	.25	.073	.116	.306	.091	.223	.25
	$\tau_{n,8,\gamma}$	.164	.178	.181	.159	.159	.143	.145	.153	.171	.123	.129	.137
-0.5	$\tau_{n,0,\gamma}$	.116	.126	.13	.081	**.075**	.092	.074	.073	.09	.061	**.052**	.075
	$\tau_{n,5,\gamma}$	.108	.115	.139	.091	.091	.105	.071	.073	.088	.063	.054	.078
	$\tau_{n,6,\gamma}$	**.053**	.061	.097	.084	.091	.156	**.057**	.067	.084	.072	.065	.101
	$\tau_{n,7,\gamma}$	.054	.063	.097	.082	.093	.157	.058	.069	.084	.073	.066	.102
	$\tau_{n,8,\gamma}$	.11	.118	.121	.109	.111	.123	.101	.109	.126	.089	.096	.109
0	$\tau_{n,0,\gamma}$	.1	.114	.113	.069	.077	.089	.088	.087	.103	.079	.072	.089
	$\tau_{n,5,\gamma}$	.102	.124	.135	.08	.086	.1	.085	.091	.096	.084	.07	.082
	$\tau_{n,6,\gamma}$	**.042**	**.042**	.023	.074	**.066**	.038	.062	.056	.045	.078	.072	**.04**
	$\tau_{n,7,\gamma}$	.041	.04	.02	.075	.067	.037	.063	.056	**.046**	.078	.071	**.04**
	$\tau_{n,8,\gamma}$	.094	.109	.116	.1	.112	.13	.082	.091	.102	.079	.083	.094

Table 17: Relative frequency of a false alarm under $H_0^{(2)}$ and constant parameters

Tables 18, 19, and 20 present the relative frequencies of false alarms under the following variance settings:

$$\sigma_{1,i}^2 = \begin{cases} 1, & \text{if } i \le n + nm/2, \\ 2, & \text{else,} \end{cases} \qquad \text{and} \qquad \sigma_{2,i}^2 \equiv 1, \tag{$\mathrm{I}_\sigma^{(c)}$}$$

$$\sigma_{1,i}^2 = \begin{cases} 1, & \text{if } i \le n + nm/4, \\ 2, & \text{else,} \end{cases} \qquad \text{and} \qquad \sigma_{2,i}^2 = \begin{cases} 1, & \text{if } i \le n + nm/4, \\ 0.5, & \text{else,} \end{cases} \tag{$\mathrm{II}_\sigma^{(c)}$}$$

or

$$\sigma_{1,i}^2 = \begin{cases} 1, & \text{if } i \le n + 3nm/4, \\ 2, & \text{else,} \end{cases} \qquad \text{and} \qquad \sigma_{2,i}^2 \equiv 1. \tag{$\mathrm{III}_\sigma^{(c)}$}$$

Under these settings only $\tau_{n,0,\gamma}$, $\tau_{n,7,\gamma}$, and $\tau_{n,8,\gamma}$ fulfill the convergence in distribution towards a maximum of the absolute value of a weighted Brownian motion. In contrast to the non-constant mean settings, here the unsuitable stopping times can also converge to a Gaussian process in each presented variance setting, cf. Remark 5.2.2. Nevertheless, $\tau_{n,7,\gamma}$ and $\tau_{n,8,\gamma}$ behave acceptably for small or moderate correlations and for $\gamma = 0$. Overall, the influence is unimportant for $\rho = 0$, and $\tau_{n,7,\gamma}$ performs better than $\tau_{n,8,\gamma}$.

		$n = 200$						$n = 2000$					
		$m = 0.5$			$m = 1.5$			$m = 0.5$			$m = 1.5$		
ρ_0	γ	0	.25	.45	0	.25	.45	0	.25	.45	0	.25	.45
-0.9	$\tau_{n,0,\gamma}$	**.132**	.153	.151	**.087**	.089	.083	**.07**	.071	.094	.069	**.065**	.08
	$\tau_{n,5,\gamma}$	.364	.355	.314	.176	.162	.139	.912	.897	.841	.479	.379	.233
	$\tau_{n,6,\gamma}$	.178	.306	.312	.206	.243	.243	.464	.451	.483	.261	.335	.298
	$\tau_{n,7,\gamma}$	.18	.311	.321	.221	.254	.251	.166	.213	.356	.114	.237	.26
	$\tau_{n,8,\gamma}$	.179	.191	.191	.155	.16	.145	.159	.173	.183	.129	.13	.136
-0.5	$\tau_{n,0,\gamma}$	.116	.126	.13	.081	.075	.092	.074	.073	.09	.061	**.052**	.075
	$\tau_{n,5,\gamma}$	.28	.265	.231	.124	.126	.116	.635	.611	.514	.263	.215	.137
	$\tau_{n,6,\gamma}$	.065	.071	.099	**.071**	.082	.154	.087	.087	.102	.076	.067	.1
	$\tau_{n,7,\gamma}$	**.061**	.072	.101	.076	.092	.159	**.065**	.071	.09	.074	.068	.102
	$\tau_{n,8,\gamma}$	.115	.122	.126	.109	.111	.124	.107	.112	.127	.091	.096	.11
0	$\tau_{n,0,\gamma}$	.1	.114	.113	.069	.077	.089	.088	.087	.103	.079	.072	.089
	$\tau_{n,5,\gamma}$	.177	.19	.183	.086	.094	.104	.152	.144	.133	.093	.078	.089
	$\tau_{n,6,\gamma}$	.055	**.047**	.027	.071	.065	**.036**	.074	.071	**.052**	.072	.066	.037
	$\tau_{n,7,\gamma}$	**.047**	.043	.024	.073	.065	.035	.066	.061	.045	.077	.07	**.038**
	$\tau_{n,8,\gamma}$	.096	.113	.117	.102	.113	.129	.081	.092	.102	.08	.083	.094

Table 18: Relative frequency of a false alarm under $H_0^{(2)}$ and variance setting ($\mathrm{I}_\sigma^{(c)}$).

		$n = 200$						$n = 2000$					
		$m = 0.5$			$m = 1.5$			$m = 0.5$			$m = 1.5$		
ρ_0	γ	0	.25	.45	0	.25	.45	0	.25	.45	0	.25	.45
-0.9	$\tau_{n,0,\gamma}$	**.132**	.153	.151	.087	.089	**.083**	**.07**	.071	.094	.069	.065	.08
	$\tau_{n,5,\gamma}$	.134	.15	.165	.095	.104	.11	.076	.073	.086	.058	**.057**	.078
	$\tau_{n,6,\gamma}$	.466	.547	.505	.501	.502	.407	1	.999	1	1	1	1
	$\tau_{n,7,\gamma}$	.381	.493	.482	.382	.407	.339	.439	.553	.643	.289	.382	.346
	$\tau_{n,8,\gamma}$	.254	.263	.259	.225	.217	.193	.221	.247	.253	.172	.172	.165
-0.5	$\tau_{n,0,\gamma}$	.116	.126	.13	.081	**.075**	.092	.074	.073	.09	.061	.052	.075
	$\tau_{n,5,\gamma}$	.108	.115	.139	.091	.091	.105	.071	.073	.088	.063	.054	.078
	$\tau_{n,6,\gamma}$	.061	.07	.101	.091	.092	.158	.141	.133	.145	.16	.145	.124
	$\tau_{n,7,\gamma}$	**.06**	.068	.1	.092	.096	.159	.071	.075	.095	.098	.076	.105
	$\tau_{n,8,\gamma}$	.104	.11	.121	.105	.108	.126	.107	.115	.126	.098	.097	.11
0	$\tau_{n,0,\gamma}$	.1	.114	.113	.069	.077	.089	.088	.087	.103	.079	.072	.089
	$\tau_{n,5,\gamma}$	.102	.124	.135	.08	.086	.1	.085	.091	.096	.084	.07	.082
	$\tau_{n,6,\gamma}$	.036	.038	.023	.067	.056	.033	.049	.044	.034	.067	.063	.037
	$\tau_{n,7,\gamma}$	.038	.033	.024	.076	.064	.038	.058	.054	.045	.078	.073	.041
	$\tau_{n,8,\gamma}$	.092	.107	.115	.098	.113	.13	.081	.09	.104	.077	.081	.095

Table 19: Relative frequency of a false alarm under $H_0^{(2)}$ and variance setting ($\mathrm{II}_\sigma^{(c)}$).

		$n = 200$						$n = 2000$					
		$m = 0.5$			$m = 1.5$			$m = 0.5$			$m = 1.5$		
ρ_0	γ	0	.25	.45	0	.25	.45	0	.25	.45	0	.25	.45
-0.9	$\tau_{n,0,\gamma}$	**.132**	.153	.151	.087	.089	**.083**	**.07**	.071	.094	.069	**.065**	.08
	$\tau_{n,5,\gamma}$	.214	.216	.207	.103	.112	.112	.442	.397	.328	.093	.078	.088
	$\tau_{n,6,\gamma}$	.155	.278	.293	.193	.235	.236	.342	.341	.411	.12	.241	.257
	$\tau_{n,7,\gamma}$	.181	.299	.311	.215	.259	.248	.184	.207	.348	.106	.234	.256
	$\tau_{n,8,\gamma}$	.178	.185	.188	.159	.161	.145	.168	.172	.184	.127	.13	.136
-0.5	$\tau_{n,0,\gamma}$	.116	.126	.13	.081	**.075**	.092	.074	.073	.09	.061	**.052**	.075
	$\tau_{n,5,\gamma}$	.164	.159	.167	.096	.101	.106	.245	.214	.186	.08	.065	.082
	$\tau_{n,6,\gamma}$	**.062**	.071	.099	.081	.088	.155	.082	.08	.097	.075	.066	.1
	$\tau_{n,7,\gamma}$	**.062**	.07	.1	.081	.091	.157	.074	**.069**	.09	.076	.068	.101
	$\tau_{n,8,\gamma}$	.114	.121	.124	.109	.11	.123	.101	.109	.127	.09	.096	.109
0	$\tau_{n,0,\gamma}$	.1	.114	.113	.069	.077	.089	.088	.087	.103	.079	.072	.089
	$\tau_{n,5,\gamma}$	.131	.146	.149	.077	.087	.1	.114	.106	.111	.083	.075	.082
	$\tau_{n,6,\gamma}$	**.048**	.042	.025	.075	.064	**.038**	.075	.068	**.049**	.079	.069	**.04**
	$\tau_{n,7,\gamma}$	.045	.042	.021	.073	.067	.037	.068	.059	**.049**	.078	.069	**.04**
	$\tau_{n,8,\gamma}$	.097	.113	.119	.098	.112	.13	.083	.091	.103	.08	.083	.094

Table 20: Relative frequency of a false alarm under $H_0^{(2)}$ and variance setting ($\mathrm{III}_\sigma^{(c)}$).

7.2.3 Influence of the Combination of Mean and Variance Estimates under $H_0^{(2)}$

In this paragraph, we focus on the behavior of stopping times using mean and variance estimates. Table 21 presents the relative frequencies of false alarms under $H_0^{(2)}$ and constant parameters. If the correlation or γ is not high, each stopping approximates the given level α. If the correlation is high, only $\gamma = 0$ is useful.

Tables 22, 23, and 22 present the relative frequency of false stops under parameter setting $(\text{III}_\mu^{(c)} \& \text{I}_\sigma^{(c)})$, $(\text{II}_\mu^{(c)} \& \text{II}_\sigma^{(c)})$, and $(\text{I}_\mu^{(c)} \& \text{III}_\sigma^{(c)})$, respectively. If the correlation is high (such as -0.9), each stopping time using estimates for the parameter (even $\tau_{1,20}^{\gamma}$) nearly always stops falsely. If, however, the correlation is moderate and small, $\tau_{1,20}^{\gamma}$ approximates the given level acceptably and well, respectively.

		$n = 200$						$n = 2000$					
		$m = 0.5$			$m = 1.5$			$m = 0.5$			$m = 1.5$		
ρ_0	γ	0	.25	.45	0	.25	.45	0	.25	.45	0	.25	.45
-0.9	$\tau_{i,0,\gamma}$	**.132**	.153	.151	.087	.089	**.083**	.07	.071	.094	.069	.065	.08
	$\tau_{i,9,\gamma}$	.141	.159	.165	.094	.104	.108	.077	.074	.084	**.055**	.059	.079
	$\tau_{i,14,\gamma}$	.212	.283	.317	.212	.242	.24	**.057**	.221	.313	.076	.216	.247
	$\tau_{i,20,\gamma}$	.219	.359	.392	.275	.314	.294	.086	.159	.347	.103	.246	.275
-0.5	$\tau_{i,0,\gamma}$	.116	.126	.13	.081	**.075**	.092	.074	.073	.09	.061	**.052**	.075
	$\tau_{i,9,\gamma}$	.114	.116	.148	.089	.093	.108	.073	.071	.09	.061	.055	.079
	$\tau_{i,14,\gamma}$	**.057**	.069	.171	.085	.099	.111	**.059**	.064	.109	.073	.064	.096
	$\tau_{i,20,\gamma}$	**.057**	.07	.103	.086	.094	.161	.06	.068	.086	.071	.063	.099
0	$\tau_{i,0,\gamma}$	.1	.114	.113	.069	.077	.089	.088	.087	.103	.079	.072	.089
	$\tau_{i,9,\gamma}$	.097	.121	.139	.077	.086	.101	.086	.088	.096	.083	.071	.085
	$\tau_{i,14,\gamma}$	.042	.041	.026	.075	.067	**.041**	.06	**.053**	.047	.074	.072	.039
	$\tau_{i,20,\gamma}$	.043	**.044**	.034	.076	.072	.037	.063	**.053**	.047	.074	.071	**.042**

Table 21: Relative frequency of a false alarm under $H_0^{(2)}$ and constant parameters.

		$n = 200$						$n = 2000$					
		$m = 0.5$			$m = 1.5$			$m = 0.5$			$m = 1.5$		
ρ_0	γ	0	.25	.45	0	.25	.45	0	.25	.45	0	.25	.45
-0.9	$\tau_{i,0}^{\gamma}$	**.132**	.153	.151	.087	.089	**.083**	**.07**	.071	.094	.069	**.065**	.08
	$\tau_{i,9}^{\gamma}$	.392	.383	.347	.175	.167	.14	.898	.876	.826	.49	.384	.245
	$\tau_{i,14}^{\gamma}$	1	1	.998	.867	.799	.557	1	1	1	1	1	1
	$\tau_{i,20}^{\gamma}$	.779	.803	.772	.355	.374	.333	.869	.875	.878	.345	.392	.339
-0.5	$\tau_{i,0}^{\gamma}$	.116	.126	.13	.081	**.075**	.092	.074	**.073**	.09	.061	**.052**	.075
	$\tau_{i,9}^{\gamma}$	.347	.329	.299	.139	.13	.122	.671	.642	.568	.267	.222	.147
	$\tau_{i,14}^{\gamma}$	.2	.199	.231	.102	.102	.117	.879	.845	.767	.243	.192	.137
	$\tau_{i,20}^{\gamma}$	**.102**	.112	.119	.079	.094	.166	.117	.119	.122	.081	.069	.097
0	$\tau_{i,0}^{\gamma}$	.1	.114	.113	.069	.077	.089	.088	.087	.103	.079	.072	.089
	$\tau_{i,9}^{\gamma}$	.271	.271	.237	.098	.104	.106	.3	.282	.225	.112	.092	.096
	$\tau_{i,14}^{\gamma}$	.057	.06	.036	.072	.065	**.039**	.09	.083	.054	.07	.067	.037
	$\tau_{i,20}^{\gamma}$	**.054**	.055	.036	.071	.07	.036	.075	.071	**.052**	.072	.068	**.039**

Table 22: Relative frequency of a false alarm under $H_0^{(2)}$ and parameter setting $(\text{III}_\mu^{(c)} \& \text{I}_\sigma^{(c)})$.

		$n = 200$						$n = 2000$					
		$m = 0.5$			$m = 1.5$			$m = 0.5$			$m = 1.5$		
ρ_0	γ	0	.25	.45	0	.25	.45	0	.25	.45	0	.25	.45
-0.9	$\tau_{i,0,\gamma}$	**.132**	.153	.151	.087	.089	**.083**	**.07**	.071	.094	.069	**.065**	.08
	$\tau_{i,9,\gamma}$	.759	.73	.607	.4	.299	.158	1	1	1	1	1	1
	$\tau_{i,14,\gamma}$	.396	.446	.443	.379	.377	.306	.996	.994	.987	.99	.985	.919
	$\tau_{i,20,\gamma}$	.838	.879	.869	.665	.672	.572	1	1	1	1	1	.999
-0.5	$\tau_{i,0,\gamma}$	.116	.126	.13	.081	**.075**	.092	.074	**.073**	.09	.061	**.052**	.075
	$\tau_{i,9,\gamma}$	.871	.854	.75	.558	.457	.236	1	1	1	1	1	1
	$\tau_{i,14,\gamma}$	**.046**	.069	.17	.102	.104	.107	.351	.346	.317	.441	.375	.237
	$\tau_{i,20,\gamma}$	.085	.098	.117	.087	.099	.167	.146	.145	.15	.17	.155	.126
0	$\tau_{i,0,\gamma}$	.1	.114	.113	.069	.077	.089	.088	.087	.103	.079	.072	.089
	$\tau_{i,9,\gamma}$	.94	.931	.846	.678	.549	.277	1	1	1	1	1	1
	$\tau_{i,14,\gamma}$	.241	.231	.149	.309	.261	.121	.98	.979	.959	.988	.981	.93
	$\tau_{i,20,\gamma}$	**.051**	.065	.054	.066	.068	**.037**	**.055**	.081	.084	.066	**.06**	.038

Table 23: Relative frequency of a false alarm under $H_0^{(2)}$ and parameter setting $(\text{II}_\mu^{(c)} \& \text{II}_\sigma^{(c)})$.

		$n = 200$						$n = 2000$					
		$m = 0.5$			$m = 1.5$			$m = 0.5$			$m = 1.5$		
ρ_0	γ	0	.25	.45	0	.25	.45	0	.25	.45	0	.25	.45
-0.9	$\tau_{i,0,\gamma}$	**.132**	.153	.151	.087	.089	**.083**	**.07**	.071	.094	.069	**.065**	.08
	$\tau_{i,9,\gamma}$	.242	.245	.232	.111	.113	.112	.447	.415	.348	.102	.086	.096
	$\tau_{i,14,\gamma}$	.904	.902	.862	.768	.715	.554	1	1	1	1	1	1
	$\tau_{i,20,\gamma}$	.648	.705	.676	.379	.39	.35	.737	.759	.767	.179	.291	.293
-0.5	$\tau_{i,0,\gamma}$	.116	.126	.13	.081	**.075**	.092	.074	**.073**	.09	.061	**.052**	.075
	$\tau_{i,9,\gamma}$	.21	.199	.197	.099	.099	.107	.301	.282	.238	.096	.083	.088
	$\tau_{i,14,\gamma}$	.124	.134	.209	.115	.117	.121	.439	.409	.356	.26	.199	.154
	$\tau_{i,20,\gamma}$	**.099**	.105	.115	.087	.096	.164	.09	.094	.104	.074	.064	.1
0	$\tau_{i,0,\gamma}$	.1	.114	.113	.069	.077	.089	.088	.087	.103	.079	.072	.089
	$\tau_{i,9,\gamma}$	.184	.193	.185	.095	.094	.102	.208	.192	.165	.11	.09	.088
	$\tau_{i,14,\gamma}$	.048	**.052**	.032	.07	.065	**.038**	.073	.067	.055	.077	.068	.035
	$\tau_{i,20,\gamma}$	.048	.048	.035	.073	.067	.034	.076	.061	**.051**	.075	.069	**.039**

Table 24: Relative frequency of a false alarm under $H_0^{(2)}$ and parameter setting $(\mathrm{I}_\mu^{(c)}\&\mathrm{III}_\sigma^{(c)})$.

7.2.4 Stopping behavior under some Alternatives

In this sub-subsection, we focus on the stopping behavior under some alternatives. In doing so, we focus on the following setups:

$$\rho_i = \rho_0 + \Delta_\rho \mathbb{1}_{\{i > n + [n\theta_\rho]\}} \tag{7.2.1}$$

with $\rho_0 \in \{-0.9, -0.5, 0\}$, $\Delta_\rho \in \{0.1, 0.25, 0.5\}$, and $\theta_\rho \in \{0.25, 0.5, 0.75\}$.

Mean Influence In this paragraph, we focus on the behaviors of stopping times under the alternatives and different mean estimates. Figure 23 presents relative frequency of alarms under constant parameters. The relative frequency of stops increases with growing n, Δ_ρ, and m. Furthermore, $\tau_{n,0,\gamma}$ and $\tau_{n,1,\gamma}$ produce more alarms than $\tau_{n,2,\gamma}$, $\tau_{n,3,\gamma}$, and $\tau_{n,4,\gamma}$, where the latter two have nearly the same the same relative frequencies.

Now, we focus on the alternatives under mean settings $(\mathrm{I}_\mu^{(c)})$ and $(\mathrm{II}_\mu^{(c)})$, cf. Figure 24 and 25. The difference between the constant parameter setting and mean setting $(\mathrm{I}_\mu^{(c)})$ is not really observable. However, in the mean setting $(\mathrm{II}_\mu^{(c)})$ the unusable stopping times $\tau_{n,1,\gamma}$ (red) and $\tau_{n,2,\gamma}$ (green) possess no monotone power in Δ_ρ. In most of the cases the alarms are caused by the simultaneous structural breaks in the means of X and Y. Here, the useful stopping times $\tau_{n,3,\gamma}$ (yellow) are slightly better than $\tau_{n,4,\gamma}$ (blue), because $\tau_{n,4,\gamma}$ uses a sequential mean estimate which corrects false early alarms for the mean change detection.

Variance Influence In this paragraph, we focus on the behavior of stopping times under the alternatives and the different variance estimates. Figure 26 presents their relative frequency of alarms under constant parameters. The relative frequency of stops increases with growing n, Δ_ρ and m. Furthermore, $\tau_{n,6,\gamma}$ (green) and $\tau_{n,7,\gamma}$ (yellow) produce more alarms than $\tau_{n,1,\gamma}$ (black), $\tau_{n,5,\gamma}$ (red), and $\tau_{n,8,\gamma}$ (blue), where the relative frequencies of $\tau_{n,6,\gamma}$ and $\tau_{n,7,\gamma}$ are nearly the same. In addition, the alarm appears more frequently in the case of high correlation. Secondly, the sequential estimation of variances after a false detection of a change in the variance yields a much conservative procedure $\tau_{n,8,\gamma}$.

Now, we focus on the alternatives under variance setting $(\mathrm{I}_\sigma^{(c)})$ and $(\mathrm{II}_\sigma^{(c)})$, cf. Figure 27 and 28. Again, the unusable stopping times $\tau_{n,5,\gamma}$ and $\tau_{n,6,\gamma}$ profit by the structure changes in some cases. The two useful stopping times $\tau_{n,7,\gamma}$ and $\tau_{n,8,\gamma}$ behave as in the constant parameter setting. Since the changes in the variances of X and Y cancel each other out in case $(\mathrm{II}_\sigma^{(c)})$, $\tau_{n,5,\gamma}$ (red) behaves as in a constant parameter setting.

Finally, we consider the behavior under $(\mathrm{II}_\mu^{(c)}+\mathrm{II}_\sigma^{(c)})$, cf. Figure 29, where only the stopping times $\tau_{n,20,\gamma}$ stay useful. We observe, that these stopping times and the ones using the exact parameters are the only ones which have monotone power in each case.

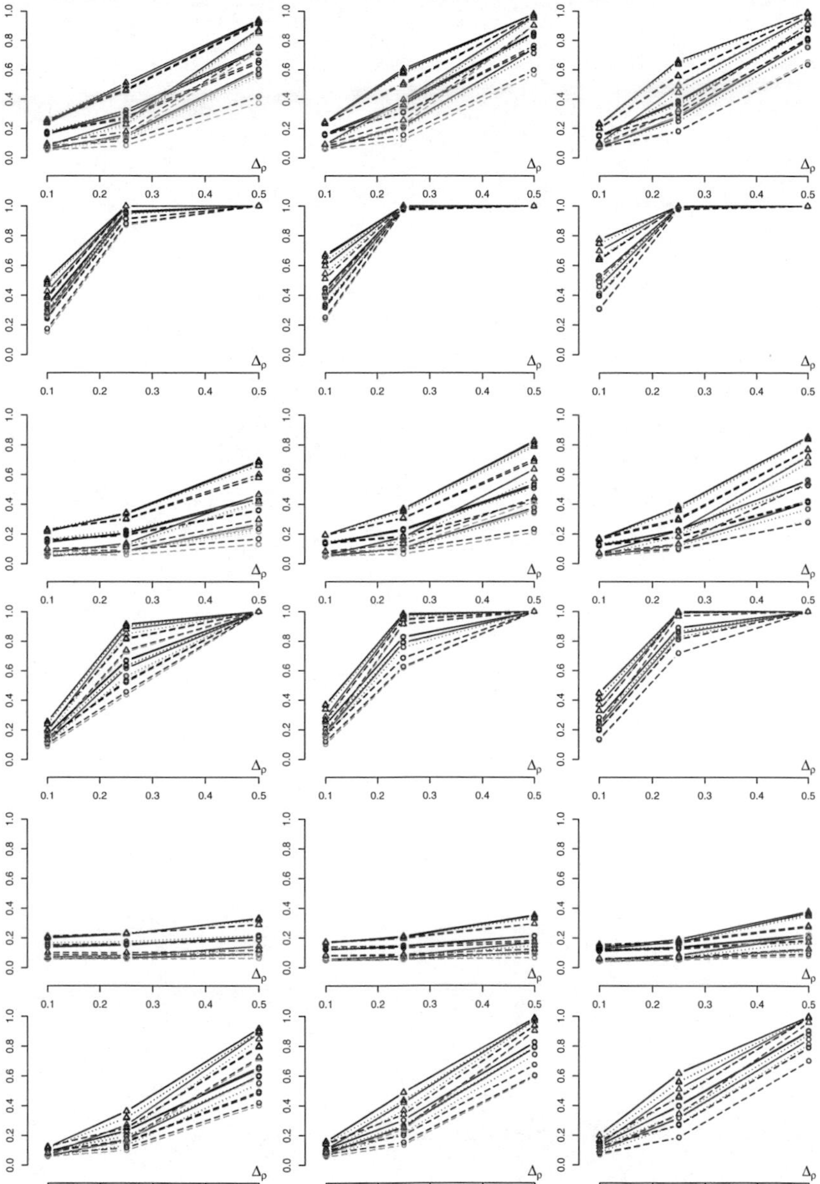

Figure 23: Each graph contains the relative frequency of stops for $\Delta_\rho = 0.1,\, 0.25,\, 0.5$ under constant parameters. In the first till the third column $\rho = -0.9,\ -0.5,\ 0$ and in the odd rows $n = 200$ and in even ones $n = 2000$. In the first, second, and third two rows $\theta_\rho = 0.25,\, 0.5,\, 0.75$.

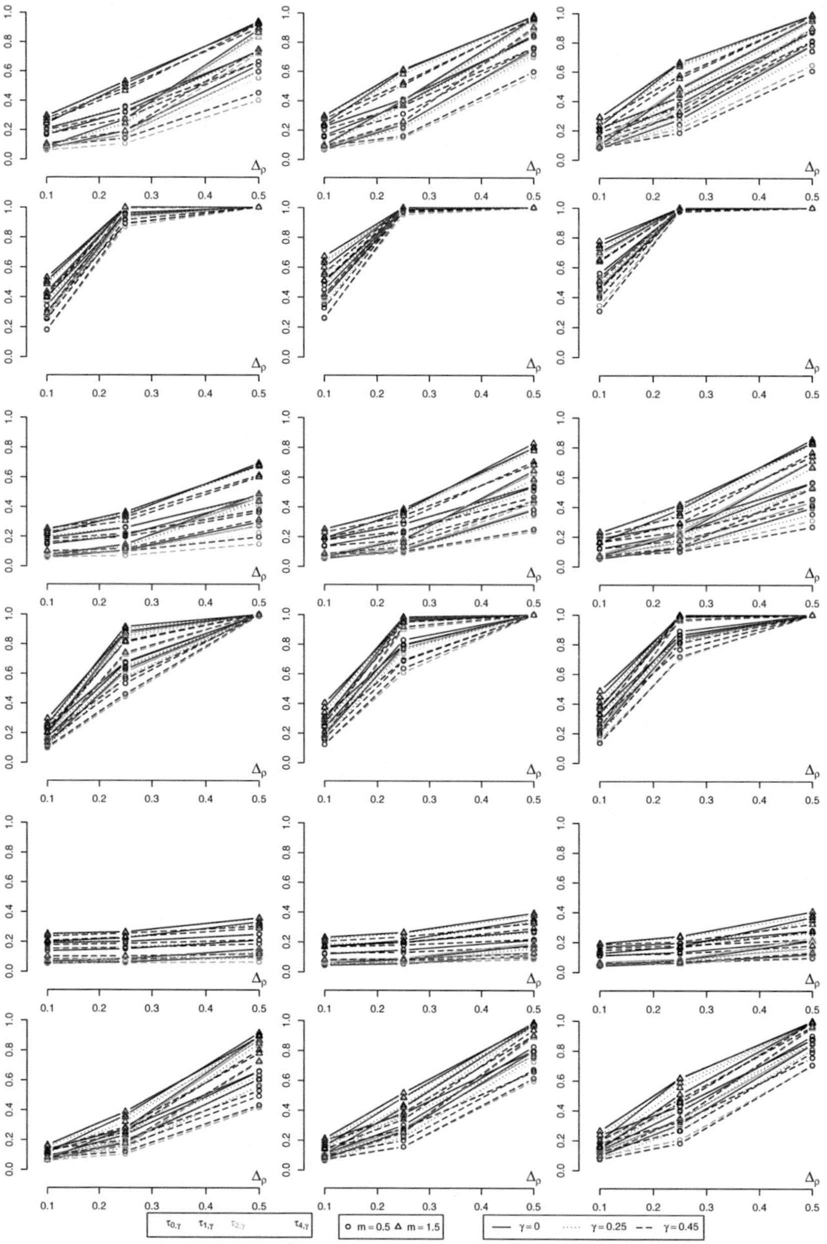

Figure 24: Each graph contains the relative frequency of stops for $\Delta_\rho = 0.1, 0.25, 0.5$ under mean setting $(\mathrm{I}_\mu^{(c)})$. In the first till the third column $\rho = -0.9, -0.5, 0$ and in the odd rows $n = 200$ and in even ones $n = 2000$. In the first, second, and third two rows $\theta_\rho = 0.25, 0.5, 0.75$.

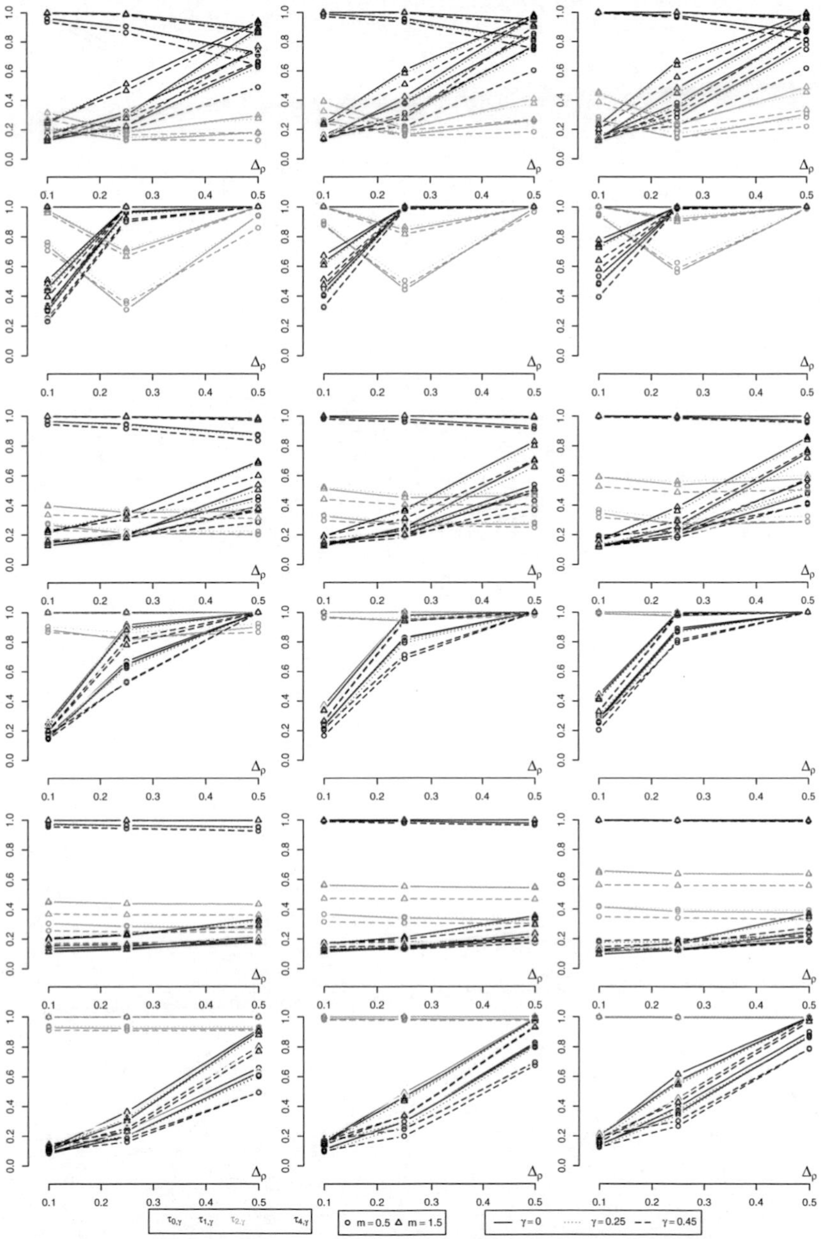

Figure 25: Each graph contains the relative frequency of stops for $\Delta_\rho = 0.1, 0.25, 0.5$ under mean setting $(\mathrm{II}_\mu^{(c)})$. In the first till the third column $\rho = -0.9, -0.5, 0$ and in the odd rows $n = 200$ and in even ones $n = 2000$. In the first, second, and third two rows $\theta_\rho = 0.25, 0.5, 0.75$.

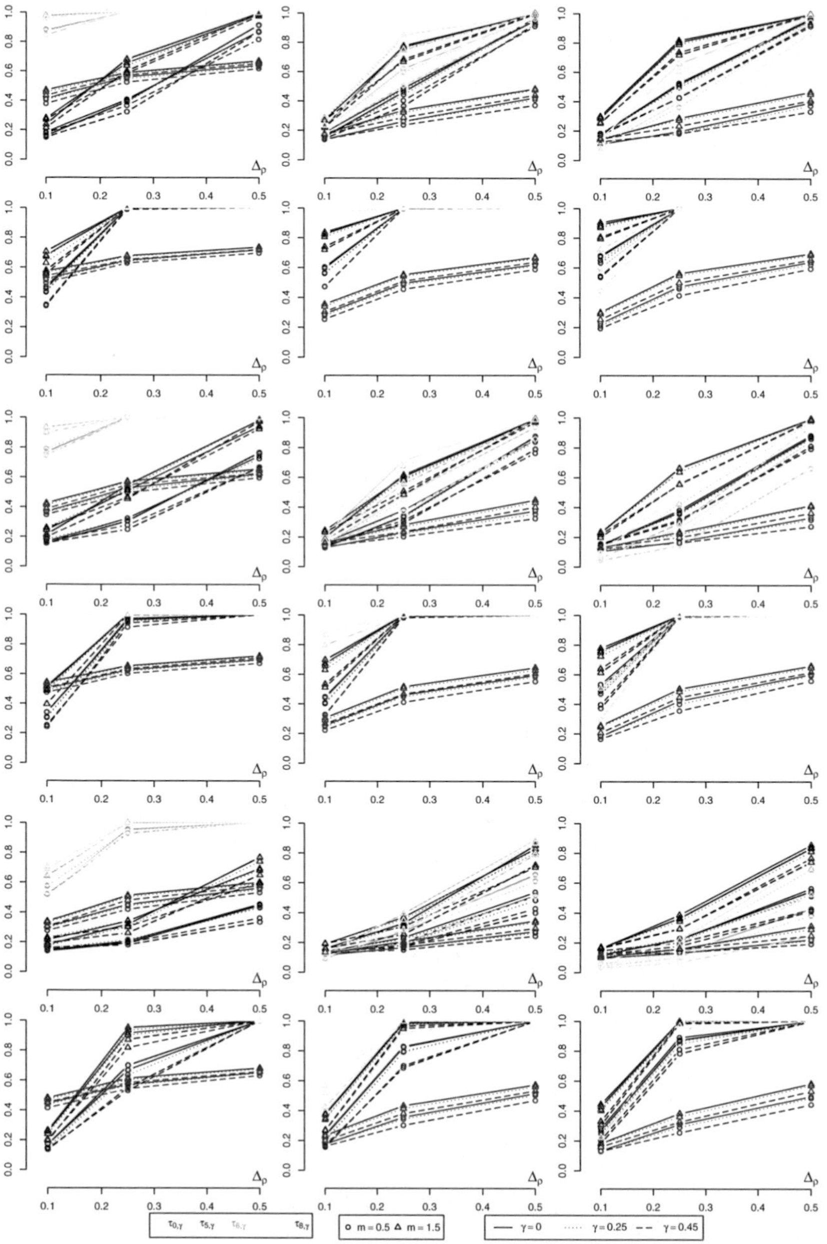

Figure 26: Each graph contains the relative frequency of stops for $\Delta_\rho = 0.1, 0.25, 0.5$ under constant parameters. In the first till the third column $\rho = -0.9, -0.5, 0$ and in the odd rows $n = 200$ and in even ones $n = 2000$. In the first, second, and third two rows $\theta_\rho = 0.25, 0.5, 0.75$.

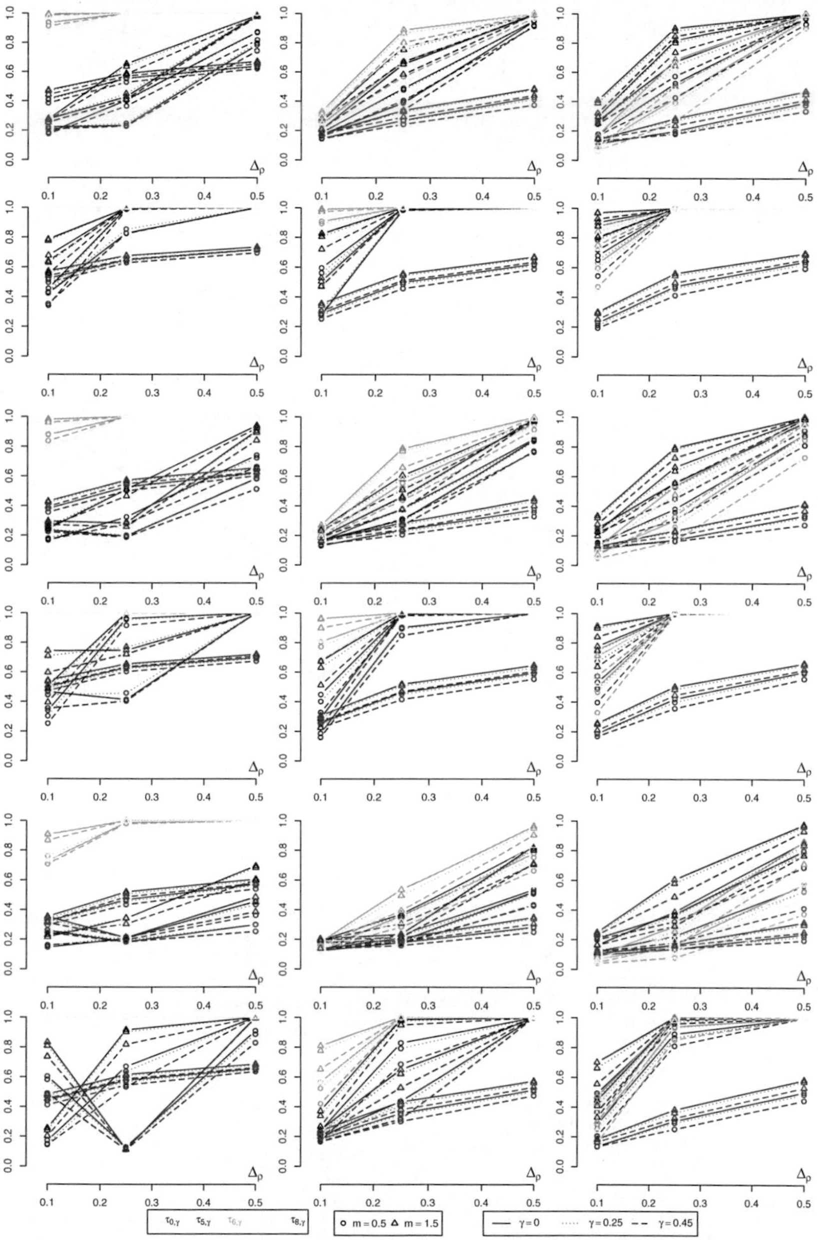

Figure 27: Each graph contains the relative frequency of stops for $\Delta_\rho = 0.1, 0.25, 0.5$ under variance setting $(\mathrm{I}_\sigma^{(c)})$. In the first till the third column $\rho = -0.9, -0.5, 0$ and in the odd rows $n = 200$ and in even ones $n = 2000$. In the first, second, and third two rows $\theta_\rho = 0.25, 0.5, 0.75$.

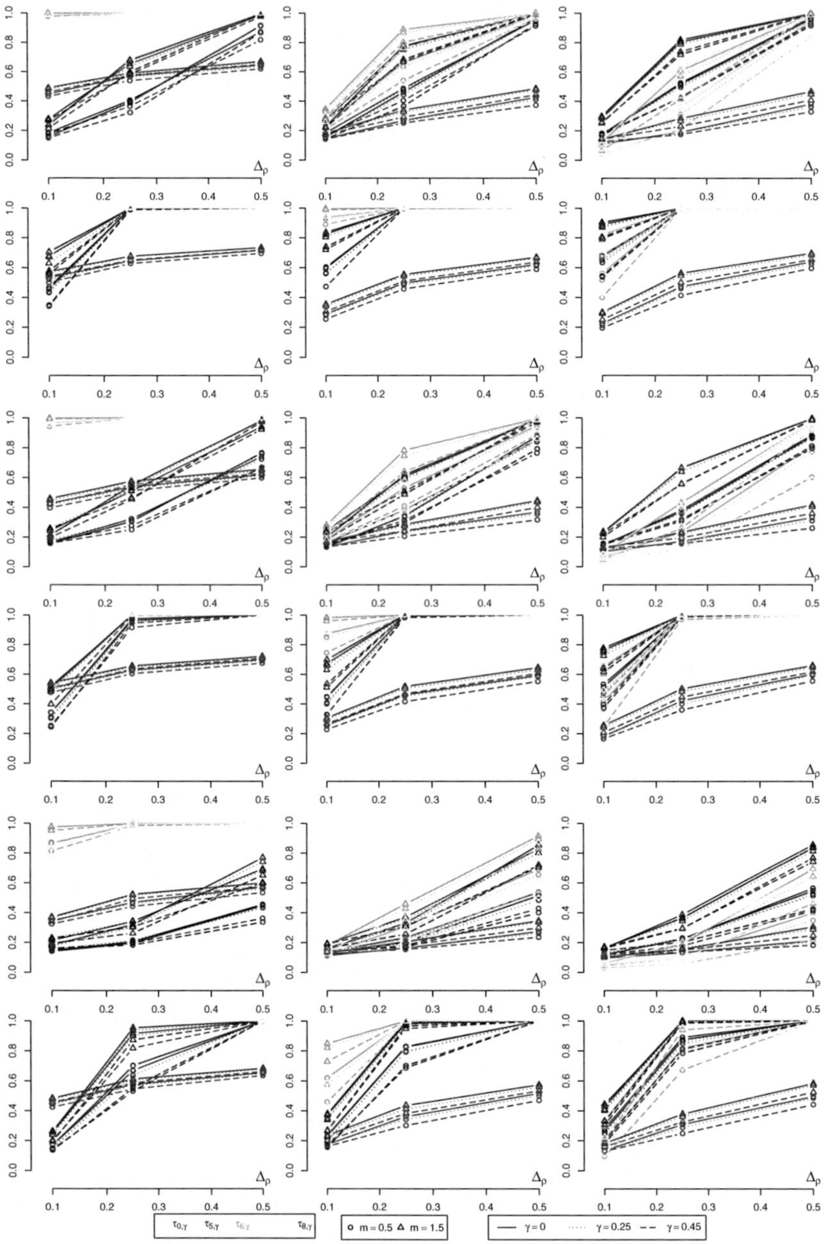

Figure 28: Each graph contains the relative frequency of stops for $\Delta_\rho = 0.1,\ 0.25,\ 0.5$ under variance setting $(\mathrm{II}_\sigma^{(c)})$. In the first till the third column $\rho = -0.9,\ -0.5,\ 0$ and in the odd rows $n = 200$ and in even ones $n = 2000$. In the first, second, and third two rows $\theta_\rho = 0.25,\ 0.5,\ 0.75$.

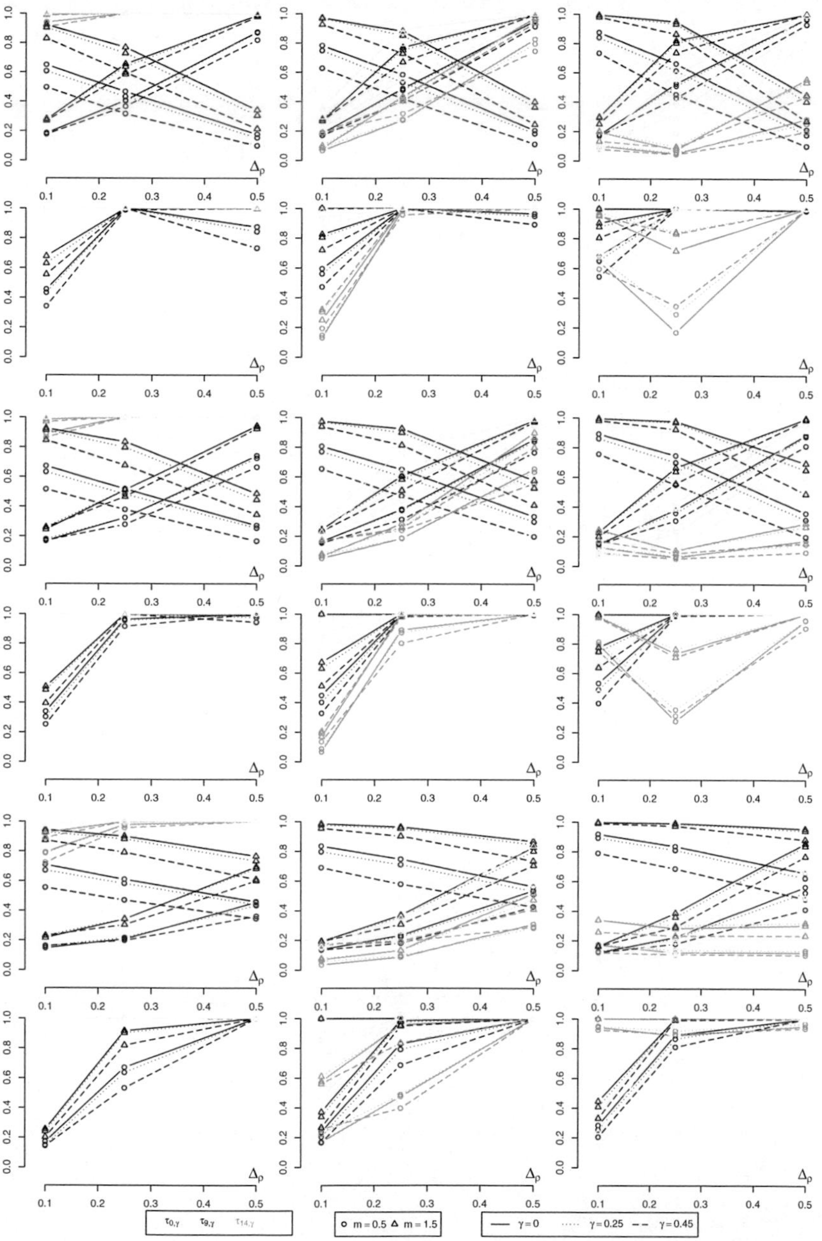

Figure 29: Each graph contains the relative frequency of stops for $\Delta_\rho = 0.1, 0.25, 0.5$ under mean and variance setting $(\mathrm{II}_\mu^{(c)}+\mathrm{II}_\sigma^{(c)})$. In the first till the third column $\rho = -0.9, -0.5, 0$ and in the odd rows $n = 200$ and in even ones $n = 2000$. In the first, second, and third two rows $\theta_\rho = 0.25, 0.5, 0.75$.

7.3 Change-Point Estimation

This subsection presents the behavior of estimates for the change-points in the correlation $\hat{\theta}^{(\psi)} = \hat{k}_1^{(\psi)}/n$ with

$$\hat{k}(Z^{(\psi)}) = \hat{k}_1^{(\psi)} \in \arg\min \Big\{ \sum_{r=1}^{2} \sum_{i=k_{r-1}+1}^{k_r} (Z^{(\psi)}_{i,k_{r-1},k_r} - \overline{Z^{(\psi)}}^{k_r}_{k_{r-1}})^2 \, : \, 0 < k_1 < n \Big\},$$

where

$$\overline{Z^{(\psi)}}^{k_r}_{k_{r-1}} = (k_r - k_{r-1})^{-1} \sum_{i=k_{r-1}+1}^{r_r} Z^{(\psi)}_{i,k_{r-1},k_r}$$

with

$$Z^{(\psi)}_{i,k_{r-1},k_r} = \begin{cases} \frac{(X_i - \mu_{1,i})\cdot(Y_i - \mu_{2,i})}{\sigma_{1,i}\sigma_{2,i}}, & \text{if } \psi = 0, \\ \frac{(X_i - \hat{\mu}^{(\psi)}_{1,i,k_{r-1},k_r})\cdot(Y_i - \hat{\mu}^{(\psi)}_{2,i,k_{r-1},k_r})}{\sigma_{1,i}\sigma_{2,i}}, & \text{if } \psi = 1,\ldots,4, \\ \frac{(X_i - \mu_{1,i})\cdot(Y_i - \mu_{2,i})}{\hat{\sigma}^{(\psi)}_{1,i,k_{r-1},k_r}\hat{\sigma}^{(\psi)}_{2,i,k_{r-1},k_r}}, & \text{if } \psi = 5,\ldots,8 \end{cases}$$

and

$$Z^{(\psi)}_{i,k_{r-1},k_r} = Z^{(4+\psi_1+4\psi_2)}_{i,,k_{r-1},k_r} = \frac{(X_i - \hat{\mu}^{(\psi_1)}_{1,i,,k_{r-1},k_r}) \cdot (Y_i - \hat{\mu}^{(\psi_1)}_{2,i,,k_{r-1},k_r})}{\hat{\sigma}^{(\psi_2)}_{1,i,,k_{r-1},k_r}\hat{\sigma}^{(\psi_2)}_{2,i,,k_{r-1},k_r}} \quad \text{for } \psi_1,\psi_2 = 1,\ldots,4.$$

Parameter Estimates Here, we use the following estimates

$$\hat{\mu}^{(1)}_{1,n,k_{r-1},k_r} \equiv n^{-1}\sum_{i=1}^{n} X_i, \quad \hat{\mu}^{(2)}_{1,n,k_{r-1},k_r} = (k_r - k_{r-1})^{-1} \sum_{i=k_{r-1}+1}^{k_r} X_i$$

$$(\hat{\sigma}^{(1)}_{1,n,k_{r-1},k_r})^2 \equiv n^{-1}\sum_{i=1}^{n}(X_i - \mu_{1,i})^2, \quad (\hat{\sigma}^{(2)}_{1,n,k_{r-1},k_r})^2 = (k_r - k_{r-1})^{-1} \sum_{i=k_{r-1}+1}^{k_r} (X_i - \mu_{1,i})^2,$$

where the exact means are replaced by the mean estimate with index (ψ_1) if the means are assumed to be unknown. $\hat{\mu}^{(3)}_{1,n,k_{r-1},k_r}$ is the piecewise sample mean which is based on change-point estimate $\hat{k}_{\mu,1}$, which is defined by

$$\hat{k}_{\mu,1} = \begin{cases} n, & \text{if } \hat{D}_X^{-1/2} \| \frac{[n\cdot]}{\sqrt{n}}(\overline{X}_{[n\cdot]} - \overline{X}_n) \| \leq c_{0.05}, \\ \hat{k}(X), & \text{else.} \end{cases}$$

Here, $\hat{D}_X$ is the classic LRV estimate with bandwidth $q_n = \log(n)$, $c_{0.05}$ is the 0.95%-quantile, and $\hat{k}(X)$ is the least square estimate defined in the first display of this page. $(\hat{\sigma}^{(3)}_{1,n,k_{r-1},k_r})^2$ is similarly defined with $(X_i - \mu_{1,i})^2$ instead of X_i, whereas the exact means are replaced by the mean estimate with index (ψ_1) in the case of unknown means. Analogously, we define the mean and variance estimates of Y.

Simulation Setup In this subsection, we consider the following change-point model

$$\rho_i = \rho_0 + \Delta_\rho \mathbb{1}_{\{i > [n\theta_\rho]\}},$$

with the following change-point setup:

n	ρ_0	θ_ρ	Δ_ρ
200, 1000, 2000	−0.9, −0.5, 0	0.1, 0.25, 0.5	0.1, 0.25, 0.5

In addition, we allow structural breaks in the parameters and consider the following parameter setup:

(1) $\mu_{1,i} \equiv 0$ and $\mu_{2,i} \equiv 0$;

(2) $\mu_{1,i} = 4 \cdot \mathbb{1}_{\{i \leq 5/8 \cdot n\}}$ and $\mu_{2,i} = \mathbb{1}_{\{i \leq 0.75 \cdot n\}}$;

(3) $\sigma_{1,i} \equiv 1$ and $\sigma_{2,i} \equiv 1$;

(4) $\sigma^2_{1,i} = 1 + 2 \cdot \mathbb{1}_{\{i > 3/4 \cdot n\}}$ and $\sigma^2_{2,i} = 1 + 1 \cdot \mathbb{1}_{\{i > 4/5 \cdot n\}}$.

Firstly, we consider the change-point estimates for the structural breaks of the correlation. Figure 30 shows the boxplots of the estimates in the case of constant means and known variances, no matter whether a test rejects the null hypothesis or not. The presented estimates $\hat{\theta}^{(\psi)}$ behave quite similarly for $\psi = 0, 1\,2, ,3, 5, 6, 14, 19$. Obviously, the estimates become more precise for larger sample sizes n and change sizes Δ_ρ. Furthermore, change-points in the middle of the observations can be estimated better than the ones which occur at the beginning or at the end of the observations. However, for $n = 1000$ the arithmetic mean is near zero in each case. The estimates $\hat{\theta}^{(2)}$, $\hat{\theta}^{(6)}$, and $\hat{\theta}^{(14)}$, which use the sequence of sample means and variances, i.e. $\hat{x}_{l,i,1,k}$ and $\hat{x}_{l,i,1+k,n}$ for $k = 1, \ldots ,n$ and $x \in \{\mu,\sigma^2\}$, to estimate the mean and variance, are a little less precise than the estimates $\hat{\theta}^{(1)}$, $\hat{\theta}^{(3)}$, $\hat{\theta}^{(5)}$, and $\hat{\theta}^{(19)}$ using the whole sample to estimate the parameter.

Now, we consider the cases where the means are non-constant and fulfill (2) but the variance remains constant to one. Then, the estimates $\hat{\theta}^{(0)}$, $\hat{\theta}^{(5)}$, and $\hat{\theta}^{(6)}$, which use the exact means, as well as the estimates $\hat{\theta}^{(3)}$ and $\hat{\theta}^{(19)}$, which use change-point estimates, remain useful. However, to analyze the influence we plot the other estimates in Figure 31, too. The change-point estimate of the correlation $\hat{\theta}^{(1)}$ untruly estimates the structural break of the means of Y, which is equal to 0.75. In contrast, $\hat{\theta}^{(2)}$ untruly estimates the structural break of the means of X, which is equal to 0.625. Furthermore, it seems that $\hat{\theta}^{(3)}$ and $\hat{\theta}^{(19)}$, which behave similarly, are slightly better estimates for the correct change-point in the correlations if these change-points lie at the beginning, i.e., for $\theta_\rho \in \{0.1, 0.25\}$.

Now, we focus on the case where the means are constantly zero but the variances possess structural breaks. Therefore, we suppose the variances setup following (4). In this setting only $\hat{\theta}^{(0)}$ and $\hat{\theta}^{(19)}$ remain useful. Nevertheless, to indicate the behavior of the other estimates we present them in Figure 32 as well. In most of the cases the unusable estimates approximate the variance change-points of Y, which is 0.75. If the change size is large ($\Delta_\rho = 0.5$) and the change-point of the correlation lies in the middle, then even $\hat{\theta}^{(1)}$, $\hat{\theta}^{(3)}$, and $\hat{\theta}^{(5)}$ correctly estimate the change-point of the correlations. The estimate $\hat{\theta}^{(19)}$ only works well if the sample size, change size, and location are appropriate. If two of these parameters are "bad" regarding a small sample size, change size, or that the change location is too early, this estimate falsely approximates the change-points of the variances.

Finally, we consider the case where the means and the variances are non-constant. Figure 32 presents the boxplots of $\hat{\theta}^{(0)}$ and $\hat{\theta}^{(19)}$ for $n \in \{200, 1000, 2000\}$. Overall, we can summarize the behavior as a combination of the previous notes.

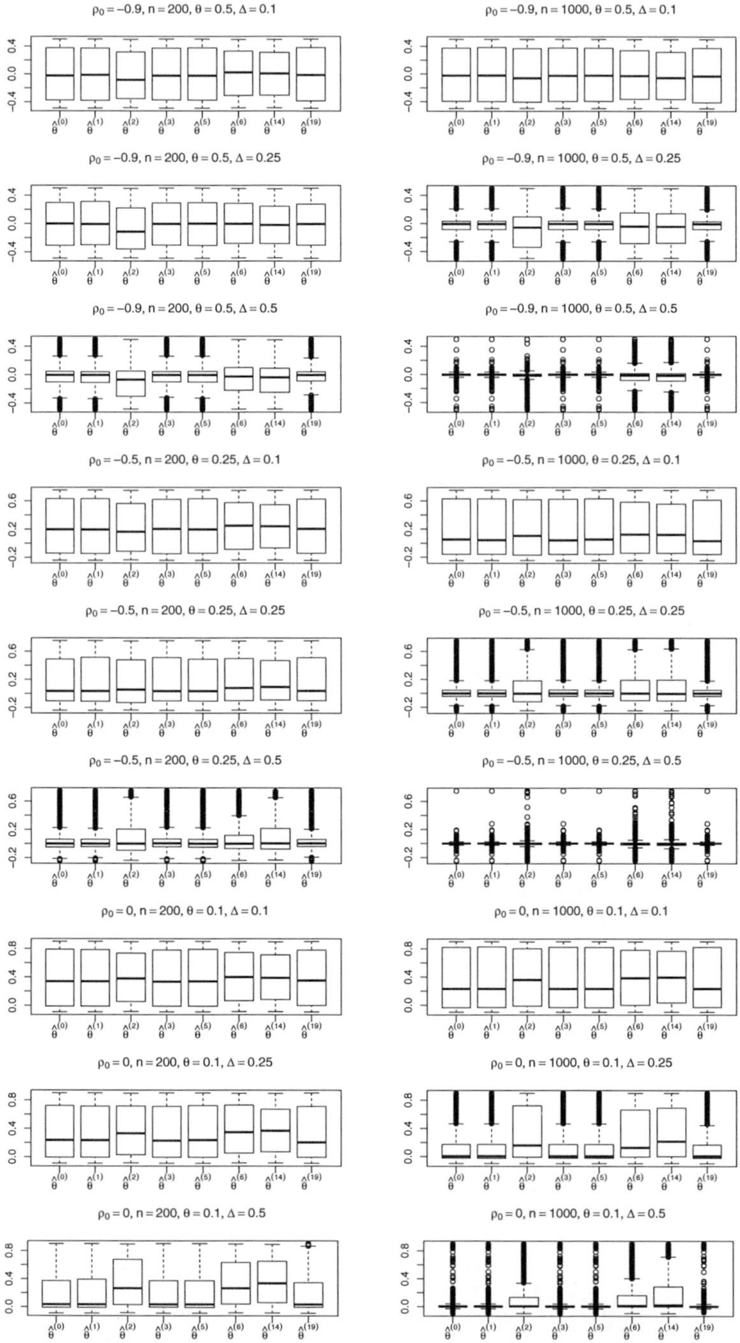

Figure 30: Boxplots of change-point estimation error $\hat{\theta}^{(\psi)} - \theta$ under parameter setup (1)

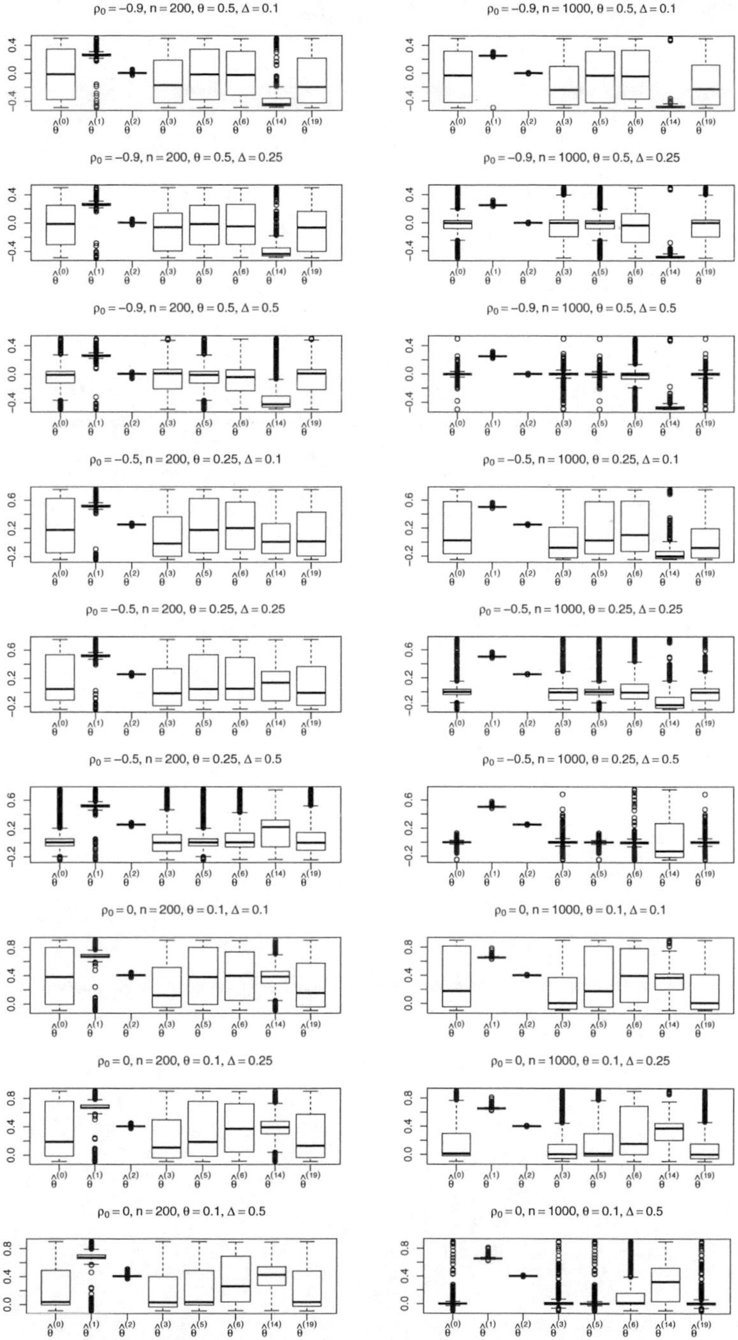

Figure 31: Boxplots of change-point estimation error $\hat{\theta}^{(\psi)} - \theta$ under mean setup (2)

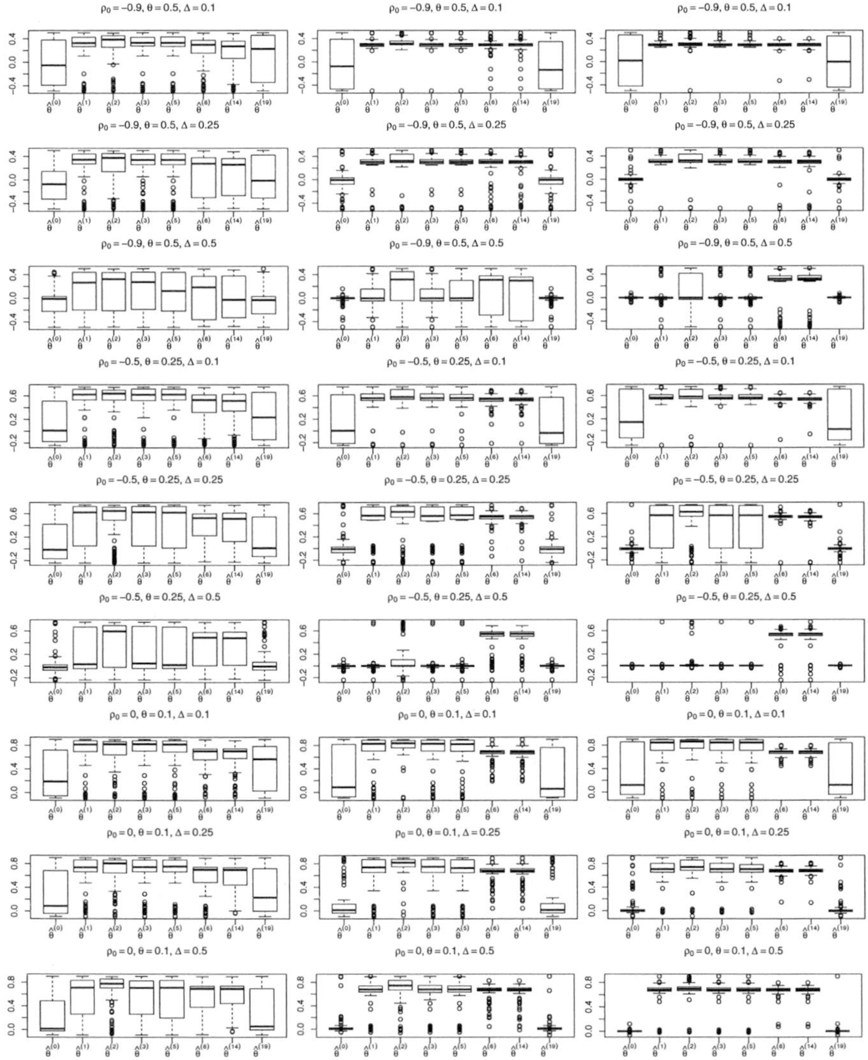

Figure 32: Boxplots of change-point estimation error $\hat{\theta}^{(\psi)} - \theta$ under parameter setup (4)

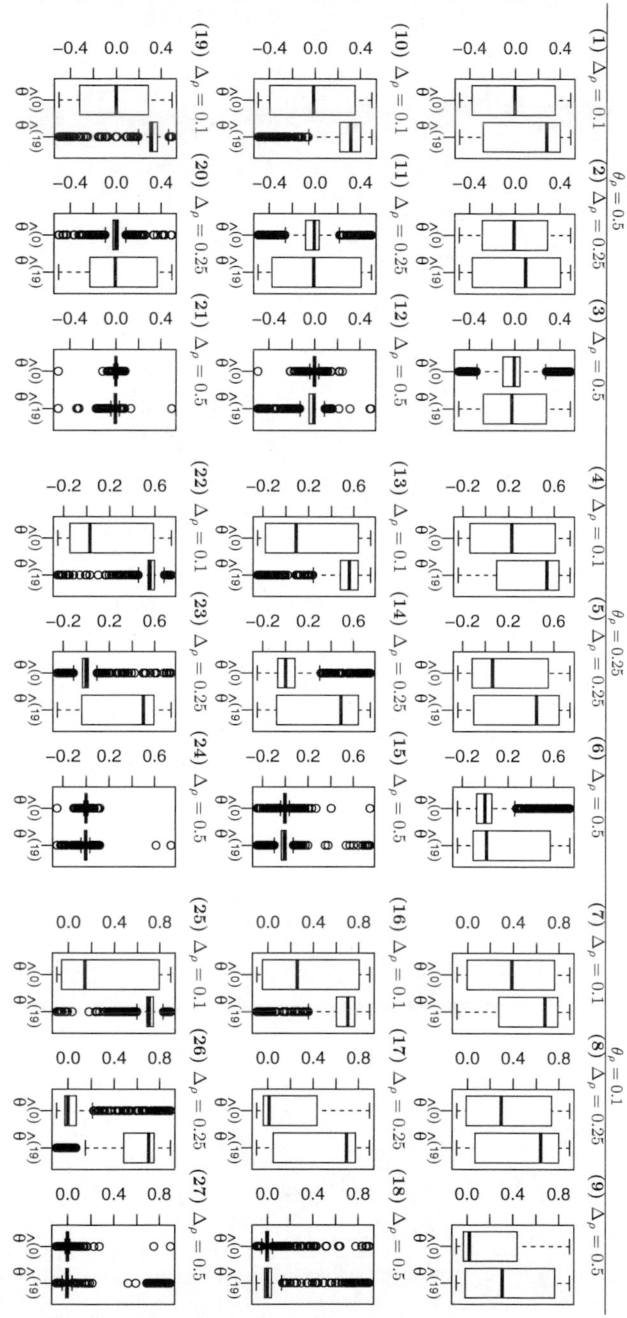

Figure 32: Boxplots of change-point estimation error $\hat{\theta}^{(\psi)} - \theta$ under mean setup (2) and variance setup (4). In the columns (1,4,7) $n = 200$, in (2,5,8) $n = 1000$, and in (3,6,9) $n = 2000$. In each column the first three rows are for $\theta_\rho = 0.5$, the second three for $\theta_\rho = 0.25$, and the last three for $\theta_\rho = 0.1$.

7.4 Real Data Application

In this subsection, we adapt our presented procedures to real financial data, taken from Finanzen (2016). We consider the correlations between the log-returns of the daily closed index DAX30 and stock VW from January 2, 2015, to March 2, 2016. The curves and the log-returns are plotted in Figure 33. (Non-trading days at Börse Frankfurt are omitted in the time axes.)

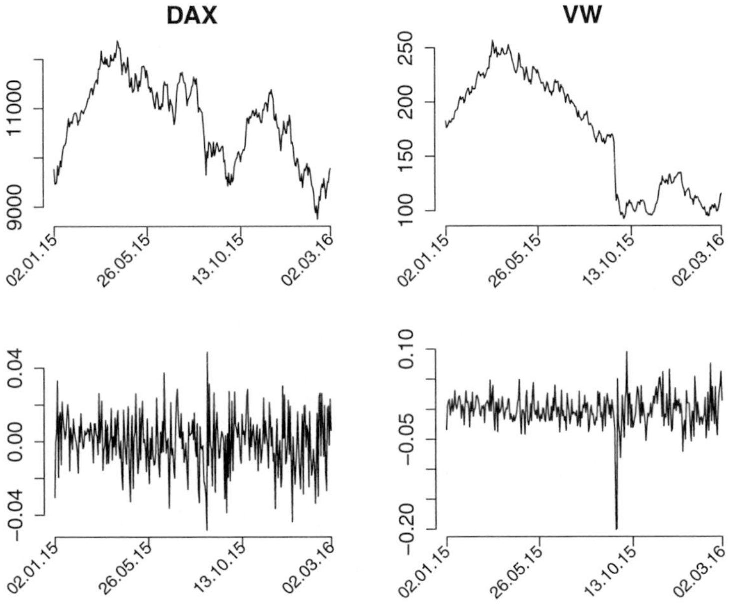

Figure 33: Chart of DAX and VW

In the selected time period, the weight of the VW stocks in the DAX index has varied as follows:

Period	01.01.2015–12.04.15	13.04.15–22.12.2015	23.12.2015–02.03.16
Amount	3,36%	3,9%	2,359%

Table 25: Amount of the VW stock in the DAX, cf. DAX-Gewichtung (2016)

Since the composition of the DAX index changed on April 13, 2015, and December 23, 2015, the correlation between these data might have changed on these two days. Regardless of the facts that we cannot state whether or not a change in the correlations has actually occurred, and that we cannot prove whether the assumptions on our results are fulfilled, we can still apply and evaluate the different procedures.

Figure 34 presents the processes $B^{9,\gamma}$, $B^{14,\gamma}$, $B^{23,\gamma}$, and $B^{24,\gamma}$ which are weighted by the critical value $c_{0.05,\gamma}$. Only $B^{14,\gamma}$, $B^{23,\gamma}$, and $B^{24,\gamma}$ reject H_0 while $B^{9,\gamma}$ does not exceed $c_{0.05,\gamma}$ for each γ.

Table 26 presents the different change-point estimates which use the restriction that there is only one change-point. Figure 35 illustrates the moving correlation based on Pearson's correlation coefficient. On the one hand, we ascertain that the correlation seems to be lower after September 16, 2015, on the other hand, we observe that the moving correlation rises back to a correlation of 0.8. Thus, the period from September 16, 2015, to around February 10, 2016, could be a random anomaly.

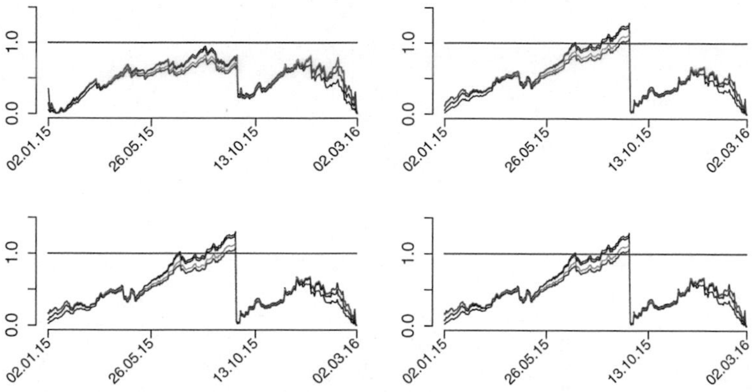

Figure 34: The left column contains the processes $B^{9,\gamma}/c_{0.05,\gamma}$ and $B^{14,\gamma}/c_{0.05,\gamma}$ and the right column contains the processes $B^{23,\gamma}/c_{0.05,\gamma}$ and $B^{24,\gamma}/c_{0.05,\gamma}$ for $\gamma = 0$ (black), $\gamma = 0.25$ (red), $\gamma = 0.45$ (green), and $\gamma = 0.49$ (blue).

$\hat{\theta}^{(1)}$	$\hat{\theta}^{(2)}$	$\hat{\theta}^{(3)}$	$\hat{\theta}^{(14)}$	$\hat{\theta}^{(19)}$	$\arg\max B^{\psi,\gamma}$
04.02.16	05.08.15	04.02.16	18.09.15	04.02.16	18.09.16

Table 26: Change-point estimates of the correlations between log returns of DAX and VW

Without claiming that there is coherence between the emissions scandal of VW and a change in the DAX/VW correlation, we take note that the Environmental Protection Agency accused Volkswagen on September 18, 2015, and VW published their apologized on September 20, 2015, cf. EPA (2015) and Pressemitteilung der Volkswagen AG (20.09.15).

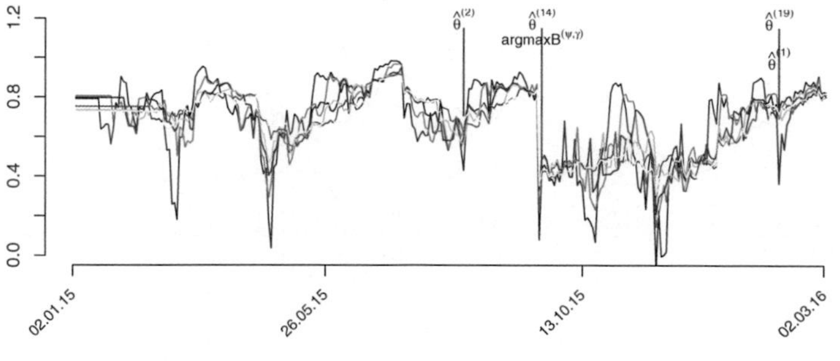

Figure 35: Moving Pearson's correlation coefficient with backward lag size of 10 (black), 15 (green), 20 (blue), 25 (cyan), 30 (pink), and 35 (yellow).

A Proofs of Section 2

In this section, we present, on the one hand, proofs of well-known results, proofs of some slightly new results, and, on the other hand, technical lemmas for the main results.

Proof of Theorem 2.1.1. Firstly, we obtain with the CTM that

$$\frac{[n\cdot]}{\sqrt{n}}\left(\hat{\rho}_{[n\cdot]}-\hat{\rho}_n\right)-R_{n,\rho}(\cdot)\overset{D[0,1]}{\longrightarrow}D^{1/2}B(\cdot) \tag{A.0.1}$$

with

$$R_{n,\rho}(\cdot)=\frac{[n\cdot]}{\sqrt{n}}\left(\overline{\rho}_{[n\cdot]}-\overline{\rho}_n\right)\quad\text{and}\quad\overline{\rho}_k=\frac{1}{k}\sum_{i=1}^{k}\rho_i. \tag{A.0.2}$$

Under H_0 it holds that $R_{n,\rho}(\cdot)\equiv 0$, since $\rho_i=\rho_0$ for all i, whereas under Assumption $\mathbf{H_{LA}}$ it holds that

$$R_{n,\rho}(\cdot)=\frac{1}{n}\sum_{i=1}^{[n\cdot]}g_\rho(i/n)-\frac{[n\cdot]}{n^2}\sum_{i=1}^{n}g_\rho(i/n)\longrightarrow D^{1/2}h(\cdot)\ \text{ as }\ n\to\infty$$

uniformly on $[0,1]$. Since $\hat{D}$ is a consistent estimate for D, Slutsky's Theorem implies the first and second result.

In the third case, we obtain that

$$\|R_{n,\rho}(\cdot)\|=\max_{1\le k\le n}|\Delta_\rho|\sqrt{n}\left|\frac{1}{n}\sum_{i=1}^{k}\mathbb{1}_{R_{k^*}}(i)-\frac{k\lambda(R_{k^*})}{n^2}\right|\ge 2^{-1}\sqrt{n}|\Delta_\rho|\frac{\lambda(R_{k^*})\lambda(R^c_{k^*})}{n^2}, \tag{A.0.3}$$

where we get the underestimation directly by choosing $k=k_1^*$ and $k=k_2^*$. With a suitable constant $c_1>0$, we therefore get that

$$\begin{aligned}&\frac{a_n}{\sqrt{n}}\|\hat{D}_n^{-1/2}\frac{[n\cdot]}{\sqrt{n}}\left(\overline{Z^{(0)}}_{[n\cdot]}-\overline{Z^{(0)}}_n\right)\|\ge\frac{a_n}{\sqrt{n}}\|\hat{D}_n^{-1/2}R_{n,\rho}(\cdot)\|-\frac{a_n}{\sqrt{n}}\|\hat{D}_n^{-1/2}(B_n(\cdot)-R_{n,\rho}(\cdot))\|\\&\ge\frac{a_n}{|\hat{D}_n^{1/2}|}\frac{(\|R_{n,\rho}(\cdot)\|-\|(B_n(\cdot)-R_{n,\rho}(\cdot))\|)}{\sqrt{n}}\ge\frac{a_n}{|\hat{D}_n^{1/2}|}\frac{\left(c_1|\Delta_\rho|\sqrt{n}\frac{\lambda(R_{k^*})\lambda(R^c_{k^*})}{n^2}-O_P(1)\right)}{\sqrt{n}},\end{aligned}$$

where we use (A.0.1). Hence, if $|\Delta_\rho|\sqrt{n}\frac{\lambda(R_{k^*})\lambda(R^c_{k^*})}{n^2}\to\infty$, which is given by $\mathbf{H_A}$, the right–hand side above is asymptotically positive. Furthermore, with $\hat{D}_n^{1/2}=o_P(a_n|\Delta_\rho|\lambda(R_{k^*})\lambda(R^c_{k^*})n^{-2})$ it follows for each $c>0$ and f_ι with $\lim_{\|x\|\to\infty}f_\iota(x)=\infty$ that $P(f_\iota(a_nn^{-1/2}B_n^{0,0,0})\ge c)\to 1$. □

Proof of Theorem 2.1.15. This proof exactly follows the proof of Bucchia and Heuser (2015, Th. 1) with the difference that here we deal with the special one-dimensional case ($d=1$) and use the notations $\tilde{Q}([ns],[nt])=\tilde{Q}_n(s,t,)$, $\tilde{Z}_i=Z_i^{(0)}-\mathbb{E}\left[Z_i^{(0)}\right]$, and $R_k=(0,k]$ in the AMOC setting as well as $R_k=(k_1,k_2]$ in the epidemic setting. Using the same decomposition of Bucchia and Heuser (2015, Th. 1) implies

$$\begin{aligned}&\mathbb{P}\left(n\|\tilde{\boldsymbol{\theta}}-\boldsymbol{\theta}^0\|\ge N+1\right)\le\mathbb{P}\left(\|\hat{k}-k^*\|\ge N\right)\\&=\mathbb{P}\left(\max_{\substack{k_1<k_2\\\|k-k^*\|\ge N}}(\tilde{Q}(k_1,k_2)-\tilde{Q}(k_1^*,k_2^*))\ge\max_{\substack{k_1<k_2\\\|k-k^*\|<N}}(\tilde{Q}(k_1,k_2)-\tilde{Q}(k_1^*,k_2^*))\right)\\&\le\mathbb{P}\left(\max_{\substack{k_1<k_2\\\|k-k^*\|\ge N}}(\tilde{Q}(k_1,k_2)-\tilde{Q}(k_1^*,k_2^*))\ge 0\right)\\&\le\mathbb{P}\left(\max_{\substack{k_1<k_2\\\|k-k^*\|\ge N}}L_{n,k_1,k_2}\left(\frac{A^{(1)}_{k_1,k_2}A^{(2)}_{k_1,k_2}}{L_{n,k_1,k_2}}+\frac{A^{(1)}_{k_1,k_2}}{B^{(1)}_{k_1,k_2}}\Delta_\rho+\Delta_\rho\frac{A^{(2)}_{k_1,k_2}}{B^{(2)}_{k_1,k_2}}+\Delta_\rho^2\right)\ge 0\right),\end{aligned}$$

where

$$\begin{aligned}
A^{(1)}_{k_1,k_2} &= \sum_{i\in R_k\backslash R_{k^*}} \tilde{Z}_i - \sum_{j\in R_{k^*}\backslash R_k} \tilde{Z}_j - \frac{\lambda(R_k\backslash R_{k^*}) - \lambda(R_{k^*}\backslash R_k)}{n} \sum_{\mathbf{1}\le i\le \mathbf{n}} \tilde{Z}_i,\\
B^{(1)}_{k_1,k_2} &= -\lambda(R_{k^*}\backslash R_k) - (\lambda(R_k\backslash R_{k^*}) - \lambda(R_{k^*}\backslash R_k))\frac{\lambda(R_{k^*})}{n},\\
A^{(2)}_{k_1,k_2} &= \sum_{i\in R_k} \tilde{Z}_i + \sum_{i\in R_{k^*}} \tilde{Z}_i - \frac{\lambda(R_k)+\lambda(R_{k^*})}{n}\sum_{\mathbf{1}\le i\le \mathbf{n}} \tilde{Z}_i,\\
B^{(2)}_{k_1,k_2} &= \lambda(R_k\cap R_{k^*}) + \lambda(R_{k^*}) - (\lambda(R_k)+\lambda(R_{k^*}))\frac{\lambda(R_{k^*})}{n},\\
L_{n,k_1,k_2} &= B^{(1)}_{k_1,k_2}B^{(2)}_{k_1,k_2}.
\end{aligned}$$

Under the assumption that R_{k^*} is an interval and $\tilde{Z}_i$ fulfills the Kolmogorov-type inequalities, it follows that $\|\sum_{i=1}^{[n\cdot]}\tilde{Z}_i\| = O_P(n^{1/r_z})$. Hence, just as in Bucchia and Heuser (2015) we get

$$0 < c \le \min_{\substack{k_1<k_2\\ \|k-k^*\|\ge N}} \frac{B^{(2)}_{k_1,k_2}}{n} < \max_{\substack{k_1<k_2\\ \|k-k^*\|\ge N}} \frac{B^{(2)}_{k_1,k_2}}{n} \le 4$$

and

$$\max_{\substack{k_1<k_2\\ \|k-k^*\|\ge N}} \frac{|A^{(2)}_{k_1,k_2}|}{n} = O_P(n^{1/r_z-1}).$$

Defining $T_{k_1,k_2} := |k_1-k_1^*| \vee |k_2-k_2^*| = \|k-k^*\|$ yields

$$0 < c \le \min_{\substack{k_1<k_2\\ \|k-k^*\|\ge N}} \frac{-B^{(1)}_{k_1,k_2}}{T_{k_1,k_2}} \le \max_{\substack{k_1<k_2\\ \|k-k^*\|\ge N}} \frac{-B^{(1)}_{k_1,k_2}}{T_{k_1,k_2}} \le C,$$

$$0 < c \le \min_{\substack{k_1<k_2\\ \|k-k^*\|\ge N}} \frac{|L_{n,k_1,k_2}|}{nT_{k_1,k_2}} \quad \text{and} \quad \max_{\substack{k_1<k_2\\ \|k-k^*\|\ge N}} L_{n,k_1,k_2} \le -c < 0,$$

for suitable constants $0 < c \le C$. It remains to be shown that

$$\max_{\substack{k_1<k_2\\ \|k-k^*\|\ge N}} \frac{|A^{(1)}_{k_1,k_2}|}{T_{k_1,k_2}} = o_P(1) + a_N O_P(1)$$

for $a_N = o(1)$ as $N\to\infty$. The assumption of $\lambda(R_{k^*})\sim n$ implies

$$\begin{aligned}
\max_{\substack{k_1<k_2\\ \|k-k^*\|\ge N}} \frac{|A^{(1)}_{k_1,k_2}|}{T_{k_1,k_2}} &\le \max_{\substack{k_1<k_2\\ \|k-k^*\|\ge N}} \frac{|\sum_{i\in R_k}\tilde{Z}_i - \sum_{i\in R_{k^*}}\tilde{Z}_i|}{T_{k_1,k_2}} + O_P(n^{1/r_z-1})\\
&= a_N O_P(1) + O_P(n^{1/r_z-1}),
\end{aligned}$$

where the last equality follows by Lemma B.0.2, for which the Kolmogorov-type inequalities are required.

Combining the above results implies

$$\begin{aligned}
&\mathbb{P}\left(n\|\hat{\boldsymbol{\theta}}-\boldsymbol{\theta}^0\| \ge N+1\right) \le \mathbb{P}\left(\|\hat{k}-k^*\|\ge N\right)\\
&\le \mathbb{P}\left(\max_{\substack{k_1<k_2\\ \|k-k^*\|\ge N}} L_{n,k_1,k_2}\left(\frac{A^{(1)}_{k_1,k_2}A^{(2)}_{k_1,k_2}}{L_{n,k_1,k_2}} + \frac{A^{(1)}_{k_1,k_2}}{B^{(1)}_{k_1,k_2}}\Delta_\rho + \Delta_\rho\frac{A^{(2)}_{k_1,k_2}}{B^{(2)}_{k_1,k_2}} + \Delta_\rho^2\right) \ge 0\right)\\
&\le \mathbb{P}\left(\max_{\substack{k_1<k_2\\ \|k-k^*\|\ge N}} L_{n,k_1,k_2}\left(o_P(1) + a_N\mathcal{O}_P(1) + \Delta_\rho^2\right) \ge 0\right)\\
&\le \mathbb{P}\left(o_P(1) + a_N\mathcal{O}_P(1) + \Delta_\rho^2 \le 0\right).
\end{aligned}$$

□

Proof of Theorem 2.1.16. This proof essentially follows the proof of Theorem 2.1.15. Using the notation $\tilde{Z}_i = Z_i^{(0)} - \rho_i$ and $a_n = n\Delta_{\rho,n}$ implies

$$\mathbb{P}\left(a_n\|\tilde{\boldsymbol{\theta}} - \boldsymbol{\theta}^0\| \geq N+1\right) \leq \mathbb{P}\left(\|\hat{k} - k^*\| \geq Nn/a_n\right)$$
$$\leq \mathbb{P}\left(\max_{\substack{k_1<k_2 \\ \|k-k^*\|\geq Nn/a_n}} L_{n,k_1,k_2}\left(\frac{A^{(1)}_{k_1,k_2}A^{(2)}_{k_1,k_2}}{L_{n,k_1,k_2}} + \frac{A^{(1)}_{k_1,k_2}}{B^{(1)}_{k_1,k_2}}\Delta_{\rho,n} + \Delta_{\rho,n}\frac{A^{(2)}_{k_1,k_2}}{B^{(2)}_{k_1,k_2}} + \Delta^2_{\rho,n}\right) \geq 0\right).$$

Since R_{k^*} is an interval and its cardinality rises with rate n, it holds with the arguments of the proof of Theorem 2.1.15 that

$$\max_{\substack{k_1<k_2 \\ \|k-k^*\|\geq Nn/a_n}} \left|\frac{A^{(2)}_{k_1,k_2}}{B^{(2)}_{k_1,k_2}}\right| = O_P(n^{1/r_z-1}) \quad \text{and} \quad \max_{\substack{k_1<k_2 \\ \|k-k^*\|\geq Nn/a_n}} \left|\frac{A^{(1)}_{k_1,k_2}}{B^{(1)}_{k_1,k_2}}\right| = O_P(\frac{a_n}{nN}),$$

where we use the Kolmogorov-type inequalities for the last equality. Combining the preceding rates implies

$$\mathbb{P}\left(a_n\|\hat{\boldsymbol{\theta}} - \boldsymbol{\theta}^0\| \geq N+1\right) \leq \mathbb{P}\left(\|\hat{k} - k^*\| \geq Nn/a_n\right)$$
$$\leq \mathbb{P}\left(\max_{\substack{k_1<k_2 \\ \|k-k^*\|\geq nN/a_n}} L_{n,k_1,k_2}\Delta^{-2}_{\rho,n}\left(\frac{A^{(1)}_{k_1,k_2}A^{(2)}_{k_1,k_2}}{L_{n,k_1,k_2}}\Delta^2_{\rho,n} + \frac{A^{(1)}_{k_1,k_2}}{B^{(1)}_{k_1,k_2}}\Delta^{-1}_{\rho,n} + \frac{A^{(2)}_{k_1,k_2}}{B^{(2)}_{k_1,k_2}}\Delta^{-1}_{\rho,n} + 1\right) \geq 0\right)$$
$$\leq \mathbb{P}\left(\max_{\substack{k_1<k_2 \\ \|k-k^*\|\geq nN/a_n}} L_{n,k_1,k_2}\left(O_P\left(n^{1/r_z-1}\Delta^{-1}_{\rho,n}N^{-1}\right) + N^{-1}O_P(1) + O_P\left(n^{1/r_z-1}\Delta^{-1}_{\rho,n}\right) + 1\right) \geq 0\right)$$
$$\leq \mathbb{P}\left(o_P(1) + N^{-1}\mathcal{O}_P(1) + 1 \leq 0\right) \to 0$$

as $n \to \infty$, followed by $N \to \infty$. Here, we insert the definition of $a_n = n\Delta_{\rho,n}$ and use that $L_{n,k_1,k_2} < -\epsilon$ for some $\epsilon > 0$.

□

Proof of Theorem 2.1.18. Firstly, for a suitable array $a_{n,N}$ we obtain that

$$P(|\hat{k} - k^*_n| \geq a_{n,N}+1) \leq P\left(\min_{k;|k-k^*|\geq a_{n,N}} Q^{(0)}_n(k) \leq Q^{(0)}_n(k^*)\right)$$

and with $\tilde{Z}_i = Z_i^{(0)} - \rho_i$ that

$$Q^{(0)}_n(k) - Q(k^*)$$
$$= \Delta^2_{\rho,n}(k^*-k)\frac{n-k^*}{n-k}\left(1 + \frac{\Delta^{-2}_{\rho,n}}{k^*-k}\sum_{i=k+1}^{k^*}\tilde{Z}_i\rho_i + \frac{\Delta^{-2}_{\rho,n}}{n-k^*}\sum_{i=k^*+1}^{n}\tilde{Z}_i\rho_i\right)\mathbb{1}_{\{k<k^*\}}$$
$$+ \Delta^2_{\rho,n}(k-k^*)\frac{k^*}{k}\left(1 + \frac{\Delta^{-2}_{\rho,n}}{k^*}\sum_{i=1}^{k^*}\tilde{Z}_i\rho_i + \frac{\Delta^{-2}_{\rho,n}}{k-k^*}\sum_{i=k^*+1}^{k}\tilde{Z}_i\rho_i\right)\mathbb{1}_{\{k\geq k^*\}}.$$

With the Hájek–Rényi–type inequalities

$$\max_{n\geq k>k^*+a_{n,N}}\left|\frac{1}{k-k^*}\sum_{i=k^*+1}^{k}\tilde{Z}_i\rho_i\right| = O_P\left(\left(\sum_{k=a_{n,N}}^{n-k_1*}k^{-r_z}\right)^{1/r_z}\right) = O_P(a_{n,N}^{-(r_z-1)/r_z}),$$
$$\max_{1\leq k\leq k^*-a_{n,N}}\left|\frac{1}{k^*-k}\sum_{i=k+1}^{k^*}\tilde{Z}_i\rho_i\right| = O_P\left(\left(\sum_{k=a_{n,N}}^{k_1*}k^{-r_z}\right)^{1/r_z}\right) = O_P(a_{n,N}^{-(r_z-1)/r_z}),$$

we obtain with (2.1.26) that the insides of the two brackets of the second display are asymptotically positive if $a_{n,N} = N\Delta_{\rho,n}^{-\frac{2r_z}{r_z-1}}$. Hence, we get $|\hat{k} - k^*| = O_P(|\Delta_{\rho,n}|^{\frac{2r_z}{r_z-1}})$.

□

Proof of Lemma 2.1.21. Base case $m = 1$: $g_2 \equiv c$ is constant. Hence, it holds that

$$\int_0^n (g_1(x) - c)^2 dx = \int_1^{k_1} (\rho_0 - c)^2 dx + \int_{k_1}^n (\rho_0 + \Delta_1 - c)^2 dx \geq \frac{k_1 \wedge (n - k_1)}{2} \Delta_1^2$$

for any $c \in \mathbb{R}$.

Induction step: Let the assumption hold for some $m \in \mathbb{N}$, $m \ll n$. Let now $g_1 \in D_{m+1,n}$ and $g_2 \in D_{m,n}$. Then, we obtain that

$$\int_0^n (g_1(x) - g_2(x))^2 dx = \int_1^{k_m} (g_1(x) - g_2(x))^2 dx + \int_{k_m}^n (g_1(x) - g_2(x))^2 dx.$$

Now, we distinguish between the cases:

1. g_2 has at most $m - 1$ jumps on $(1,k_m]$;
2. g_2 has m jumps on $(1,k_m]$.

In the first case, we obtain by the use of the induction hypothesis that

$$\begin{aligned}\int_1^{k_m} (g_1(x) - g_2(x))^2 dx &= \frac{k_m}{n} \int_1^n (g_1(xk_m/n) - g_2(xk_m/n))^2 dx \\ &\geq \frac{1}{2} \min_{1\leq i\leq m} \Delta_i^2 \min_{1\leq i\leq m} (k_i - k_{i-1})\end{aligned}$$

and

$$\int_{k_m}^n (g_1(x) - g_2(x))^2 dx \geq 0.$$

In the second case, we obtain by the same estimations as under the induction base that

$$\int_1^{k_m} (g_1(x) - g_2(x))^2 dx \geq 0$$

and

$$\int_{k_m}^n (g_1(x) - g_2(x))^2 dx \geq \frac{(k_{m+1} - k_m) \wedge (n - k_{m+1})}{2} \Delta_{m+1}^2.$$

Hence, by combining both cases we get that

$$\int_0^n (g_1(x) - g_2(x))^2 dx \geq \frac{1}{2} \min_{1\leq i\leq m+1} \Delta_i^2 \min_{1\leq i\leq m+2} (k_i - k_{i-1}).$$

□

Proof of Theorem 2.1.33. Set $-d_z = (1 - r_z)/r_z$. Applying the Kolmogorov's inequality we observe that

$$\frac{1}{n} \max_{1\leq k\leq n} \left| \sum_{i=1}^k (Z_i^{(0)} - \rho_i) \right| = O_P(n^{-d_z}),$$

which will be used below. We obtain that

$$\tilde{R}_n = \frac{1}{n} \sum_{i,j=1}^n \mathfrak{f}\left(\frac{i-j}{q_n}\right) (T_{1,i,j} + T_{2,i,j} + T_{3,i,j})$$

with

$$T_{1,i,j} = \hat{R}_{\rho,n}(i)\left(Z_j^{(0)} - \rho_j\right)^T, \quad T_{2,i,j} = \left(Z_i^{(0)} - \rho_i\right)\hat{R}_{\rho,n}(j)^T, \quad T_{3,i,j} = \hat{R}_{\rho,n}(i)\hat{R}_{\rho,n}(j)^T.$$

We define $\tilde{R}_n^{(W)}$ for $W \in \{A,B,C,D,E,F,G,H\}$ as $\tilde{R}_n$ in the corresponding cases (A)–(H).

In case of (A) we get

$$\begin{aligned}
\tilde{R}_n^{(A)} &= O_P(n^{-\delta_1})\frac{1}{n}\sum_{i,j=1}^{n}\mathfrak{f}\left(\frac{i-j}{q_n}\right)(Z_i^{(0)} - \rho_i) + O_P(n^{-2\delta_1})\frac{1}{n}\sum_{i,j=1}^{n}\mathfrak{f}\left(\frac{i-j}{q_n}\right) \\
&= O_P(n^{-\delta_1})\sum_{i=-n}^{n}\mathfrak{f}\left(\frac{i}{q_n}\right)\frac{1}{n}\sum_{j\in N_i}(Z_j^{(0)} - \rho_j) + O_P(q_n n^{-2\delta_1}) \\
&= O_P(q_n n^{-(d_z+\delta_1)}) + O_P(q_n n^{-2\delta_1}),
\end{aligned}$$

where we use that N_i is a union of two intervals and $\mathfrak{f}$ is integrable. In the case of (B) and, analogously, of (C) we obtain that

$$\begin{aligned}
\tilde{R}_n^{(B)} &= \tilde{R}_n^{(A)} + O_P(n^{-\delta_1})\frac{2}{n}\sum_{i,j=1}^{n}\mathfrak{f}\left(\frac{i-j}{q_n}\right)d_{i,n}^{(k)} + \frac{1}{n}\sum_{i,j=1}^{n}\mathfrak{f}\left(\frac{i-j}{q_n}\right)d_{i,n}^{(k)}d_{j,n}^{(k)} \\
&\quad + \frac{2}{n}\sum_{i,j=1}^{n}\mathfrak{f}\left(\frac{i-j}{q_n}\right)d_{i,n}^{(k)}(Z_j^{(0)} - \rho_j) \\
&= O_P(q_n n^{-(d_z+\delta_1)}) + O_P(q_n n^{-2\delta_1}) + O_P(q_n n^{-(1/2+\delta_1)}) + O(q_n n^{-1}) + O_P(a_n^{(k)})
\end{aligned}$$

for $k = 1, 2$, where we use the absolute boundary of $\mathfrak{f}$ and $\sum_{i=1}^{n}|d_{i,n}^{(1)}| = o(\sqrt{n})$ in the situation of case (B) and additionally, $\|d_{i,n}^{(2)}\| = o(n^{-1/2})$ in the case of (C). In case of (D), where we only assume an uniform bound for $d_{i,n}^{(3)}$, we get that

$$\tilde{R}_n^{(D)} = O_P(q_n n^{-(d_z+\delta_1)}) + O_P(q_n n^{-2\delta_1}) + O_P(q_n n^{-\delta_1}) + O(q_n\|d_{in}^{(3)}\|^2) + O_P(a_n^{(3)}).$$

In case of (E) we obtain that

$$\begin{aligned}
\left|\frac{1}{n}\sum_{i,j=1}^{n}\mathfrak{f}\left(\frac{i-j}{q_n}\right)T_{1,i,j}\right| &\leq n^{-1}\sum_{k=1}^{m}O_P(n^{-\delta_k})\left|\sum_{j=-n}^{n}\mathfrak{f}\left(\frac{j}{q_n}\right)\sum_{i\in N_j\cap\hat{C}_k}(Z_{i+j}^{(0)} - \rho_{i+j})\right| \\
&\leq n^{-1}\sum_{k=1}^{m}O_P(n^{-\delta_k})\left|\sum_{j=-n}^{n}\mathfrak{f}\left(\frac{j}{q_n}\right)\right|\max_{1\leq j_1\leq j_2\leq n}\left|\sum_{i=j_1}^{j_2}(Z_i^{(0)} - \rho_i)\right| \\
&= O_P(q_n n^{-(1-1/r_z+\min_k\delta_k)}),
\end{aligned}$$

where we use the Kolmogorov-type inequality $(\mathcal{K}_{\mathrm{r}}^{(1)})$. In a similar way, we obtain the rate for $T_{2,i,j}$ by switching i and j. Furthermore, we get that

$$\begin{aligned}
\left|\frac{1}{n}\sum_{i,j=1}^{n}\mathfrak{f}\left(\frac{i-j}{q_n}\right)T_{3,i,j}\right| &\leq n^{-1}\sum_{k_1=1}^{m}\sum_{k_2=1}^{m}O_P(n^{-\delta_{k_1}-\delta_{k_2}})\left|\sum_{j=-n}^{n}\mathfrak{f}\left(\frac{j}{q_n}\right)\right|\max_{1\leq j\leq n}\left|\sum_{i\in N_j\cap\hat{C}_{k_1}, i+j\in\hat{C}_{k_2}}1\right| \\
&= O_P\left(q_n\sum_{k_1=1}^{m}\sum_{k_2=1}^{m}n^{-1-\delta_{k_1}-\delta_{k_2}}\#(C_{k_1})\wedge\#(C_{k_1})\right) + o_P(1) \\
&= O_P\left(q_n\max_{1\leq k_1,k_2\leq m}n^{-1-\delta_{k_1}-\delta_{k_2}}\#(C_{k_1})\wedge\#(C_{k_1})\right) + o_P(1).
\end{aligned}$$

In cases of (F)–(G) we use the triangle inequality so that we get the rates as in case of (E) plus types of the form:

$$\frac{1}{n}\left|\sum_{i,j=1}^{n} \mathfrak{f}\left(\frac{i-j}{q_n}\right) d_{jn}^{(k)}(Z_i^{(0)}-\rho_i)\right|, \; \frac{1}{n}\left|\sum_{i,j=1}^{n} \mathfrak{f}\left(\frac{i-j}{q_n}\right) d_{jn}^{(k)} d_{in}^{(k)}\right|,$$

$$\frac{1}{n}\left|\sum_{l=1}^{m}\sum_{i,j=1}^{n} \mathfrak{f}\left(\frac{i-j}{q_n}\right) d_{jn}^{(k)} \mathbb{1}_{\hat{C}_l}(j) O_P(n^{-\delta_l})\right|,$$

where the first appears in case of $T_{1,i,j}$ and of $T_{2,i,j}$ and is equal to $O_P(a_n^{(k)})$. The two other summands appear in case of $T_{3,i,j}$, where the first one is equal to $o(1)$, $O(q_n n^{-1})$, or $O(q_n)$ in cases of (F), (G), or (H), respectively. The last term is equal to $o_P(q_n n^{-1/2-\min_k \delta_k}) + o_P(1)$, $O_P(q_n n^{-1-1/2} \max_k \#(C_k) n^{-\delta_k}) + o_P(1)$, and $O_P(q_n n^{-1} \max_k \#(C_k) n^{-\delta_k}) + o_P(1)$ in cases of (F), (G), and (H), respectively. □

B Lemmas for the Main Results

Lemma B.0.1. *For any arbitrary $r^* \in [1,2]$ let $\{Z_n\}$ be a sequence of real–valued random variables with uniformly bounded r^*th moments and $\{d_n\}$ a deterministic sequence with $\sum_{i=1}^n |d_i| = o(\sqrt{n})$. Then, $\{d_n Z_n\}$ satisfies the first and second Kolmogorov inequality with $r = r^*$.*

Proof. Set $\|\cdot\|_r = (\mathbb{E}[[] \,|\cdot|^r])^{1/r}$. Per definition we have to show that there is a constant $C \in \mathbb{R}$ for each $n \in \mathbb{N}$ and $\eta > 0$ so that

$$P\left(\max_{1\le k\le n}\left|\sum_{i=1}^{k} d_i Z_i\right| \ge \eta\right) \le \frac{C}{\eta^{r^*}}\sum_{i=1}^{n}\alpha_i,$$

where $\{\alpha_n\}$ is a uniformly bounded sequence. We obtain with Markov's inequality that

$$\begin{aligned}P\left(\max_{1\le k\le n}\left|\sum_{i=1}^{k} d_i Z_i\right| \ge \eta\right) &\le \frac{1}{\eta^{r^*}}\left\|\max_{1\le k\le n}\sum_{i=1}^{k} d_i Z_i\right\|_{r^*}^{r^*} \le \frac{1}{\eta^{r^*}}\left\|\sum_{i=1}^{n}|d_i Z_i|\right\|_{r^*}^{r^*}\\ &\le \frac{1}{\eta^{r^*}}\left(\sum_{i=1}^{n}\|d_i Z_i\|_{r^*}\right)^{r^*} \le \frac{C_1}{\eta^{r^*}}\left(\sum_{i=1}^{n}|d_i|\right)^{r^*} \le \frac{C_2}{\eta^{r^*}}\sum_{i=1}^{n}1,\end{aligned}$$

where the first constant C_1 is an upper bound of the uniformly bounded second moments and the second derives from the property of $\sum_{i=1}^n |d_i| = o(\sqrt{n})$. □

Lemma B.0.2. *Let $\|\cdot\|$ denote the Euclidean norm. Set $k^* = (k_1^*,k_2^*)$ and $k = (k_1,k_2)$.*

1. *Let $\{d_n\}$ fulfill $\sum_{i=1}^n |d_i| = o(\sqrt{n})$ and $d_n = o(1)$, as $n \to \infty$. Let $\sqrt{n} = O(k_2^* - k_1^*)$ as $n \to \infty$. Then, it holds that*

$$\max_{\substack{1\le k_1<k_2\le n\\ \|k-k^*\|\ge N}} \frac{1}{\|k-k^*\|}\left[\sum_{i=1+k_1}^{k_2} d_i - \sum_{i=1+k_1^*}^{k_2^*} d_i\right] = o(1).$$

2. *Let $\{Z_n\}$ be a real random sequence fulfilling the second and shifted Kolmogorov inequalities $(\mathcal{K}_r^{(2)})$ and $(\mathcal{K}_r^{(3)})$ for any $r > 1$. Let $k_1^* \vee (n-k_2^*) = o((k_2^*-k_1^*)^r)$. Then, it holds that*

$$\max_{\substack{1\le k_1<k_2\le n\\ \|k-k^*\|\ge N}} \frac{1}{\|k-k^*\|}\left[\sum_{i=1+k_1}^{k_2} Z_i - \sum_{i=1+k_1^*}^{k_2^*} Z_i\right] = a_N O_P(1),$$

as $n \to \infty$, where $a_N = o(1)$, as $N \to \infty$.

3. *Under the second conditions let $\hat{I}_n$ a random set with $P(\hat{I}_n \subset I) \to 1$ as $n \to \infty$, where $I \subset \{1,\dots,n\}$. Then, it holds that*

$$\begin{aligned}&P\left(\max_{\substack{1\le k_1<k_2\le n\\ \|k-k^*\|\ge N}} \frac{1}{\|k-k^*\|}\left[\sum_{i\in(k_1,k_2]\cap\hat{I}_n} Z_i - \sum_{i\in(k_1^*,k_2^*]\cap\hat{I}_n} Z_i\right] \ge \eta\right)\\ &= O\left(\frac{\mathbb{1}_{\{k_1^*\ge\min I\}}}{(N\vee(k_1^*-\max I))^{r-1}} + \frac{\mathbb{1}_{\{k_2^*\ge\min I\}}}{(N\vee(k_2^*-\max I))^{r-1}}\right.\\ &\quad\left. + \frac{\mathbb{1}_{\{k_1^*\le\max I\}}}{(N\vee(\min I-k_1^*))^{r-1}} + \frac{\mathbb{1}_{\{k_2^*\le\max I\}}}{(N\vee(\min I-k_2^*)))^{r-1}}\right).\end{aligned}$$

Proof. **1.** We obtain that

$$\sum_{i=1+k_1}^{k_2} d_i - \sum_{i=1+k_1^*}^{k_2^*} d_i = \begin{cases} \sum_{i=1+k_1}^{k_2} d_i - \sum_{i=1+k_1^*}^{k_2^*} d_i, & \text{if } k_1 < k_2 \leq k_1^* < k_2^*, \\ \sum_{i=1+k_1}^{k_1^*} d_i - \sum_{i=1+k_2}^{k_2^*} d_i, & \text{if } k_1 \leq k_1^* \leq k_2 < k_2^*, \\ -\sum_{i=1+k_1^*}^{k_1} d_i - \sum_{i=1+k_2}^{k_2^*} d_i, & \text{if } k_1^* < k_1 < k_2 < k_2^*, \\ \sum_{i=1+k_1}^{k_1^*} d_i + \sum_{i=1+k_2^*}^{k_2} d_i, & \text{if } k_1 < k_1^* < k_2^* < k_2, \\ -\sum_{i=1+k_1^*}^{k_1} d_i + \sum_{i=1+k_2^*}^{k_2} d_i, & \text{if } k_1^* < k_1 \leq k_2^* < k_2, \\ \sum_{i=1+k_1}^{k_2} d_i - \sum_{i=1+k_1^*}^{k_2^*} d_i, & \text{if } k_1^* < k_2^* \leq k_1 < k_2 \end{cases}$$

$$= \begin{cases} o(n^{1/2}), & \text{if } k_1 < k_2 \leq k_1^* < k_2^*, \\ o(\|k-k^*\| \wedge n^{1/2}), & \text{else}, \\ o(n^{1/2}), & \text{if } k_1^* < k_2^* \leq k_1 < k_2, \end{cases}$$

$$= o(\|k-k^*\| \wedge n^{1/2}),$$

where we use in the first and last case that $\|k-k^*\| = O(k_2^* - k_1^*)$ and that for $j = 1, 2$

$$\max_{\substack{k_j \leq k_j^* \\ \|k-k^*\| \geq N}} \frac{1}{\|k-k^*\|} \sum_{i=k_j}^{k_j^*} |d_i| = \max_{\substack{k_j \leq k_j^* \\ N \leq |k_j - k_j^*| \leq \sqrt{n}}} \frac{1}{\|k-k^*\|} \sum_{i=k_j}^{k_j^*} |d_i| \tag{B.0.4}$$

$$+ \max_{\substack{k_j \leq k_j^* \\ \|k-k^*\| \geq \sqrt{n}}} \frac{1}{\|k-k^*\|} \sum_{i=k_j}^{k_j^*} |d_i| = o(1). \tag{B.0.5}$$

Here, we use $d_{a_n} = o(1)$ for any $a_n \to \infty$ as $n \to \infty$ for the first summand. Analogously, the other case follows.

2. Now, we prove the second conclusion and use the same segmentation of the set, over which the maximum is taken, as at the beginning. Since the maximum over the whole set is smaller than the sum of the maximum over each segment area, it is sufficient to show that each maximum is equal to $O_P(1)$. For the first segment it follows by Kolmogorov's type inequalities that

$$P\left(\max_{\substack{k_1 < k_2 \leq k_1^* < k_2^* \\ \|k-k^*\| \geq N}} \frac{1}{\|k-k^*\|} \left| \sum_{i=1+k_1}^{k_2} Z_i - \sum_{i=1+k_1^*}^{k_2^*} Z_i \right| \geq \epsilon \right)$$

$$\leq P\left(\max_{\substack{k_1 < k_2 \leq k_1^* < k_2^* \\ \|k-k^*\| \geq N}} \frac{1}{(k_2^* - k_1^*)} \left| \sum_{i=1+k_1}^{k_2} Z_i - \sum_{i=1+k_1^*}^{k_2^*} Z_i \right| \geq \epsilon \right)$$

$$\leq \frac{c}{\epsilon^r (k_2^* - k_1^*)^r} \sum_{i=1}^{k_1^*} c_i + \frac{c}{\epsilon^r (k_2^* - k_1^*)^r} \sum_{i=1+k_1^*}^{k_2^*} c_i = o(1),$$

as $n \to \infty$, where we use $k_1^* = o((k_2^* - k_1^*)^r)$. Similarly, we obtain the rate for the last segment, where we use $n - k_2^* = o((k_2^* - k_1^*)^r)$. For the other segments we use the Hájek-Rényi type inequalities, which are equivalent to the Kolmogorov's type inequalities, and get for example

$$P\left(\max_{\substack{k_1 \leq k_1^* \\ \|k-k^*\| \geq N}} \frac{1}{\|k-k^*\|} \left| \sum_{i=1+k_1}^{k_1^*} Z_i \right| \geq \epsilon \right) \leq \frac{c}{\epsilon^r} \sum_{i=1}^{k_1^*} \frac{c_i}{(i \vee N)^r} = a_N O(1),$$

as $n \to \infty$, where $a_N \to 0$ as $N \to \infty$.

3. In the third case, just as in the second, we use the decomposition of the set over which the maximum is taken. On the first set $k_1 < k_2 \leq k_1^* < k_2^*$ we obtain:

$$\max_{\substack{1 \leq k_1 < k_2 \leq k_1^* \\ \|k-k^*\| \geq N}} \frac{1}{\|k-k^*\|} \left| \sum_{i \in (k_1, k_2] \cap \hat{I}_n} Z_i - \sum_{\in (k_1^*, k_2^*] \cap \hat{I}_n} Z_i \right| \leq \frac{c}{k_2^* - k_1^*} \max_{1 \leq k_1 < k_2 \leq k_2^*} \left| \sum_{i \in (k_1, k_2] \cap \hat{I}_n} Z_i \right|.$$

Now, we use $P(\hat{I}_n \subset I) \to 1$ as $n \to \infty$ so that the right–hand side above is estimated by

$$\frac{c}{k_2^* - k_1^*} \max_{k_1,k_2 \in I} \left| \sum_{i\in(k_1,k_2]} Z_i \right| + o_P(1) \le \frac{2c}{k_2^* - k_1^*} \max_{\min I \le k \le \max I} \left| \sum_{i=k}^{\max I} Z_i \right| + o_P(1)$$
$$= O_P(\#I(k_2^* - k_1^*)^{-r}) + o_P(1) = o_P(1),$$

where we use the Kolmogorov-type inequalities for the rates. Analogously, we deal with the last set $k_1^* < k_2^* \le k_1 < k_2$. Hence, we now consider the term on the second set $k_1 \le k_1^* \le k_2 < k_2^*$ and obtain:

$$\max_{\substack{1\le k_1 \le k_1^* \le k_2 \le k_2^* \\ \|k-k^*\| \ge N}} \frac{1}{\|k - k^*\|} \left| \sum_{i\in(k_1,k_2]\cap \hat{I}_n} Z_i - \sum_{\in(k_1^*,k_2^*]\cap \hat{I}_n} Z_i \right|$$
$$\le \max_{1\le k_1 \le k_1^*} \frac{1}{(k_1^* - k_1) \vee N} \left| \sum_{i\in(k_1,k_1^*]\cap \hat{I}_n} Z_i \right| + \max_{k_1^* \le k_2 \le k_2^*} \frac{1}{(k_2^* - k_2) \vee N} \left| \sum_{i\in(k_2,k_2^*]\cap \hat{I}_n} Z_i \right|.$$

We treat the first and second summand analogously so that we only consider the first one. Using $P(\hat{I}_n \subset I) \to 1$ as $n \to \infty$, we obtain that the first summand is equal to $o_P(1)$ if $\min I > k_1^*$. In the other case we estimate the summand by

$$\max_{k_1,k_2 \in I} \frac{1}{(k_1^* - k_1) \vee N} \left| \sum_{i\in(0,k_1^*]\cap(k_1,k_2]} Z_i \right| + o_P(1) \le 2 \max_{k \in I} \frac{1}{(k_1^* - k) \vee N} \left| \sum_{i=k}^{\max I \wedge k_1^*} Z_i \right| + o_P(1)$$

Now, we apply a Kolmogorov-type inequality, which yields that

$$P\left(\max_{k\in I} \frac{1}{(k_1^* - k) \vee N} \left| \sum_{i=k}^{\max I \wedge k_1^*} Z_i \right| \ge \eta \right) \le \frac{c}{\eta^r} \sum_{i=\min I}^{k_1^* \wedge \max I} \frac{1}{(k_1^* - i)^r \vee N^r}$$
$$= O\left(((k_1^* - \max I) \vee N)^{1-r} \right).$$

Analogously, we handle the maximum over the other sets so that we get:

$$P\left(\max_{\substack{1\le k_1 < k_2 \le n \\ \|k-k^*\| \ge N}} \frac{1}{\|k - k^*\|} \left[\sum_{i\in(k_1,k_2]\cap \hat{I}_n} Z_i - \sum_{\in(k_1^*,k_2^*]\cap \hat{I}_n} Z_i \right] \ge \eta \right)$$
$$= O\left(\frac{\mathbb{1}_{\{k_1^* \ge \min I\}}}{(N \vee (k_1^* - \max I))^{r-1}} + \frac{\mathbb{1}_{\{k_2^* \ge \min I\}}}{(N \vee (k_2^* - \max I))^{r-1}} \right.$$
$$\left. + \frac{\mathbb{1}_{\{k_1^* \le \max I\}}}{(N \vee (\min I - k_1^*))^{r-1}} + \frac{\mathbb{1}_{\{k_2^* \le \max I\}}}{(N \vee (\min I - k_2^*)))^{r-1}} \right).$$

□

Lemma B.0.3. *Let* $m \in \mathbb{N}$ *be finite and* $C_n = \cup_{i=1}^m (n\theta_{2i-1}, n\theta_{2i}]$*, where* $\theta_i \in (0,1)$ *and* $\theta_i \ne \theta_j$ *for* $i \ne j$*. In addition let* $\hat{C}_n = \cup_{i=1}^{\hat{m}} (n\hat{\theta}_{2i-1}, n\hat{\theta}_{2i}]$ *with* $\lambda(\hat{C}_n \triangle C_n) = O_P(n^{1-\delta})$ *and let* $\{Z_n\}$ *be a random sequence, which fulfills the Kolmogorov's inequalities* $(\mathcal{K}_r^{(1)})$ *and* $(\mathcal{K}_r^{(2)})$ *with a* $r > 1$*. Then, it holds that*

$$\max_{C\in R_n} \left| \sum_{i\in C\cap(\hat{C}_n \triangle C_n)} Z_i \right| = \begin{cases} O_P(mn^{1/2}), & \text{if } |\hat{m} - m| = O_P(1) \\ O_P(mn^{(1-\delta)(1/r+\tilde{\delta})} \wedge n^{1/2}) & \text{if } |\hat{m} - m| = o_P(1) \end{cases}$$

for any arbitrarily small $\tilde{\delta} > 0$*, where* $R_n = \{\bigcup_{i=1}^m (ns_i, nt_i] \;:\; s_i, t_i \in (0,1), s_i \le t_i < s_{i+1}\}$

Proof. Let $\eta > 0$ then it holds that

$$\begin{aligned}
&P\left(\max_{C\in R_n}\left|\sum_{i\in C\cap(\hat{C}_n\triangle C_n)} Z_i\right| > \eta\right)\\
&= P\left(\max_{C\in R_n}\left|\sum_{i\in C\cap(\hat{C}_n\triangle C_n)} Z_i\right| > \eta, |\hat{m}-m| > N_1\right)\\
&\quad + P\left(\max_{C\in R_n}\left|\sum_{i\in C\cap(\hat{C}_n\triangle C_n)} Z_i\right| > \eta, |\hat{m}-m| \le N_1\right)\\
&\le P\left(|\hat{m}-m| > N_1\right) + P\left(\max_{C\in R_n}\left|\sum_{i\in C\cap(\hat{C}_n\triangle C_n)} Z_i\right| > \eta, |\hat{m}-m| \le N_1\right)
\end{aligned}$$

as $n \to \infty$. Since $\hat{C}_n$ is a union of at most $m + N_1$ disjoint intervals, $\hat{C}_n\triangle C_n$ and $C\cap\hat{C}_n\triangle C_n$ are a union of at most $2m + N_1$ intervals under the condition $|\hat{m}-m| \le N_1$. Hence, it holds that

$$\max_{C\in R_n}\left|\sum_{i\in C\cap(\hat{C}_n\triangle C_n)} Z_i\right| \le \max_{C\in R_n}\left|\sum_{i\in C} Z_i\right| \le 2(2m+N_1)\max_{1\le k\le n}|S_n - S_{n-k}| = O_P(mn^{1/r}),$$

where we just assume that $|\hat{m}-m| = O_P(1)$. If we assume $|\hat{m}-m| = o_P(1)$, we replace N_1 by an $\epsilon_1 > 0$ and consider

$$\begin{aligned}
&P\left(\max_{C\in R_n}\left|\sum_{i\in C\cap(\hat{C}_n\triangle C_n)} Z_i\right| > \eta, |\hat{m}-m| \le \epsilon_1, \lambda(\hat{C}_n\triangle C_n) \le N_2 n^{1-\delta}\right)\\
&\le P\left(\sum_{j=1}^{m}\max_{1\le k_1,k_2\le N_2 n^{1-\delta}}\left|\sum_{i=n\theta_{2j-1}}^{n\theta_{2j}} Z_i - \sum_{i=n\theta_{2j-1}-k_1}^{n\theta_{2j}+k_2} Z_i\right| > \eta\right),
\end{aligned}$$

where by applying Lemma B.0.2 it follows that

$$\max_{1\le k_1,k_2\le N_2 n^{1-\delta}}\left|\sum_{i=n\theta_{2j-1}}^{n\theta_{2j}} Z_i - \sum_{i=n\theta_{2j-1}-k_1}^{n\theta_{2j}+k_2} Z_i\right| = O_P(N_2^{1/r+\tilde{\delta}} n^{(1-\delta)(1/r+\tilde{\delta})} \wedge n^{1/2})$$

for any arbitrary $\tilde{\delta} > 0$. Choosing $\eta = N_2^{1+1/r+\tilde{\delta}}$ makes clear that the result follows. □

Lemma B.0.4. *Under the assumption that g is piecewise continuous, it holds on $[0,1]$ that*

$$\left\|\frac{1}{n}\sum_{i=1}^{[n\cdot]} g(i/n) - \int_0^{\cdot} g(x)dx\right\|_\infty \to 0, \quad \textit{as } n \to \infty. \tag{B.0.6}$$

Proof. Let $\epsilon > 0$ and $z_1,\ldots,z_m \in [0,1]$, $m \in \mathbb{N}$, the discontinuous points of g. Then, we estimate the maximum over $[0,1]$ by the sum of the maxima over $[z_i,z_{i+1}]$, $i = 1,\ldots,m-1$. Therefore, it is sufficient to show that each of those summands are smaller than $\epsilon' = \epsilon/m > 0$ for each $n \ge N$ for a suitably large N. Since

$$\begin{aligned}
\sup_{z\in[z_j,z_{j+1}]}\left|\frac{1}{n}\sum_{i=1}^{[nz]} g(i/n) - \int_0^z g(x)dx\right| &\le \sum_{k=1}^{j-1}\left|\frac{1}{n}\sum_{i=[nz_k]}^{[nz_{k+1}]} g(i/n) - \int_{z_k}^{z_{k+1}} g(x)dx\right|\\
&\quad + \sup_{z\in[z_i,z_{i+1}]}\left|\frac{1}{n}\sum_{i=[z_j]}^{[nz]} g(i/n) - \int_{z_j}^{z} g(x)dx\right|
\end{aligned}$$

and $\frac{1}{n}\sum_{i=[nz_k]}^{[nz_{k+1}]} g(i/n)$ pointwise converges towards $\int_{z_k}^{z_{k+1}} g(x)dx$, it is sufficient to consider

$$\begin{aligned}
&\sup_{z\in[z_i,z_{i+1}]}\left|\frac{1}{n}\sum_{i=[z_j]}^{[nz]} g(i/n) - \int_{z_j}^{z} g(x)dx\right| \\
&\qquad\leq \sup_{z\in[z_i,z_{i+1}]}\left|\frac{z-z_j}{[nz]}\sum_{i=1}^{n} g(i/n(z-z_j)) - \int_{z_j}^{z} g(x)dx\right| \\
&\qquad\quad + \sup_{z\in[z_i,z_{i+1}]}\left|\frac{z-z_j}{n}\sum_{i=1}^{n} g(i(z-z_j)/n) - \frac{1}{n}\sum_{i=[z_j]}^{[nz]} g(i/n)\right|,
\end{aligned}$$

where the first summand converges towards zero, since $f_n(z) = \frac{z-z_j}{[nz]}\sum_{i=1}^{n} g(i/n(z-z_j))$ is equicontinuous (g continuous) and converges pointwise towards the integral. □

References

Antoch, J. and Hušková, M. (1996). Tests and estimators for epidemic alternatives. *Tatra Mt. Math. Publ.,* **7**, 311-329.

Anderson, T.W. (1984). *An Introduction to Multivariate Statistical Analysis.* 2nd ed. Wiley publications in statistics. Mathematical statistics, New York.

Andrews, D.W. (1991). Heteroskedasticity and autocorrelation consistent covariance matrix estimation. *Econometrica,* **59**, 817-858.

Aue, A. and Hörmann, S. and Horváth, L. and Reimherr, M. (2009) .Break detection in the covariance structure of multivariate time series models. *The Annals of Statistics,* **37**, 4046-4087.

Aue, A. and Horváth, L. (2013). Structural breaks in time series. *Journal of Time Series Analysis,* **34**, 1-16.

Bahr, B. and Esseen, C. (2013). Inequalities for the rth absolute moment of a sum of random variables, $1 \leq r \leq 2$. *The Annals of Mathematical Statistics,* **36**, 299-303.

Bai, J. (1994). Least Square estimation of a shift in linear processes. *Journal of Time Series Analysis,* **15**, 5, 453 -472.

Bai, J. (1997). Estimating multiple breaks one at time. *Econometric Theory,* **13**, 315 -352.

Bai, J. and Perron, P. (1998). Estimating and testing linear models with multiple structural changes. *Econometrica,* 47-78.

Billingsley, P. (1968). *Convergence of Probability Measures Wiley.* New York.

Bradley, R. C. (2009). *Introduction to Strong Mixing Conditions, Kendrick Press.* Voll 2nd printing.

Brodsky, E. and Darkhovsky, B. S. (2013). Nonparametric Methods in Change-Point Problems. Vol. 243. *Springer Science & Business Media.*

Bucchia, B. and Heuser, C. (2015). Long-run variance estimation for spatial data under change-point alternatives. *Journal of Statistical Planning and Inference,* **165**, 104-126.

Csörgö, M. and Horváth, L. (1997). *Limit Theorems in Change-Point Analysis.* Wiley New York.

Darling, D. A. and Erdös, P. (1956). A limit theorem for the maximum of normalized sums of independent random variables. *Duke Math. J.* **23**, 143–155.

Davidson, J. (1994) *Stochastic Limit Theory: An Introduction for Econometricians.* Oxford university press.

Davidson, J. and De Jong, R. M. (2000). Consistency of kernel estimators of heteroscedastic and autocorrelated covariance matrices. *Econometrica,* **68**, 407-423.

Davidson, J. (2002). Establishing conditions for the functional central limit theorem in nonlinear and semiparametric time series processes. *Elsevier,* **106**, 243–269.

Davidson, J. (2004). Moment and memory properties of linear conditional heteroscedasticity models, and a new model. *Journal of Business and Economic Statistics,* **22**, 16-29.

DAX-Gewichtung (2016). Accessed on 05.03.2016. URL: *http://www.finitmat.de/DAX-Gewichtung.html.*

De Jong, R. M. (2000). A strong consistency proof for heteroskedasticity and autocorrelation consistent covariance matrix estimators. *Econometric Theory,* **16**, 262-268.

Dehling, H., Vogel, D., Wendler, M., and Wied, D. (2015). Testing for Changes in the Rank Correlation of Time Series. URL: *http://arxiv.org/abs/1203.4871v4*, 1-26.

Donsker, M. (1951). An invariance principle for certain probability limit theorems, Four Papers on Probability. *Mem. Amer. Math. Soc.* **6**.

Eckley, I., Fearnhead, P., and Killick, R. (2011). *Analysis of changepoint models. Bayesian Time Series Models, Cambridge University Press* Chapter: 10.

California Environmental Protection Agency. (18.09.15). In-use Compliance Letter.

Fazekas, I. and Klesov, O. (2001). A general approach to the strong law of large numbers. *Theory of Probability & Its Applications*, **45**, 436-449.

Ferger, D. (1994). Change-point estimators in case of small disorders. *Journal of Statistical Planning and Inference*, **40**, 33-49.

Ferger, D. (2001). Analysis of Change-Point Estimators under the Null Hypothesis . *Bernoulli Society for Mathematical Statistics and Probability*, **7**, 487-506.

Finanzen (2016).Accessed on 06.03.2016. URL: *http://www.finanzen.net/historische-kurse/Volkswagen* and *http://www.finanzen.net/index/DAX/Historisch.*

Herrndorf, N. (1984). A Functional central limit theorem for weakly dependent sequences of random variables. *Ann. Probab.* **12**, 141-153.

Heuser, C. (2013). Testprozeduren zum Aufdecken von Strukturbrüchen in den Korrelationen von Zeitreihen. Masterarbeit, Universität zu Köln.

Hill, J B. (2010). A New Moment Bound and Weak Laws for Mixingale Arrays without Memory or Heterogeneity Restrictions, with Applications to Tail Trimmed Arrays. *Dept. of Economics,* University of North Carolina - Chapel Hill.

Hušková, M., 1995. Estimators for epidemic alternatives. Comment. *Math. Univ. Carolinae,* **36**, 279-291.

Hušková, M. and Kirch, C. (2010). A note on studentized confidence intervals for the change-point. *Computational Statistics,* **25**, 269-289.

Horváth, L. and Hušková, M. and Kokoszka, P. and Steinebach, J.G. (2004). Monitoring changes in linear models. *Journal of Statistical Planning and Inference,* **126**, 225-251.

Horváth, L. and Steinebach, J.G. (2004). Testing for changes in the mean or variance of a stochastic process under weak invariance. *Journal of Statistical Planning and Inference,* **91**, 365-376.

Ibragimov, I. A. and Linnik, J. V. (1971). Independent and stationary sequences of random variables. Wolters-Noordhoff, Groningen.

Juhl, T. and Xiao, Z. (2009). Tests for changing mean with monotonic power. *Journal of Econometrics,* **148**, 14-24.

Kim, J. and Pollard, D. (1990). The analysis of change-point data with dependent errors. *Annals of Statistics,* **18**, 191-219.

Levin, Bruce and Kline, Jennie, 1985: The cusum test of homogeneity with an application in spontaneous abortion epidemiology. *Statistics in Medicine,* **4**, 469 488.

Lombard, F. and Hart, J. D. (1994). Cube root asymptotics. *Lecture Notes–Monograph Series,* **23**, 194-209.

Qiu, J. and Lin, Z. (2011). The functional central limit theorem for linear processes with strong near-epoch dependent innovations. *Journal of Mathematical Analysis and Applications,* **376**, 373-382.

Qu, Z. and Perron, P. (2007). Estimating and testing structural changes in multivariate regressions. *Econometrica,* **75**, 459-502.

Lavielle, M. and Moulines, E. (2000). Least squares estimation of an unknown number of shifts in a time Series. *Journal of time series analysis,* **21**, 33-59.

McLeish, D. L. (1975). A maximal inequality and dpendent strong laws. *Annals of Probability,* **5**, 329-339 .

Müller, H. G., and Siegmund, D. (1994). *Change-Point Problems.* IMS Lecture Notes, **23**.

Newey, W.K. and West, K. D. (1987). A simple, positive semi-definite, heteroskedasticity and autocorrelationconsistent covariance matrix. *Econometrica,* **55**, 703708.

Tomacs, T. and Líbor, Z. (2006). A Hájek–Rényi type inequality and its applications. *Eszterházy Károly College, Institute of Mathematics and Computer Science,* **33**, 141-149 .

van der Vaart, A.W. (1998). *Asymptotic Statistics.* Cambridge: New York.

Wied, D. and Galeano, P. (2013). Monitoring correlation change in a sequence of random variables. *Journal of Statistical Planning and Inference,* **143**, 186-196.

Wied, D. (2010). Ein Fluktuationstest auf konstante Korrelation. Dissertation, TU Dortmund.

Wied, D., Krämer, W. and Dehling, H. (2012). Testing for a change in correlation at an unknown point in time using an extended functional delta method. *Econometric Theory,* **28**, 570-589.

Galeano, P. and Wied, D. (2014). Multiple break detection in the correlation structure of random variables. *Computat. Statist. Data Analysis,* **76**, 262-282.

Winterkorn, M. (2015). Pressemitteilung der Volkswagen AG, 20.09.15. Accessed on 09.03.2016. URL: *https://www.volkswagen-media-services.com/detailpage/-/detail/Erklrung-des-Vorstandsvorsitzenden-der-Volkswagen-AG-Professor-Dr-Martin-Winterkorn/view/2709299/7a5bbec13158edd433c6630f5ac445da?p_p_auth=0qKdNca7.*

Yao, Q. Tests for change-points with epidemic alternatives. *Biometrika,* **80**, 179-191.